Integrated Energy System Resilience

Zhi Wu · Qirun Sun · Wei Gu · Suyang Zhou ·
Pengxiang Liu · Yue Qiu

Integrated Energy System Resilience

Modeling, Evaluation, and Enhancement

Springer

Zhi Wu
Southeast University
Nanjing, Jiangsu, China

Wei Gu
Southeast University
Nanjing, Jiangsu, China

Pengxiang Liu
City University of Hong Kong
Kowloon Tong, Kowloon, Hong Kong

Qirun Sun
Hohai University
Nanjing, Jiangsu, China

Suyang Zhou
Southeast University
Nanjing, Jiangsu, China

Yue Qiu
Nanjing Normal University
Nanjing, Jiangsu, China

This work was supported by the Natural Science Foundation of China (Grant No. 52522704), Natural Science Foundation of China (Grant No. 52177077), National Key R&D Program of China (Grant No. 2023YFE0108600), Natural Science Foundation of Jiangsu Province (Grant No. BK20240076), and Carbon Peak and Neutrality Science and Technology Innovation Project in Jiangsu Province (BE2023093-1).

ISBN 978-981-95-2362-7 ISBN 978-981-95-2363-4 (eBook)
https://doi.org/10.1007/978-981-95-2363-4

This Springer imprint is published by the registered company Springer Nature Singapore Pte Ltd.
The registered company address is: 152 Beach Road, #21-01/04 Gateway East, Singapore 189721, Singapore

Preface

Against the backdrop of increasingly frequent extreme disasters, a diversified energy structure, and the growing demand for urban energy security, the safe operation of integrated energy systems (IES) faces new challenges. On the one hand, the increasing frequency and intensity of extreme disaster events pose serious threats to energy supply security and system operational stability. On the other hand, the deepening coupling of multiple energy carriers—including electricity, heat, and gas—in urban energy systems results in greater operational complexity and vulnerability. Building a resilient IES capable of withstanding and rapidly recovering from disasters has become essential to safeguarding urban energy security and promoting the sustainable development of the economy and society.

This book presents key technologies for enhancing the resilience of multi-energy coupled IES in response to extreme disaster events. It focuses on three core areas: resilience assessment, resilience planning, and resilience recovery, and discusses the corresponding fundamental theories and experimental methodologies, striving for a balance between theoretical rigor and practical relevance. The book emphasizes systematic and practical approaches, offering valuable insights for energy industry planners, researchers, and government decision-makers, while also serving as a useful reference for faculty and students in higher education institutions engaged in related research and studies.

The book is organized as follows: Chaps. 1 and 2 review the destructive impacts of recent global extreme events on energy systems and analyze the policy background, fundamental properties, and core characteristics of IES resilience. Chapter 3 investigates the cross-system failure propagation mechanisms in combined heat and power IES through dynamic simulation, exploring how multi-energy coupling structures influence fault evolution. Chapter 4 proposes a resilience assessment methodology that considers gas-thermal inertia and multi-energy interdependency and establishes a multi-energy interdependency assessment index system for IES. Chapters 5 and 6 focus on rapid resilience assessment methods considering non-convex energy flow constraints, introducing acceleration strategies based on second-order cone equality reformulation, spatial branching, and combinatorial cutting-plane strategies. Chapter 7 proposes a resilience planning approach that balances economic and risk factors, accompanied by efficient solution algorithms for large-scale models. Chapter 8 explores modeling and scheduling optimization of electricity–hydrogen-enriched natural gas systems under multiple uncertainties, covering production, storage, blending, transmission, and end-use stages, and presents assessment indicators and efficient solution strategies. Chapter 9 focuses on post-disaster recovery processes, proposing an energy supply restoration method for IES based on multi-stage and multi-energy flow coordination.

Nanjing, China
June 2025

Zhi Wu
Qirun Sun
Wei Gu
Suyang Zhou

Acknowledgments This book summarizes some of our research on energy system resilience in recent years. We are deeply grateful to all those who have contributed their insights, expertise, and encouragement throughout the development of this book. We extend our sincere thanks to Dr. Pengxiang Liu from City University of Hong Kong, Prof. Haifeng Qiu and Mr. Aobo Guan from Southeast University, and Dr. Yue Qiu from Nanjing Normal University. We also thank Ms. Yuxin Liu, Mr. Ziang Lu, Mr. Zijian Wang, and Ms. Xinran Jia for their careful examination of the entire book. In addition, we acknowledge the innovative research and practice contributed by others in this increasingly important field. Finally, we appreciate the assistance and help of the staff at Springer.

This book has been supported by the Natural Science Foundation of China (Grant No. 52522704), Natural Science Foundation of China (Grant No. 52177077), National Key R&D Program of China (Grant No. 2023YFE0108600), Natural Science Foundation of Jiangsu Province (Grant No. BK20240076), and Carbon Peak and Neutrality Science and Technology Innovation Project in Jiangsu Province (BE2023093-1). We do appreciate their supports.

Zhi Wu
Qirun Sun
Wei Gu
Suyang Zhou
Pengxiang Liu
Yue Qiu

Competing Interests The authors have no competing interests to declare that are relevant to the content of this manuscript.

Contents

Chapter 1
Introduction of IES Resilience

1.1 Background

In the context of the carbon peaking and carbon neutrality goals, integrated energy systems (IESs) have emerged as an innovative energy utilization pattern, garnering increasing international attention. By integrating key equipment such as combined heat and power (CHP) units, gas turbines, and heat pumps, IESs effectively combine various energy subsystems, including electricity, natural gas, and heating systems, into a unified and coordinated framework. The multi-energy complementarity of IESs facilitates the cascade utilization of energy, leading to significant improvements in energy efficiency, reductions in greenhouse gas emissions, and enhanced system stability. As a key strategy for achieving China's "30–60" targets, electric-gas-thermal coupled IESs have been incorporated into the national "14th Five-Year Plan" for energy development, with substantial progress achieved in both research and practical applications [1].

As a critical infrastructure supporting the stable operation of modern society, the safe and reliable operation of IESs plays an irreplaceable role in ensuring the normal functioning of social and economic activities. Energy supply security is also a crucial component of national security, and safeguarding energy supply security is of paramount importance to national security. However, extreme weather events caused by global climate change, such as hurricanes, heatwaves, and cold waves, have posed unprecedented shocks and challenges to global energy systems [2]. Figure 1.1 illustrates the frequency of various natural disasters from 1960 to 2023.

For example, in September 2019, Australia experienced months of wildfires triggered by extreme heat. These catastrophic fires destroyed more than 10 million hectares of land and caused significant damage to energy infrastructure. The fires directly destroyed large amounts of power infrastructure, including substations, power lines, and transmission towers, resulting in widespread power outages. Particularly in remote and disaster-stricken areas, the recovery time for electricity was prolonged, severely impacting residents' lives and economic activities. At the same

Z. Wu et al., *Integrated Energy System Resilience*,
https://doi.org/10.1007/978-981-95-2363-4_1

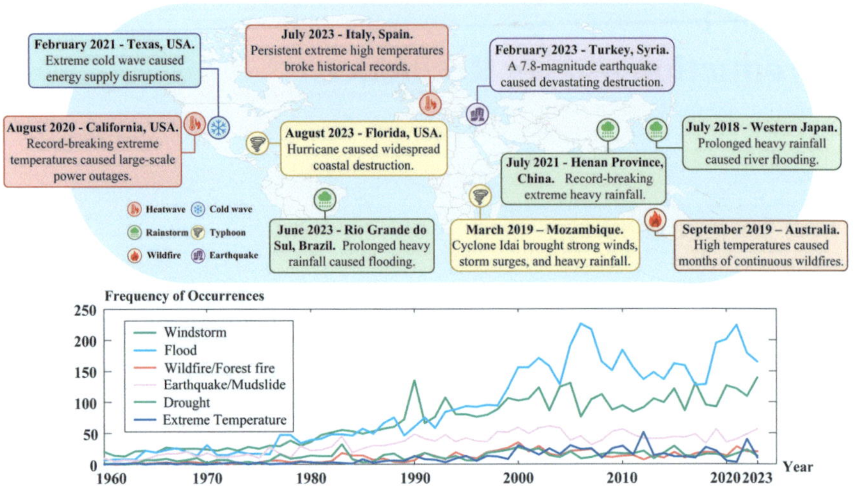

Fig. 1.1 Global typical extreme disaster events and historical records. *Source* International disaster database EM-DAT [3]

time, the fires disrupted transportation infrastructure, particularly affecting the transportation and distribution of natural gas, with the disruption severely hampering the natural gas supply process. Furthermore, the fires caused significant damage to heating systems, primarily through the destruction of heating equipment and pipelines, leading to widespread heating interruptions.

In August 2020, California, USA, faced rare extreme heat conditions that brought significant challenges to the local energy system. The extreme temperatures led to a surge in electricity demand, putting unprecedented pressure on the power grid. In response to the power shortage, California had to implement rolling blackouts, affecting more than 800,000 people and leaving over 3.3 million households without power. Additionally, the extreme heat exacerbated the drought, leading to a sharp decline in hydropower generation and further weakening the power supply. Simultaneously, the extreme heat posed a threat to natural gas supply facilities, as gas-fired power plants shut down under the high temperatures, leading to a failure in peak shaving and intensifying the power supply shortage.

In February 2021, Texas, USA, encountered an extreme cold wave, causing a severe energy supply disruption. About 60% of Texas residents rely on electric heating systems. During the cold wave, electricity demand surged to 69.22 million kilowatts, while the grid's maximum generating capacity was only 48 million kilowatts, leading to a 20 million kilowatt electricity shortfall. At the same time, natural gas pipelines were interrupted due to freezing and blockages, causing gas-fired power plants to shut down, which led to a sudden drop in gas-generated electricity by approximately 25 million kilowatts. With widespread frozen gas pipelines and slow repair work, municipal heating systems collapsed on a large scale. This cold wave led to an overload of the Texas power grid, severe power shortages, interruptions in gas supply, shutdown of gas units, and a massive failure of municipal heating systems, ultimately resulting in an economic loss of approximately $130 billion.

In July 2021, Henan Province in China experienced a historically rare torrential rainstorm, with precipitation in Zhengzhou breaking historical records, leading to severe flooding. This extreme weather event caused significant damage to the energy supply system, affecting 1.2663 million households. About 45% of the base stations in Zhengzhou were shut down due to power outages or damage to equipment, and at least three water plants ceased operations due to power interruptions, leaving multiple neighborhoods without running water. The heavy rain also caused several accidents in the oil and gas pipelines in Henan, particularly the rupture of the Yizheng gas pipeline and leakage of the Wugang Tianlun gas pipeline.

On February 6, 2023, a 7.8 magnitude earthquake struck the Turkey-Syria border region, followed by a 7.5 magnitude aftershock. During the earthquake, power infrastructure was severely damaged, with large-scale transmission line breakages and catastrophic damage to substations and other key power facilities, leading to widespread power outages. The earthquake also triggered multiple natural gas pipeline accidents, disrupting gas supplies and exacerbating the energy shortage. Furthermore, severe fires and explosions occurred in multiple regions following the earthquake, complicating rescue efforts. The earthquake also caused widespread ruptures in district heating network pipes, particularly in central heating areas, leading to heating disruptions and leaving many residents struggling with insufficient warmth in cold weather, further increasing the risk to post-disaster social stability.

According to statistics from the United Nations Office for Disaster Risk Reduction (UNDRR), the frequency of extreme natural disasters has significantly increased in recent years. From 2000 to 2019, there were 350 to 500 major and medium-sized disaster events globally each year, with a total of 7,348 disaster events recorded over ten years, causing approximately $2.97 trillion in economic losses. In comparison, only 4,212 extreme events were recorded between 1980 and 1999. It is expected that by 2030, the annual number of extreme disaster events will rise to 560, or about 1.5 events per day. This trend presents increasing challenges and demands for the security and reliability of energy systems.

As a critical support for the stability and development of modern society, IESs expose a large amount of infrastructure and equipment to extreme disasters [4], making them highly susceptible to multi-point cascading faults, as shown in Fig. 1.2. The electricity, gas, and heating systems are tightly coupled through devices such as power-to-gas equipment, gas turbines, and CHP units. When a failure occurs in one of these systems, the deep coupling characteristics of electricity, gas, and heating provide pathways for the propagation of faults, easily triggering multi-point cascading faults. This can lead to severe energy supply disruptions and even cause broader social unrest and economic losses. To address the increasing frequency and severity of future extreme climate events, countries urgently need to invest in disaster response and energy resilience to ensure the security and stability of energy supply in the face of frequent extreme weather events.

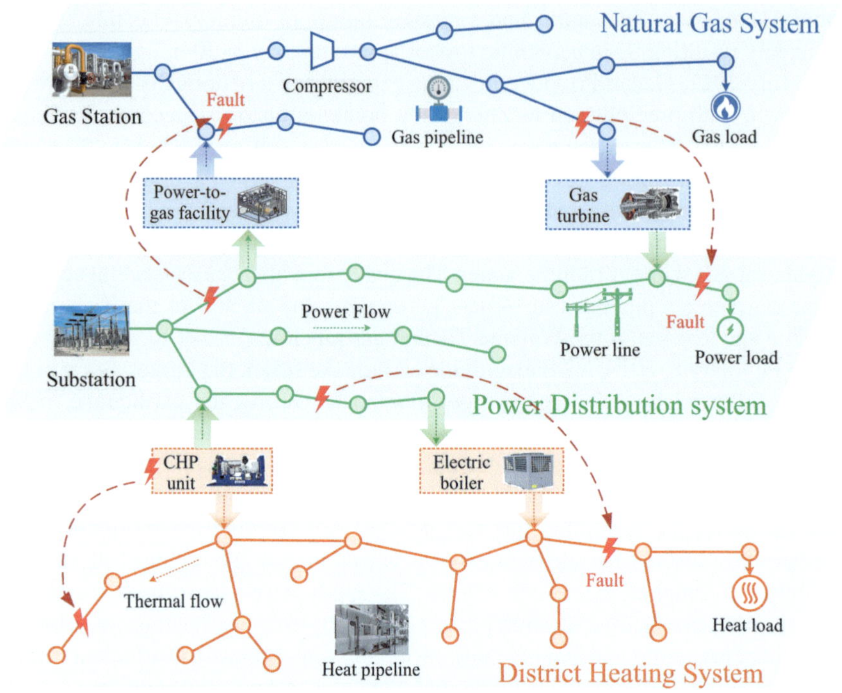

Fig. 1.2 Schematic diagram of multi-energy coupled cascading faults

1.2 Concepts Related to IES Resilience

1.2.1 Policies Related to IES Resilience

Energy security, as the cornerstone of national development, is essential for national prosperity and social stability. In the face of the increasing frequency of extreme disaster events recently, building a highly resilient energy system has become a strategic priority for governments worldwide. In recent years, many countries have introduced a series of policy documents and guiding frameworks regarding the construction of resilient energy systems, as shown in Table 1.1. This reflects the global attention to the construction of resilient IES.

With the increasing frequency of extreme weather events worldwide, governments and international organizations have elevated the construction of energy system resilience to the level of national strategy. The resilience of IESs has become a crucial indicator for assessing national energy security and social stability. Since the United States' Presidential Policy Directive 21 in 2013 first incorporated critical infrastructure resilience into policy frameworks, and culminating in the NATO-EU Joint Critical Infrastructure Resilience Final Evaluation Report released in 2023, a series

Table 1.1 Policies on constructing resilient energy systems worldwide

Year	Issuing authority	Policy	Relevant content
2013	U.S. presidential policy directive 21	"Critical Infrastructure Security and Resilience"	Introduced the concept of energy system resilience for the first time, and emphasized the importance of strengthening the security and resilience of critical infrastructure under extreme conditions [5]
2017	National academy of engineering (USA)	"Enhancing the Resilience of the Nation's Electricity System"	Proposed strategies to enhance the resilience of national power systems against potential large-scale, long-duration outages [6]
2018	Energy Research Partnership (UK)	"Future Resilience of the UK Electricity System"	Promoted collaboration with government agencies to assess risks to infrastructure and adopt protection measures against potential threats [7]
2018	U.S. department of energy	"Multiyear Plan for Energy Sector Cybersecurity"	Established offices for cybersecurity, energy security, and emergency response [8]
2021	State council of the People's Republic of China	"Regulations on the security protection of critical information infrastructure"	Strengthened the security system for critical information infrastructure to ensure the stable and continuous operation of critical information facilities [9]
2022	National energy administration (China)	"14th five-year modern energy system plan"	Enhanced the stability and safety of energy supply, continuously improving risk response capabilities [10]
2023	North Atlantic treaty organization and European union	"NATO-EU task force on the resilience of critical infrastructure final assessment report"	Strengthened the construction of critical infrastructure and supply chain resilience while implementing measures to reduce potential risks [11]

of policies and guiding documents aimed at enhancing energy system resilience have been issued.

These documents not only reflect the international community's heightened focus on building energy system resilience but also provide a solid policy foundation and assurance for the sustainable development of energy systems. By formulating strategies to improve energy system resilience, assessing infrastructure security risks, implementing necessary protection measures, and enhancing the stability and security of energy supply chains, governments worldwide are actively addressing the threats and risks posed by potential extreme disaster events.

1.2.2 Basic Concepts of IES Resilience

The term "resilience" originates from the Latin word "resiliere", meaning "to rebound". In scientific research, the concept can be traced back to the 1970s when it was first introduced to ecology to evaluate an ecosystem's ability to maintain a stable population structure in the face of parameter fluctuations, which laid the foundation for the development of resilience theory [12]. Later, Simmie and Martin expanded the scope of resilience research, focusing on the system's long-term adaptability and proposing the concept of "evolutionary resilience". This concept is defined as the ability of a system to continuously evolve and self-adjust, thereby improving its capacity to withstand extreme events, maintain stable operations, and quickly recover to its original equilibrium. This evolution of the resilience concept not only deepened its theoretical understanding but also supported its application in broader fields. Over time, the concept of resilience has gradually extended from ecology to other disciplines such as environmental science and psychology. Despite varying emphases across disciplines, the core essence of resilience remains highly consistent. Generally, it is used to assess a system's ability to resist disturbances, respond rapidly, and recover efficiently when faced with external challenges.

In the power and energy sector, which is critical to national development and public welfare, governments, academia, and industries demonstrate diverse understandings and definitions of resilience, as summarized in Table 1.2.

The concept of power system resilience first gained widespread attention in the United States. Over time, as policies evolved, resilience emerged as a core principle emphasizing the ability of power systems to maintain supply continuity and restore power rapidly during disturbances. This reflects the United States' focus on the secure and stable operation of power systems, providing a foundation for further research and practice in power resilience. In contrast, the United Kingdom emphasizes resilience in response to multiple natural disasters, defining it as the ability of power systems to continue operations under various natural hazards. The European Union adopts a broader perspective, focusing on the system's adaptability to long-term changes, particularly the potential impacts of climate change on power system stability. This definition expands the scope of resilience and offers theoretical support for addressing global challenges such as climate change. Japan, a country prone to frequent natural disasters, focuses on the resilience of power systems during disasters, specifically the ability to ensure power supply to critical users.

In China, significant progress has been made in defining and implementing power system resilience through collaborative efforts between the government, grid companies, and universities. Policies from the State Council and the National Energy Administration emphasize the ability of power systems to maintain stable operation under extreme conditions. For instance, The State Grid Zhejiang Electric Power Company proposed the concept of a "multi-integrated, high-flexibility power grid" to address external environmental changes and challenges. The State Grid Shanghai Electric Power Company identified six key characteristics of resilient power grids and initiated technological advancements and innovative practices around these features.

Table 1.2 Definitions of resilience by different institutions in the power and energy sector

Country	Institution or document	Definition of resilience
The United States	Presidential policy directive 21	The resilience of critical infrastructure refers to the ability of systems to prepare in advance, adapt, absorb shocks, and recover quickly from disruptive events [5]
	National academy of engineering	Power system resilience refers to the ability of systems to identify potential risks of large-scale power outages, prepare adequately in advance, minimize the impacts of events, restore supply quickly, and incorporate lessons learned to inform future prevention and response measures [6]
	Cybersecurity and infrastructure security agency	The resilience of energy infrastructure depends not only on its physical characteristics but also on its management and organizational capabilities, including collaboration among operators, regulators, emergency responders, and stakeholders [13]
The United Kingdom	Cabinet	Resilience refers to the ability of systems to prepare for, absorb, adapt to, and rapidly recover from disruptive events [14]
	Energy research partnership	Power system resilience refers to the ability to endure and mitigate the scale and duration of disruptive natural disasters, including prediction, absorption, adaptation, and rapid recovery [7]
European Union	European commission's joint research centre	Resilience is primarily reflected in the system's ability to adapt to medium- and long-term cumulative changes (such as climate change, policy adjustments, and energy mix transitions) [15]
	North Atlantic treaty organization and European union	Resilience refers to a system's ability to predict, protect, respond to, resist, mitigate, absorb, adapt, and recover from extreme events [11]
Japan	Cabinet	Power system resilience refers to its ability to prepare for external disturbances, adapt during disturbances, and recover quickly after disturbances [16, 17]
China	State council	The resilience of critical infrastructure refers to its ability to maintain business continuity when faced with destruction or functional failure while minimizing potential adverse impacts [9]
	National energy administration	Resilience in energy systems refers to the ability to maintain stable and continuous operation when responding to climate change, extreme weather, and market shocks [10]
	State grid Zhejiang electric power company	Grid resilience refers to the ability to quickly respond, actively adjust, and restore after various disruptions [18]

(continued)

Table 1.2 (continued)

Country	Institution or document	Definition of resilience
	State grid Shanghai electric power company	Resilience refers to a power grid's ability to perceive, diagnose, and respond quickly to its operating state. It can coordinate internal and external resources, proactively deal with disruptions, and rapidly restore critical loads while improving system defense and recovery capabilities [18, 19]
	Tsinghua University	Resilience refers to the ability of the power system to adjust itself during disruptions or failures, minimizing losses caused by disturbances and quickly restoring to a normal state after failures are resolved [20]
	Xi'an Jiaotong University	Power system resilience refers to the ability to predict, resist, absorb, adapt, and quickly recover when facing small-probability but high-impact extreme events [21]
	Hohai University	Power system resilience refers to the ability to avoid collapse or disintegration under sustained disruptions, ensuring normal operation [22]

Scholars from various universities have explored power system resilience from different perspectives, particularly highlighting its importance in ensuring recovery and adaptability under extreme events to guarantee the secure and stable operation of power systems.

In the field of power system research, resilience primarily focuses on the stability of single-energy systems, such as electricity, when subjected to external shocks like natural disasters or cyberattacks. However, as global consensus deepens on energy green transitions and the carbon peak and carbon neutrality goals, the interdependence and coupling between power systems and other key energy infrastructures, such as natural gas, heating, and cooling, have become increasingly prominent. This multi-energy flow system encompasses all aspects of the power system, including generation, transmission, distribution, and consumption, shifting the research focus from power system resilience to IES resilience.

Different countries and regions have distinct understandings and priorities for the resilience of IESs, forming diverse practical approaches and theoretical explorations. The United States enhances IES resilience by promoting distributed energy resources and the development of multi-energy microgrids, improving prevention, endurance, and recovery capabilities in the face of climate change and cybersecurity threats. The European Union focuses on building modernized energy infrastructures and improving the coupling of multiple energy forms to strengthen the power system's adaptability to the impacts of climate change. Japan emphasizes the integration of distributed energy supply networks and renewable energy sources, ensuring the continuity of energy supply during emergencies. China adopts a multi-energy complementary strategy to enhance the overall resilience of the energy supply chain, improving the flexibility and recovery capability of multi-energy coupled systems under complex challenges, thus achieving sustainable energy system development.

In conclusion, the resilience of multi-energy systems has become a core issue in the context of energy transitions. The explorations and practices of different countries and regions regarding IES resilience reflect their strategic choices and policy orientations in energy transitions while offering valuable experiences and insights for the future development of global energy systems.

1.2.3 Organization of Book

Energy security is of strategic importance to national stability and development. To address extreme disaster events and ensure the security of energy supply, the establishment of "resilient energy systems" has become a national strategy for governments worldwide. Specifically, resilient IESs should possess adequate preparedness before the occurrence of extreme disasters, as well as emergency response and rapid recovery capabilities following such events. The core components for enhancing IES resilience are resilience assessment, resilient-oriented planning, and resilient-oriented restoration. These three aspects are interdependent and together form a systematic strategy for responding to sudden extreme disaster events.

Resilience assessment aims to quantify and analyze the system's capacity to resist and recover from extreme disaster events. It evaluates the system's overall performance under extreme scenarios and effectively identifies vulnerabilities and weak points. The findings provide a scientific basis for pre-disaster planning and post-disaster energy supply recovery. During the pre-disaster stage, resilient-oriented planning focuses on reinforcing critical infrastructure and optimizing emergency resource allocation. The goal is to enhance the system's disaster resistance and minimize the potential scope of failures caused by extreme events. During the post-disaster stage, resilient-oriented restoration involves the rational dispatch of mobile emergency equipment, repair teams, and other resources to gradually restore damaged infrastructure. The aim is to ensure the rapid and orderly recovery of energy supply to normal levels.

This book focuses on key foundational technologies for enhancing the resilience of IESs. It aims to serve as a reference for research and engineering applications in this field, contributing to advancements in resilience strategies and the sustainable development of modern energy systems.

This book is divided into nine chapters.

This chapter and Chap. 2 review the destructive impacts of recent global extreme events on energy systems, outline the policy context related to energy system resilience, and explore the fundamental attributes of IES resilience. These two chapters analyze the core characteristics of IESs in responding to extreme events and elaborate on the key technologies for enhancing system resilience.

Chapter 3 explores the cross-system fault propagation mechanisms in IES through dynamic simulation. First, it presents common operational models of CHP-based IES and develops a pipeline leakage model that comprehensively accounts for both thermal and hydraulic effects. Next, a simulation framework for CHP-based IES under fault conditions is proposed, enabling adaptive adjustments of simulation strategies based on different fault scenarios. Finally, the impact of system coupling configurations on fault propagation mechanisms is analyzed in detail.

Chapter 4 presents an IES resilience assessment method considering gas-thermal inertia and multi-energy interdependencies. Firstly, numerical simulation methods for multi-energy cascading faults and energy supply restoration during extreme events are introduced. Secondly, the IES multi-stage resilience curve is elaborated, and an evaluation index system for electric-gas-thermal interdependencies under extreme disasters is proposed. Lastly, the specific impacts of gas-thermal inertia and multi-energy interdependencies on system resilience are analyzed.

Chapters 5 and 6 present an accelerated resilience assessment method for electricity-gas IES considering non-convex gas flow dynamics.

Chapter 5 focuses on non-convex constraint handling. First, the original non-convex optimization model of the IES is introduced, followed by a unified reformulation of its non-convex constraints into standard second-order cone equality constraints. To efficiently solve the resulting non-convex optimal energy flow problem, a novel spatial branching strategy tailored for second-order cone equality constraints is proposed within the SB&B algorithm framework.

Chapter 6 focuses on algorithmic acceleration. A new branching strategy for conic spaces is introduced, along with a combined cutting-plane generation strategy that transforms the exhaustive fault scenario traversal into a guided single search tree generation. By structuring the solution process for each fault scenario as a subset of the search tree, this approach significantly reduces computational overhead, enabling the rapid evaluation of system resilience under multiple fault scenarios.

Chapter 7 presents a resilient-oriented IES planning method from the perspective of risk-economic coordination. First, a hierarchical fault scenario generation method is introduced to produce a representative set of fault scenarios, which can incorporate multiple crisis levels and their probabilistic characteristics. Second, a resilient planning model for IES is proposed to achieve a controllable balance between economics and risk mitigation. Finally, an enhanced Benders decomposition algorithm is introduced to improve the computational efficiency of large-scale planning models.

Chapter 8 introduces a network-based modeling and operation optimization approach for electricity hydrogen-enriched compressed natural gas IES (E-HCNG-IES) under multiple uncertainties. First, an improved HCNG network model is developed to account for the impact of hydrogen blending on gas network dynamics. Second, an optimal E-HCNG-IES dispatch model that considers the entire HCNG supply chain is proposed. Finally, a convex-concave procedure with a bound-tightening method is proposed to solve the model efficiently.

Chapter 9 introduces an IES energy supply restoration method based on multi-stage and multi-energy flow coordination. First, the multi-stage emergency response and energy restoration process for IES under extreme events is outlined. Next, a restoration model is proposed that leverages the synergistic effects of multi-stage processes and multi-energy flow coordination to enhance overall restoration efficiency. Finally, a customized progressive hedging algorithm based on scenario decoupling is introduced to efficiently solve the model.

References

1. Lv, J., Zhang, S., Cheng, H., et al. (2021). Research review on regional integrated energy system planning considering interconnection and interaction. *Proceedings of the CSEE, 41*(12), 4001–4021.
2. Van Loenhout, J., Below, R., Mcclean, D. (2020). The human cost of disasters: An overview of the last 20 Years (2000–2019). Centre for Research on the Epidemiology of Disasters (CRED) and United Nations Office for Disaster Risk Reduction (UNISDR), [2024-08-16]. https://www.preventionweb.net/files/74124_humancostofdisasters20002019reportu.pdf.
3. Wang, Y., Yang, S., Zhang, L., et al. (2022). Comparison and outlook of three global meteorological disaster databases. *Advances in Climate Change Research, 18*(2), 253–260.
4. Wan, G. J., Pinson, P., Chatzivasileiadis, S., et al. (2023). On machine learning-based techniques for future sustainable and resilient energy systems. *IEEE Transactions on Sustainable Energy*, 14(2), 1230–1243.
5. The White House, Office of the Press Secretary. (2013). Presidential Policy Directive 21: Critical Infrastructure Security and Resilience[EB/OL]. Washington, DC: The White House, Office of the Press Secretary. [2024–08–16]. https://obamawhitehouse.archives.gov/the-press-office/2013/02/12/presidential-policy-directive-critical-infrastructure-security-and-resil.
6. National Academies of Sciences, Engineering, Medicine. (2017). *Enhancing the resilience of the Nation's electricity system*. Washington, DC: The National Academies Press. [2024–08–16]. https://www.nationalacademies.org/news/2017/07/enhancing-the-resilience-of-the-nations-electricity-system.
7. The Energy Research Partnership Working Group Members. (2018). *Future resilience of the UK electricity system*. Birmingham: Energy Research Partnership. [2024–08–16]. https://erpuk.org/wp-content/uploads/2019/01/Summary_Report.pdf.
8. Office of Electricity Delivery & Energy Reliability. (2018). *Multiyear plan for energy sector cybersecurity*. Washington, DC: U. S. Department of Energy. [2024–08–16]. https://www.energy.gov/ceser/articles/doe-multiyear-plan-energy-cybersecurity.
9. State Council Decree. (2021). *Regulations on the security protection of critical information infrastructure*. Beijing: State Council. [2024–08–16]. https://www.gov.cn/gongbao/content/2021/content_5636138.htm.
10. National Development and Reform Commission. (2022). *"14th five-year plan" for modern energy system*. Beijing: National Energy Administration. https://www.gov.cn/zhengce/zhengceku/2022-03/23/5680759/files/ccc7dffca8f24880a80af12755558f4a.pdf.
11. EU-NATO Task Force. (2023). *EU-NATO task force on the resilience of critical infrastructure final assessment report*. Brussels: EU-NATO Task Force. https://commission.europa.eu/system/files/2023-06/EU-NATO_Final%20Assessment%20Report%20Digital.pdf.
12. HOLLING C S. (1973). Resilience and stability of ecological systems. *Foundations of Socio-Environmental Research*, 460–482.
13. Cybersecurity and Infrastructure Security Agency. (2024). *Infrastructure resilience planning framework*. Washington, DC: Cybersecurity and Infrastructure Security Agency. [2024–08–18]. https://www.cisa.gov/resources-tools/resources/infrastructure-resilience-planning-framework-irpf.
14. Cabinet Office. (2011). *Keeping the country running: Natural hazards and infrastructure*. London: Cabinet Office. [2024–08–18]. https://www.gov.uk/government/publications/keeping-the-country-running-natural-hazards-and-infrastructure.

15. Fulli, G. (2016). *Electricity security: Models and methods for supporting the policy decision making in the European Union.* Polytechnic University of Turin.
16. Strategic planning office of the agency for natural resources and energy, The sixth strategic energy plan. Tokyo: Japanese Cabinet. [2024–08–18]. https://www.meti.go.jp/press/2021/10/20211022005/20211022005-1.pdf.
17. Marnay, C., AkI, H., Hirose, K., et al. (2015). Japan's pivot to resilience: How two microgrids fared after the 2011 earthquake. *IEEE Power and Energy Magazine, 13*(3), 44–57.
18. Ruan, Q., Xie, W., Xu, Y., et al. (2020). Concept and key characteristics of resilient grids. *Proceedings of the CSEE, 40*(21), 6773–6784.
19. Chen, L., Deng, X., Chen, H., et al. (2022). Research review on power system resilience assessment and enhancement. *Power System Protection and Control, 50*(13), 11–22.
20. Gao, H., Chen, Y., Huang, S., et al. (2015). Research progress on distribution network resilience and related studies. *Automation of Electric Power Systems, 23*, 1–8.
21. BIe, Z., LiN, Y., Li, G., et al. (2017). Battling the extreme: A study on the power system resilience. *Proceedings of the IEEE, 105*(7), 1253–1266.
22. Ju, P., Wang, C., Xin, H., et al. (2019). Research on flexibility, resilience, and robustness of power systems. *Electric Power Automation Equipment, 39*(11), 1–7.

Chapter 2
Overview of IES Resilience

Abstract The increasing frequency and severity of extreme disaster events have exposed the limitations of traditional reliability-based approaches in ensuring the stable operation of integrated energy systems (IESs). This chapter provides a systematic overview of IES resilience from conceptual, comparative, and technological perspectives. First, the fundamental attributes and multi-stage evolution of IES resilience are introduced. Next, the similarities and differences between IES resilience and conventional power system resilience are analyzed, revealing the complementary roles of multi-energy coupling and cross-domain coordination in mitigating cascading failures. Finally, the chapter elaborates on key technologies for enhancing IES resilience.

Nomenclature

IES	Integrated energy system
SCADA	Supervisory control and data acquisition
GIS	Geographic information system
GPS	Global positioning system
CHP	Combined heat and power

2.1 Fundamental Attributes of IES Resilience

2.1.1 Core Connotations of IES Resilience

The frequent occurrence and devastating impacts of extreme disaster events highlight the inadequacies of energy system protection strategies. Traditional energy system research typically focuses on the reliability design for "high-probability, low-impact" failure events, often neglecting the challenges posed by "low-probability,

Z. Wu et al., *Integrated Energy System Resilience*,
https://doi.org/10.1007/978-981-95-2363-4_2

high-impact" extreme disaster events. Generally, such low-probability but highly destructive extreme events have the following characteristics:

1. their occurrence time and location are highly unpredictable, resulting in significant uncertainty;
2. the incidents may persist for a long time, causing irreversible damage to energy infrastructure and increasing the difficulty of system recovery;
3. if cascading faults occur, they may lead to widespread energy supply disruptions within a short time.

Given the catastrophic consequences caused by such extreme events, resilience, defined as the system's ability to effectively resist and rapidly recover after extreme disasters, has gradually attracted significant attention. According to relevant definitions of IES resilience [1, 2], resilience primarily focuses on the system's responsiveness, defensiveness, and recoverability when dealing with "low-probability, high-impact" extreme events. It emphasizes the system's ability, through the complementary integration of multiple energy forms and the dispatch of emergency resources, to respond quickly, minimize energy supply interruption time, and restore normal operation [3].

2.1.2 Comparison Between IES Resilience and Reliability

To clarify the fundamental characteristics of IES resilience, this section compares the commonalities and differences between resilience and reliability [4, 5], as summarized in Table 2.1. Both resilience and reliability serve as critical indicators for the safety and defense of IESs. Reliability focuses on the system's ability to maintain continuous energy supply when facing small-scale disturbances, whereas resilience emphasizes the resistance, adaptation, and recovery processes of the system under large-scale disturbances. The two concepts fundamentally differ in terms of their applicable scenarios, impact scope, system design, and evaluation indicators, while remaining complementary to each other.

2.1.3 IES Multi-stage Resilience Curve

The full process of an IES responding to extreme disaster events can be visually represented using a resilience curve [1, 6, 7], as shown in Fig. 2.1. The vertical axis represents the system's functional performance, typically measured as the proportion of the IES's energy supply load relative to the total load. R_0 represents the system's energy supply function under ideal operating conditions (100%). The resilience curve links the multi-stage temporal and spatial evolution of extreme disasters with the IES's corresponding response performance. The process can be divided into four stages:

Table 2.1 Comparison of reliability and resilience

Feature	Reliability	Resilience
Applicable scenarios	High-frequency and small-scale routine fault events	Low-probability and large-scale extreme disaster events
Impact scope	Primarily affects operational stability, typically causing localized supply interruptions with limited damage	Often leads to system-wide collapse or prolonged supply disruptions, resulting in large-scale socio-economic losses
System design	Emphasizes the configuration of high-quality equipment and components to ensure stable operation under normal conditions	Emphasizes multi-energy complementarity and allocation of emergency resources to improve system flexibility and self-recovery during emergencies
Target loads	Covers all loads	Under emergency conditions, prioritizes critical loads
Cost consideration	Focuses on routine system maintenance, inspection costs, fault repair, and system reliability costs	Includes redundancy design costs, pre-disaster emergency resource deployment costs, and post-disaster recovery costs
Evaluation metrics	Energy supply interruption time, interruption frequency, equipment failure rate, reliability indices	Fault propagation speed, lost load, supply recovery time, system redundancy, and adaptability

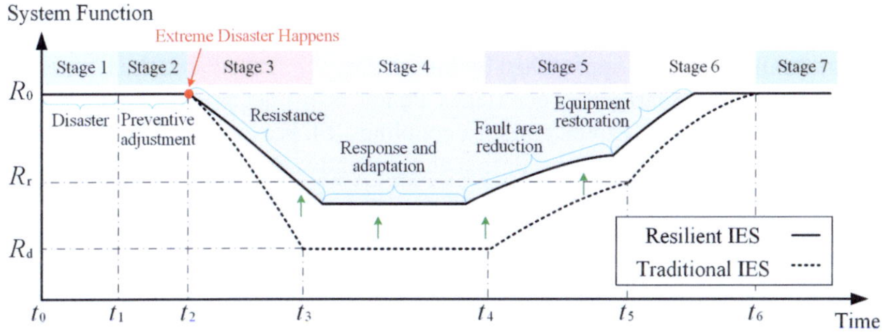

Fig. 2.1 Schematic diagram of the resilience curve

1. ***Pre-disaster Stage*** t_0–t_2: Before the disaster happens, the system's functional performance remains at R_0. By thoroughly analyzing historical disaster data, constructing predictive models for potential disaster events, and implementing disaster prevention designs, the system can be adjusted to a preventive emergency state, significantly enhancing its ability to resist sudden extreme disasters.
2. ***System Degradation Stage*** $t_2 - t_3$: After the disaster occurs, the system's energy supply function drops sharply to R_d due to damage to multiple equipment components, making normal energy supply unsustainable. At this point, emergency

isolation of faulty components can reduce fault propagation paths, while coordinated dispatch of multi-energy flow emergency resources helps maintain energy supply to critical loads.

3. ***Adaptation Stage*** $t_3 - t_4$: After a period of emergency adjustments, system performance stabilizes at R_d. During this stage, system operators can use tools such as the Supervisory Control and Data Acquisition (SCADA) system, the Geographic Information System (GIS), and the Global Positioning System (GPS) [6, 8] to comprehensively collect and analyze fault information. At the same time, by pre-deploying various types of emergency resources, corresponding emergency response and energy restoration strategies can be formulated to ensure orderly energy recovery in subsequent stages.
4. ***Restoration Stage*** t_4–t_6: Through precise fault localization and coordinated dispatch of diverse emergency resources, energy supply to non-fault areas can be restored [9, 10], gradually increasing the system's functional performance to R_r. Subsequently, maintenance personnel begin repairing damaged equipment and components, allowing the IES to progressively return to its pre-disaster operational state.

2.1.4 Key Fundamental Characteristics of IES Resilience

As shown in Fig. 2.1, the resilience level of an IES is closely related to system performance at each stage. The shaded area under the curve represents the overall resilience level of the system in responding to extreme disaster events. Enhancing system resilience can be approached by minimizing the shaded area. It is important to note that the boundaries between stages are not as distinct as depicted in Fig. 2.1, as different stages often exhibit complex couplings [6]. For example, during the pre-disaster stage, disaster prevention design and preventive emergency adjustments can be conducted simultaneously; during the post-disaster stage, emergency resource dispatch and equipment repair can be coordinated in a rolling manner. The resilience curve intuitively illustrates the fundamental characteristics of IES resilience across different stages, including predictability, resistance, adaptability, and recoverability [1, 11, 12].

1. ***Predictability***: Extreme disaster events are often sudden and difficult to predict accurately. Predictability refers to the system's ability to forecast, learn, and identify potential risks of disaster events. By enhancing the panoramic perception and disaster early-warning capabilities of IES, analyzing the interaction mechanisms between various disasters and the system, and accurately identifying risks across generation, grid, load, and storage components, the system's disaster predictability can be significantly improved [13, 14].
2. ***Resistance***: Resistance refers to the system's ability to buffer the impact of disaster events and mitigate the extent of damage, primarily determined by the robustness of infrastructure and the availability of emergency resources such

as distributed energy, multi-type energy storage, and backup energy. High resistance ensures continuous energy supply to critical loads during disasters, avoiding system collapse. By strengthening the protection of key infrastructure, such as power lines, natural gas pipelines, and heating pipelines, the system's resistance to extreme disaster events can be enhanced [15, 16].

3. ***Adaptability***: Adaptability refers to the system's ability to adjust multi-energy flow operations and optimize emergency resource allocation in response to equipment damage, enabling a gradual transition to a new equilibrium state after a disaster. This involves modifying load supply paths, dynamically dispatching distributed energy, and implementing demand response to maintain a stable energy supply for critical loads under prolonged supply interruptions and equipment failures [17, 18].
4. ***Recoverability***: Recoverability refers to the system's ability to repair damaged equipment quickly and restore normal energy supply levels after a disaster. Recoverability involves not only infrastructure repairs but also the rational prioritization of energy restoration in fault-affected areas. It ensures the system is smoothly restored to its pre-disaster state based on real-time system changes [19, 20].

The above content primarily elaborates on the characteristics of IES resilience. At the broad level, IES resilience not only encompasses the ability to respond to various types of extreme disaster events but also extends to the system's adaptability and sustainability in the face of long-term and systemic changes [1, 4]. These changes include global climate change, international geopolitical shifts, and the increasing complexity of energy infrastructure. Broad-level resilience emphasizes the system's ability to maintain stability and functionality in dynamic environments, ensuring long-term safe, reliable, and sustainable operation under complex challenges [21].

2.2 Similarities and Differences Between IES Resilience and Power System Resilience

2.2.1 Comparison Between IES Resilience and Power System Resilience

IESs encompass multiple energy networks, including electricity, heating, and gas, as well as coupling equipment. Based on their geographical distribution, IESs can be broadly categorized into regional, urban, and cross-regional levels [22]. This section compares the key characteristics of IES resilience and power system resilience, analyzing their differences and commonalities in responding to extreme events. This comparison provides a reference for developing technologies to enhance IES resilience.

Under extreme events, the cascading faults and energy supply interruptions in IESs are complex and diverse. Due to multiple coupling links between energy systems at various levels, a single point of failure in one area can easily trigger a "domino effect," propagating across different layers of energy systems and forming a series of cascading faults, as shown in Fig. 2.2. This illustrates the potential negative effects of multi-energy coupling during extreme events.

However, from a positive perspective, multi-energy coupling also plays a significant beneficial role during extreme disasters. Compared to power systems, IESs can leverage the horizontal complementarity of multiple energy forms (electricity, gas, and heat) and the vertical coordination among generation, transmission, load, and storage to establish a multi-layer collaborative joint prevention and control mechanism [23, 24]. During disasters, such multi-layer coordination facilitates mutual support among systems, reducing the risk of cascading faults across different subsystems. After disasters occur, the formulation of coordinated multi-energy emergency resource allocation strategies can significantly improve restoration efficiency.

Therefore, the development of technologies to enhance IES resilience requires effective mitigation of cascading fault risks while fully leveraging the complementarity and mutual support advantages of various energy forms.

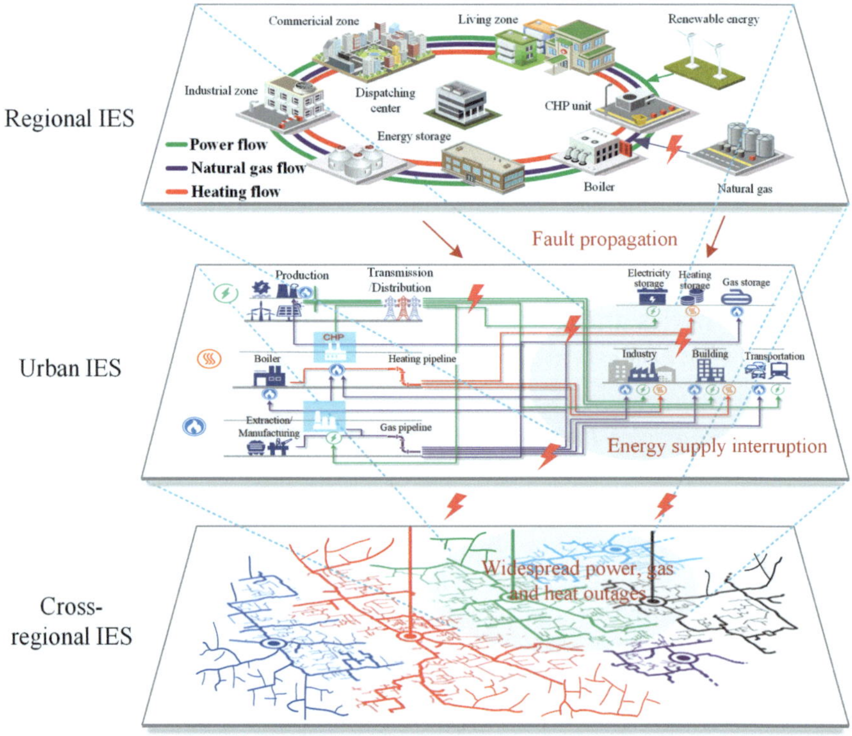

Fig. 2.2 Schematic diagram of IES energy supply interruption under extreme events

2.2.2 Advantages of IESs in Resilience Enhancement

Table 2.2 provides a detailed comparison of the key characteristics of resilience between power systems and IESs. IESs offer three major advantages in resilience enhancement [16, 25]:

1. The system integrates CHP units, gas turbines, electric heat pumps, and other coupling devices, enabling flexible energy conversion among different energy flows. This diversity in energy supply methods enhances system flexibility. Under extreme disaster events, when a single energy system fails, compensation can be provided through other energy subsystems [20].
2. The system incorporates a variety of centralized and distributed energy storage resources, including electric energy storage, thermal energy storage, and natural gas storage [26, 27]. Compared to electric energy storage, gas storage tanks and thermal storage tanks provide more sustained energy supply capabilities during disasters. This ensures that IESs exhibit stronger resilience in maintaining continuous energy supply during prolonged power outages or gas supply interruptions.

Table 2.2 Comparison of IES resilience and power system resilience

	Power system resilience	IES resilience
Energy types	Single energy flow form	Multiple energy flow forms, including electricity, natural gas, and heating
Fault propagation mechanism	Faults can propagate through generation, transmission, distribution, and user-side networks	Fault propagation paths are more complex; faults can propagate across systems and regions
Source side	Under fault scenarios, power supply comes from generators, distributed energy, and mobile power sources	Under fault scenarios, energy supply sources are more diversified, and multi-energy complementarity improves energy supply flexibility
Network side	Primarily causes damage to power lines and transmission towers	Simultaneous damage can occur to power lines, natural gas pipelines, and heating networks
Energy storage side	Power storage provides fast responses but has limited duration	Includes various types of emergency storage, such as power storage, gas storage, and thermal storage. Natural gas pipelines and heating pipelines provide effective storage capacities
Demand side	Users have low tolerance for power outages, with significant impacts from interruptions	Multi-energy demand response and thermal load inertia enhance energy supply flexibility during disasters

3. The slow dynamic characteristics of natural gas and heating systems impart equivalent storage capacities to gas pipelines and hot water pipelines [28]. When gas sources or heat sources fail, the stored energy in pipelines can continue to supply energy for a certain period. On the demand side, thermal loads possess thermal inertia characteristics, unlike the instantaneous supply–demand balance required by power systems. Following a heating supply interruption, there is a delay before indoor temperatures drop from a comfortable range to an uncomfortable level, providing a critical buffer period for system response to extreme disasters [29, 30]. Additionally, implementing multi-energy demand response [31, 32] can further enhance the energy supply flexibility of multi-energy coupled systems under extreme disasters, thereby improving the overall resilience of IESs.

2.3 Key Technologies for Enhancing IES Resilience

The technologies for enhancing the resilience of IESs aim to strengthen their prevention, resistance, response, and recovery capabilities under increasingly frequent extreme disturbances. This section provides an in-depth analysis and explanation of the core technologies for improving IES resilience from three dimensions: resilience assessment, resilience planning, and energy supply restoration.

2.3.1 IES Resilience Assessment

Resilience assessment and weak-point identification methods form the foundation of IES resilience enhancement technologies, providing critical references for resilience planning and energy supply restoration studies. The dynamic and static processes of electricity, gas, and thermal systems influence one another and may trigger cascading faults. Meanwhile, the multi-energy complementarity of IESs, along with the slow dynamic characteristics of natural gas and heating systems, adds operational flexibility under extreme conditions. To conduct resilience assessment, it is essential to start from the interdependency among electricity, gas, and thermal systems under extreme fault scenarios. By analyzing the multi-stage dynamic behavior of cascading fault propagation and energy supply restoration, the fault propagation mechanisms can be thoroughly examined, and weak links caused by multi-energy coupling relationships can be identified. Establishing a comprehensive and precise assessment index system allows exploration of effective pathways to improve IES resilience, providing critical decision-making support for resilience assessment and planning.

2.3.2 IES Resilient-Oriented Planning

The core objective of resilient-oriented planning is to enhance the ability of IESs to resist extreme disasters by strengthening infrastructure robustness and optimizing emergency resource allocation. Specific measures include improving the structural stability of critical outdoor equipment and components, increasing redundancy in energy supply paths, and rationally deploying emergency resources. Given the low-probability but high-impact nature of extreme disaster events, resilient planning often involves significant investment costs. Therefore, disaster prevention planning must account for the diminishing marginal returns of resilience measures, evaluating their actual contributions to improving IES resilience while avoiding excessive costs caused by over-redundancy. At the same time, resilient-oriented planning must consider the uncertainty of disaster scenarios, ensuring that supply interruption risks are reasonably controlled across different crisis levels. This helps achieve an optimal balance between planning investments and risk control under various extreme conditions.

From a broader perspective, resilience planning for IESs must address multi-dimensional disturbances over extended planning periods. Under climate change trends, large-scale renewable energy integration and multi-energy loads are strongly affected by multi-time scale uncertainties. These include the long-term impacts of extreme weather conditions on temperature-sensitive loads and short-term source-load fluctuations, which pose significant challenges to energy supply security. To ensure long-term energy supply security, resilience planning models must analyze operational risks across different time granularities during the planning period and develop optimal strategies to address multiple disturbances. This enhances the system's adaptability to a variety of extreme scenarios. Furthermore, the multi-time scale uncertainties increase the complexity of modeling and the difficulty of solving the models, highlighting the urgent need for high-performance algorithms that balance computational efficiency and solution accuracy.

2.3.3 IES Energy Supply Restoration

The post-disaster energy supply restoration of IESs consists of several stages, including fault localization, fault isolation, and energy supply restoration. To achieve an orderly and efficient recovery of multi-energy loads, it is essential to design detailed scheduling strategies based on a multi-stage process. In the multi-stage recovery process, the stages are interdependent, as the system state at an earlier stage directly influences decision-making in subsequent stages. Therefore, the development of energy supply restoration strategies requires organic coordination across these stages. Moreover, the coupling relationships among electricity, gas, and thermal energy flows further increase the complexity of the restoration process. In summary, to establish an orderly and efficient energy supply restoration strategy for IESs, it is necessary to fully consider the operational characteristics of multi-energy flow systems at different stages, such as network reconfiguration and coupled equipment scheduling. The priority of energy supply restoration for multi-energy loads must be determined, and the interactions among various energy flows across stages must be coordinated. This ensures the collaborative operation of multi-energy emergency resources, thereby maximizing restoration efficiency.

References

1. Huang, H., Vincent, P. H., Davis, K. R., et al. (2024). Toward resilient modern power systems: From single-domain to cross-domain resilience enhancement. *Proceedings of the IEEE, 112*(4), 365–398.
2. Jasiūnas, J., Lund, P. D., & Mikkola, J. (2021). Energy system resilience—A review. *Renewable and Sustainable Energy Reviews, 150*, Article 111476.
3. Lin, Y., & Bie, Z. (2016). Study on the resilience of the integrated energy system. *Energy Procedia, 103*, 171–176.
4. Ahmadi, S., Saboohi, Y., & Vakili, A. (2021). Frameworks, quantitative indicators, characters, and modeling approaches to analysis of energy system resilience: A review. *Renewable and Sustainable Energy Reviews, 144*, Article 111476.
5. Pahadia P., Jain P., Harit A., et al. (2023). Resilience metrics for integrated energy systems: A review. In *Proceedings of the 5th Biennial International Conference on Nascent Technologies in Engineering (ICNTE)* (pp. 1–6), Navimumbai, India.
6. Zhang G. (2021). *Research on resilient distribution network planning and restoration for natural disasters.* Shandong University, Ph.D. Dissertation, Shandong.
7. Liu J. (2021). *Research on resilience enhancement methods for distribution systems with multi-resource coordination under extreme events.* Tianjin University, Ph.D. Dissertation, Tianjin.
8. Chen, B., Chen, C., Wang, J., et al. (2018). Sequential service restoration for unbalanced distribution systems and microgrids. *IEEE Transactions on Power Systems, 33*(2), 1507–1520.
9. Li, B. D., Chen, Y., Wei, W., et al. (2022). Online coordination of LNG tube trailer dispatch and resilience restoration of integrated power-gas distribution systems. *IEEE Transactions on Smart Grid, 13*(3), 1938–1951.
10. Gazijahani, F. S., Salehi, J., & Shafie-Khah, M. (2023). A parallel fast-track service restoration strategy relying on sectionalized interdependent power-gas distribution systems. *IEEE Transactions on Industrial Informatics, 19*(3), 2273–2283.
11. Wang, J., Pinson, P., Chatzivasileiadis, S., et al. (2023). On machine learning-based techniques for future sustainable and resilient energy systems. *IEEE Transactions on Sustainable Energy, 14*(2), 1230–1243.
12. Xie, H. P., Sun, X. T., Chen, C., et al. (2022). Resilience metrics for integrated power and natural gas systems. *IEEE Transactions on Smart Grid, 13*(3), 2483–2486.
13. Wang, C., Hou, Y. H., Qiu, F., et al. (2017). Resilience enhancement with sequentially proactive operation strategies. *IEEE Transactions on Power Systems, 32*(4), 2847–2857.
14. Yu, F., Guo, Q., Wu, J., et al. (2024). Early warning and proactive control strategies for power blackouts caused by gas network malfunctions. *Nature Communications, 15*(1), 4714.
15. Zhou, Y., Li, X., Han, H., et al. (2024). Resilience-oriented planning of integrated electricity and heat systems: A stochastic distributionally robust optimization approach. *Applied Energy, 353*, Article 122053.
16. Li, T., Han, X., Wu, W., et al. (2023). Robust expansion planning and hardening strategy of meshed multi-energy distribution networks for resilience enhancement. *Applied Energy, 341*, Article 121066.

17. Li, B. D., Chen, Y., Wei, W., et al. (2022). Enhancing resilience of emergency heat and power supply via deployment of LNG tube trailers: A mean-risk optimization approach. *Applied Energy, 318*, Article 119204.
18. Amirioun, M. H., Aminifar, F., & Shahidehpour, M. (2019). Resilience-promoting proactive scheduling against hurricanes in multiple energy carrier microgrids. *IEEE Transactions on Power Systems, 34*(3), 2160–2168.
19. Wang, Y., Qiu, D., Sun, X., et al. (2024). Coordinating multi-energy microgrids for integrated energy system resilience: A multi-task learning approach. *IEEE Transactions on Sustainable Energy, 15*(2), 920–937.
20. Li, X., Du, X. X., Jiang, T., et al. (2022). Coordinating multi-energy to improve urban integrated energy system resilience against extreme weather events. *Applied Energy, 309*, Article 118455.
21. Cheng, Y., Liao, H., & Elsayed, E. A. (2024). From reliability to resilience: More than just taking one step further. *IEEE Transactions on Reliability, 73*(1), 42–46.
22. Yu, X. D., Xu, X. D., Chen, S. Y., et al. (2016). Brief overview of integrated energy systems and energy internet. *Journal of Electrical Engineering, 31*(1), 1–13.
23. Jing R., Wang X., Zhao Y., et al. (2021). Planning urban energy systems adapting to extreme weather. *Advances in Applied Energy,* 100053.
24. Saravi, V. S., Kalantar, M., & Anvari-Moghaddam, A. (2022). Resilience-constrained expansion planning of integrated power-gas-heat distribution networks. *Applied Energy, 323*, Article 119315.
25. Huang, W. J., Zhang, X., Li, K. P., et al. (2022). Resilience-oriented planning of urban multi-energy systems with generalized energy storage sources. *IEEE Transactions on Power Systems, 37*(4), 2906–2918.
26. Yan, M. Y., He, Y. B., Shahidehpour, M., et al. (2019). Coordinated regional-district operation of integrated energy systems for resilience enhancement in natural disasters. *IEEE Transactions on Smart Grid, 10*(5), 4881–4892.
27. Xuan, A., Shen, X. W., Guo, Q. L., et al. (2021). A conditional value-at-risk based planning model for integrated energy systems with energy storage and renewables. *Applied Energy, 294*, Article 116971.
28. Sun, W., Wang, Q., Ye, Y., et al. (2022). Unified modelling of gas and thermal inertia for integrated energy systems and its application to multitype reserve procurement. *Applied Energy, 305*, Article 117963.
29. Li, Z., Xu, Y., Wang, P., et al. (2024). Restoration of a multi-energy distribution system with joint district network reconfiguration via distributed stochastic programming. *IEEE Transactions on Smart Grid, 15*(3), 2667–2680.
30. Lu, S., Gu, W., Meng, K., et al. (2020). Thermal inertial aggregation model for integrated energy systems. *IEEE Transactions on Power Systems, 35*(3), 2374–2387.
31. Yin, X., Ye, C., Ding, Y., et al. (2023). Exploiting internet data centers as energy prosumers in integrated electricity-heat systems. *IEEE Transactions on Smart Grid, 14*(1), 167–182.
32. Zhang, D., Zhu, H., Zhang, H., et al. (2022). Multi-objective optimization for smart integrated energy systems considering demand responses and dynamic prices. *IEEE Transactions on Smart Grid, 13*(2), 1100–1112.

Chapter 3
Cross-System Fault Propagation Analysis in Combined Heat and Power IES Based on Dynamic Simulation

Abstract Combined heating and power integrated energy system (IES) have gained increasing attention due to their high energy efficiency. However, the stronger coupling between subsystems also introduces the risk of cascading failures, which poses a significant challenge to their widespread application. To address this issue, this chapter employs dynamic simulation to investigate the fault propagation mechanism between heating and power systems, using pipeline leakage as a case study. Insights are provided from a simulation perspective to enhance the resilience of IES. First, this chapter introduces the equipment models, network operating conditions, and pipeline leakage models that account for both hydraulic and thermal effects under normal operating conditions. Next, a dynamic simulation framework for fault conditions in IES is proposed. This framework adaptively adjusts simulation strategies based on different fault scenarios and is applicable to common fault analyses in such systems. Finally, a case study of a 22-node IES is conducted to compare the impact of pipeline leakage on strongly coupled IES (SC-IES) and weakly coupled IES (WC-IES). Simulation results indicate that in SC-IES, the effects of pipeline leakage can propagate bidirectionally between the heating and power systems via combined heat and power (CHP) units, underscoring the critical role of coupling equipment conditions in enhancing system fault tolerance.

Keywords Combined heat and power system · Cross-system fault propagation · Pipeline leakage · Dynamic simulation · Fault tolerance

Nomenclature

h	Heating power, W
c	Specific heat capacity, J/(kg °C)
m	Mass flow rate, kg/s
T	Temperature, K
v	Flow rate, m/s
R	Thermal resistance, (m K)/W

Z. Wu et al., *Integrated Energy System Resilience*,
https://doi.org/10.1007/978-981-95-2363-4_3

$\mathbf{A}_f$	Node-pipe relationship matrix
G	Volume flow rate, m^3/h
ΔQ	Net outflow at nodes, m^3/h
S	Resistance characteristic factor, $Pa/(m^3/h)^2$
Z	Available pressure at each substation, mh_2o
DH	Head pressure provided by circulation pump, mh_2o
$\mathbf{B}_f$	Basic loop matrix
H	Head pressure difference, Pa
A	Area, m^2
P	Active power, W
Q	Reactive power, Var
V	Voltage, V
θ	Phase angle, rad
k	Power coefficient in CHP
τ	Time interval, s
r	Equivalence radius
IES	Integrated energy system
SC-IES	Strongly coupled IES
WC-IES	Weakly coupled IES
DHS	District heating system
PDE	Partial differential equation
FDM	Finite difference method
CHP	Combined heat and power
PV	Photovoltaic
OPF	Optimal power flow

3.1 Introduction

Against the backdrop of the global energy transition, improving the efficiency of sustainable energy utilization, enhancing the flexibility of energy systems, and reducing carbon emissions have become core objectives of modern society [1–3]. Integrated energy systems (IES) achieve high-efficiency cascading energy utilization by integrating multiple single-carrier energy systems, such as thermal, electrical, and gas energy, thereby leveraging the synergies and complementarities between heterogeneous energy flows. This multi-energy complementary characteristic enables IES to play a crucial role in improving energy utilization efficiency and optimizing energy structures, garnering significant attention from both academia and industry [4]. Among various types of IESs, combined heating and power IESs have attracted particular interest due to their ability to simultaneously meet electricity and heating demands and have been gradually deployed in the energy infrastructures of multiple countries [5]. By tightly coupling district heating system (DHS) with power systems, combined heating and power IESs provide greater flexibility for the efficient utilization of renewable energy. However, this high degree of coupling also introduces significant operational security challenges [6, 7]. In particular, potential faults in the system may not only disrupt individual subsystems but also propagate throughout the entire system via coupling mechanisms, potentially triggering cascading failures that severely threaten system stability and security [8]. For combined heating and power IES operations, frequent and widely propagating faults pose a major obstacle to further development. These faults often lead to severe system parameter fluctuations and, in extreme cases, large-scale energy supply interruptions. For example, in a thermal power plant in Shanxi, China, a leakage fault in the DHS resulted in the disconnection of the thermal power plant from the power system, ultimately leading to a localized blackout. This typical case highlights the profound impact that local faults in DHSs can have on the secure operation of combined heating and power IES. Addressing these challenges necessitates investigating the impact of faults on combined heating and power IES operations to ensure their security. However, given the complexity and variety of faults in combined heating and power IESs, it is impractical to analyze the impact of each fault individually. Instead, selecting representative fault types for in-depth analysis offers a more feasible approach. Among various possible faults, pipeline leakage in DHSs has been identified as a key focus of research due to the following characteristics [9–11]:

(1) ***Extensive impact***: Pipeline leakage disrupts multiple subsystems within combined heating and power IESs, including hydraulic, thermal, and electrical systems, as well as coupling devices [9]. Therefore, analytical frameworks and algorithms developed for pipeline leakage can be extended to other types of faults.
(2) ***High destructiveness***: Pipeline leakage is one of the most common and destructive faults in DHSs [10]. Since leakage is typically a gradual process, it may have already caused prolonged adverse effects on system operations before detected.
(3) ***Propagation mechanisms***: Although specific models may vary across different fault types, their propagation mechanisms within combined heating and power IESs share similarities [11]. Studying the propagation mechanism of pipeline leakage can provide valuable insights for preventing and mitigating other types of faults.

This chapter focuses on comprehensively analyzing the impact of pipeline leakage on the operation of combined heating and power IES. Such analysis not only lays the foundation for a deeper understanding of fault propagation mechanisms but also provides critical insights for enhancing the operational security of combined heating and power IESs. Simulation is an essential tool for studying the operational conditions of combined heating and power IESs. Based on whether thermodynamic dynamics are considered, simulation methods can be classified into steady-state and dynamic simulations. Steady-state simulations typically rely on simplified algebraic equations to describe the system's long-term operating conditions, whereas dynamic simulations employ partial differential equations (PDEs) to provide a more detailed representation of temperature variations and parameter changes during transient processes, offering superior accuracy. Given its ability to better capture system behavior under fault conditions, this study employs dynamic simulation to analyze the operational state of combined heating and power IESs under pipeline leakage faults. Various algorithms have been developed to solve the PDEs in dynamic simulation models. For instance, Qin et al. [12] employs the finite difference method (FDM) to transform PDEs into algebraic equations and validates its superiority in terms of accuracy and convergence through numerical calculations. Subsequently, Yao et al. [13] proposes an improved finite difference scheme that achieves a better balance between stability, convergence, and computational efficiency. Further, Yao et al. [14] develops a decomposition-iteration algorithm based on a discrete model to solve complex combined heating and power IES models. Additionally, Yu [15] introduces a semi-discrete scheme that effectively mitigates the dissipation and dispersion errors inherent in the finite difference method, while [16] enhances computational efficiency through a matrix-form heat power flow model.

With the widespread adoption of IESs, increasing research attention has been directed toward their coupled operational security. For instance, Lu et al. [17] proposed a thermodynamics-based static voltage stability margin model for power systems, revealing the impact of DHSs on power system voltage security. Zhang et al. [18] analyzed the steady-state security region of combined heating and power IES to mitigate the risk of cascading failures. Pan et al. [19] alleviated power system congestion through a quasi-dynamic interaction method and validated its feasibility in a real-world system. Meanwhile, Zhao et al. [20] introduced a reliability assessment method based on a fault occurrence matrix, quantifying the impact of equipment failures on system operation. As a critical issue in the field of pipeline safety [21–23], pipeline leakage faults have become a research hotspot, and significant progress has been made in modeling and detection. For example, Zhou et al. [24] developed a pipeline leakage model and proposed a simulation method to assess its impact on the hydraulic characteristics of regional DHSs. Zheng et al. [25] further analyzed the hydraulic model of pipeline leakage and employed the particle swarm optimization algorithm to achieve precise localization of leakage points. In addition, data-driven approaches have been extensively studied. For instance, Quinones-Grueiro et al. [26] proposed a pipeline leakage detection method based on historical data, while Ying et al. [27] integrated neural networks and fuzzy algorithms to identify leakage states.

Despite extensive research on the simulation, security analysis, and modeling of pipeline leakage in combined heating and power IESs, several gaps remain in the literature: (1) Existing studies on IES security primarily focus on operational scheduling, with limited research on how faults affect system operation. (2) While dynamic simulation algorithms have been widely studied under normal operating conditions, the additional complexities of energy flow calculations under fault conditions remain underexplored. (3) Current research on pipeline leakage mainly addresses its hydraulic impact while overlooking its effects on DHSs. Given the strong coupling between subsystems in combined heating and power IESs, investigating the thermal impact of pipeline leakage and developing corresponding models is of great significance.

To address these challenges, this chapter proposes a method for accurately analyzing the operational conditions of combined heating and power IESs under fault scenarios. The key contributions include: (1) establishing a detailed pipeline leakage model that comprehensively accounts for leakage diffusion and its effects on heating temperature and heating power; (2) proposing a generalized dynamic simulation framework capable of accommodating different fault stages and applicable to various common fault types; (3) conducting an in-depth analysis of fault propagation mechanisms in WC-IESs and SC-IESs, providing practical insights for enhancing the operational security of IESs.

The remainder of this chapter is structured as follows: Sect. 3.1 introduces the dynamic IES simulation model, Sect. 3.2 presents the cross-system fault propagation mechanism analysis method, Sect. 3.3 conducts case studies to investigate the propagation mechanisms of pipeline leakage under different coupling scenarios in combined heating and power IES, and Sect. 3.4 concludes the chapter.

3.2 Cross-System Fault Propagation Simulation Model for IES

This section provides a comprehensive overview of the simulation model employed in this study, laying the foundation for subsequent analysis. Specifically, the detailed modeling of each subsystem in the combined heating and power IES is first introduced, including the DHS, hydraulic system, electrical system, and coupling device models. These subsystem models aim to accurately characterize the operational characteristics of the system, as well as the internal energy flow and conversion processes. Furthermore, to thoroughly analyze the impact of pipeline leakage faults on system operation, a precise pipeline leakage model is developed. This model not only accounts for the disruption of hydraulic characteristics caused by leakage but also further examines its effects on thermal characteristics, thereby providing a comprehensive representation of the system's dynamic behavior under fault conditions.

3.2.1 Combined Heating and Power IES Model

District Heating System: The DHS model is a critical component of the simulation of the combined heating and power IES, with its core focus on the modeling and analysis of fundamental DHS units, including heat sources, pipelines, and heat loads. In DHSs, pipelines are generally categorized into supply pipelines and return pipelines, which are typically designed symmetrically in practical engineering applications to simplify design and optimize operation. Additionally, DHS operation can be regulated using two primary modes: temperature-based regulation and flow-based regulation, both of which significantly impact the operational characteristics within the system.

In temperature-based regulation, the flow distribution in pipelines remains constant, and heat supply adjustments are achieved by varying the supply water temperature at the heat source. This mode is characterized by simple control and rapid response; however, due to the high-temperature water transport, substantial heat loss occurs, resulting in relatively low energy efficiency. In contrast, flow-based regulation balances heating power by adjusting the mass flow rate in each pipeline while maintaining a constant supply water temperature. Although this mode reduces heat losses, it involves a more complex control strategy and requires higher regulation capability of equipment. In practical engineering applications, most DHSs adopt temperature-based regulation to simplify control and meet real-time operational requirements. Therefore, this chapter employs temperature-based regulation in modeling and simulation to closely align with real-world application scenarios.

To accurately describe the dynamic behavior of the DHS, this chapter adopts a dynamic thermal energy flow calculation model to analyze the system's response to special events, such as pipeline leakage faults. This model captures the temporal and spatial variations in pipeline and node temperatures, providing theoretical support for studying system behavior under fault conditions. To reduce computational complexity while maintaining accuracy, the following assumptions are made in the model development:

(1) The hot water flowing through a given pipeline has uniform inner diameter and material properties.
(2) The influence of water density variations with temperature is neglected.

These assumptions strike a balance between computational accuracy and model complexity reduction. Based on the above assumptions, the fundamental state variables of the DHS include: temperature distribution across nodes and pipelines over time and space, mass flow rate of each pipeline, and water pressure distribution. These variables collectively determine the dynamic characteristics of the DHS. Accordingly, the proposed DHS model consists of three main components:

Nodal Power Balance Equation: Treating the supply and return water networks as an integrated system, the power balance at any node k in the DHS at any given time satisfies the following equilibrium relationship:

$$\sum\nolimits_{j}\left[m_j \cdot \left(T_j^{\text{in}} - T_a\right)\right] + \frac{h_{n,k}}{c} = \left(T_{n,k} - T_a\right) \cdot \sum_{j} m_j, \ \left(j \in E^{\text{in}}\right) \tag{3.1}$$

where $m_{b,j}$ represents the mass flow rate in pipeline j. $T_{b,j}$ denotes the hot water temperature in pipeline j. T_a is the ambient temperature. Superscripts out and in denote outflowing and inflowing physical quantities, respectively. $h_{n,k}$ represents the heat power exchange between the DHS and heat exchanger when node k acts as either a heat source or a heat load (positive when injected into the node, negative when extracted). c is the specific heat capacity of water. $T_{n,k}$ denotes the relative temperature between the outgoing hot water at node k and the ambient temperature. $E_{n,k}^{\text{in}}$ denotes the set of all pipelines flowing into node k.

Additionally, for any node k, the temperature at the upstream end of all outflowing pipelines is equal to the node temperature:

$$T_{b,j}^{\text{out}} - T_a = T_{n,k} - T_a, \quad \left(j \in E_{n,k}^{\text{out}}\right) \tag{3.2}$$

Pipeline Heat Transfer Characteristic Equation: The dynamic heat transfer process in any pipeline of the DHS follows the heat transfer characteristic equation, which in partial differential form is given by:

$$\frac{\partial T}{\partial t} + v\frac{\partial T}{\partial x} + \frac{v}{m_{\text{b}}cR}\left(T - T^{\text{a}}\right) = 0 \tag{3.3}$$

where R is the thermal resistance of the pipeline.

Load Model: Compared to constant power heat loads, the thermal load nodes modeled via water-to-water heat exchangers ensure that the heat power received on the secondary side remains constant. Under constant input temperature, the heat load exhibits constant power consumption. However, when the input temperature fluctuates, the heat load power correspondingly adjusts, making it more representative of real-world operating conditions and enhancing simulation accuracy. Therefore, this study employs a water-to-water heat exchanger as the thermal load model.

In this model, the two fluid flows on either side of the heat exchanger are treated as two thermal masses. The metal plates of the heat exchanger are considered as an additional thermal mass. The following assumptions are made: (1) Uniform flow distribution within the heat exchanger. (2) Constant heat capacity of the metal plates at different temperatures. (3) Negligible heat loss from the heat exchanger to the external environment. (4) Lumped temperature modeling, using the outlet temperatures on both sides of the heat exchanger. A simplified schematic of the heat exchanger model is illustrated in Fig. 3.1.

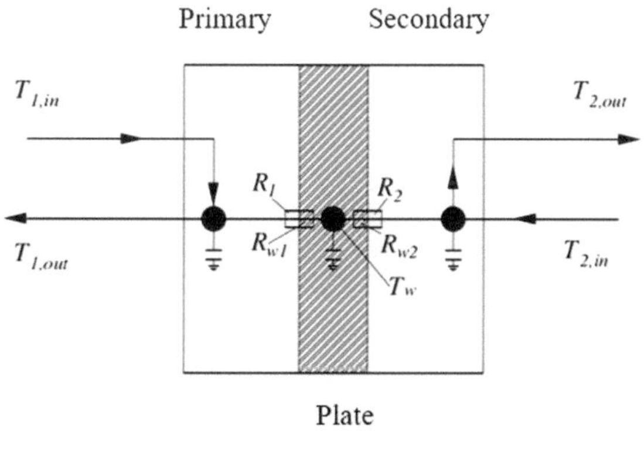

Fig. 3.1 Simplified schematic diagram of a water-water heat exchanger

For a water-water heat exchanger, expressing the hot water temperature in terms of the outlet temperature, the heat balance relationships for the primary side, secondary side, and metal plate elements can be derived based on the principle of energy conservation as follows:

$$\begin{cases} M_1 c_1 \frac{\mathrm{d}T_{1,\mathrm{out}}}{\mathrm{d}t} = m_1 c_1 \left(T_{1,\,\mathrm{in}} - T_{1,\,\mathrm{out}}\right) - k_1 A (T_{m1} - T_w) \\ M_2 c_2 \frac{\mathrm{d}T_{2,\mathrm{out}}}{\mathrm{d}t} = k_2 A (T_w - T_{m2}) - m_2 c_2 \left(T_{2,\,\mathrm{out}} - T_{2,\mathrm{in}}\right) \\ M_w c_w \frac{\mathrm{d}T_w}{\mathrm{d}t} = k_1 A (T_{m1} - T_w) + k_2 A (T_{m2} - T_w) \end{cases} \tag{3.4}$$

where M_1 denotes the total mass of the primary-side hot water in the heat exchanger. c_1 represents the specific heat capacity of the primary-side water. $T_{1,\mathrm{out}}$ is the outlet temperature of the primary-side hot water. m_1 denotes the mass flow rate of the primary-side hot water. $T_{1,\mathrm{in}}$ is the inlet temperature of the primary-side hot water. k_1 represents the heat transfer coefficient of the primary side. A is the total heat transfer area of the heat exchanger. The mean temperature of the primary-side fluid in the water-water heat exchanger is given by $T_{m1} = \frac{T_{1,in} + T_{1,\mathrm{out}}}{2}$. T_w represents the temperature of the metal plates in the plate heat exchanger. Similarly, for the secondary side: M_2 denotes the total mass of the secondary-side hot water in the heat exchanger. c_2 represents the specific heat capacity of the secondary-side water. $T_{2,\mathrm{out}}$ is the outlet temperature of the secondary-side hot water. k_2 represents the heat transfer coefficient of the secondary side. The mean temperature of the secondary-side fluid in the plate heat exchanger is given by $T_{m2} = \frac{T_{2,\mathrm{in}} + T_{2,\mathrm{out}}}{2}$. m_2 denotes the mass flow rate of the secondary-side hot water. $T_{2,\mathrm{in}}$ is the inlet temperature of the secondary-side hot water. c_w represents the specific heat capacity of the metal plates in the water-water heat exchanger.

Hydraulic System: In the simulation of the combined heating and power IES, when the DHS operates under a mass-based regulation mode in a normal state, the system's hydraulic conditions remain unchanged, and the hydraulic model is generally not considered. However, when the system enters a fault state under mass-based regulation or operates under volume-based regulation, hydraulic conditions vary, necessitating the recalculation of hydraulic conditions to assess their impact on the system. Therefore, it is essential to model the hydraulic conditions. The fundamental equations for DHS hydraulic calculations are well-established and will not be elaborated in detail here. Instead, a brief introduction to the model is provided. The hydraulic model consists of the flow conservation Eq. (3.5), the pipeline pressure drop Eq. (3.6), and the loop pressure drop Eq. (3.7) [28]. The flow conservation equation describes the relationship between flow rates in different pipelines, the pipeline pressure drop equation characterizes the relationship between pressure drop and flow rate, and the loop pressure drop equation defines the pressure drop relationships among interconnected pipelines.

$$\boldsymbol{A}_f \cdot m = \Delta Q \tag{3.5}$$

$$\Delta H = \text{diag}(\boldsymbol{S})\text{diag}(|m|)m + Z - DH \tag{3.6}$$

$$\boldsymbol{B_f} \cdot \Delta H = 0 \tag{3.7}$$

$\boldsymbol{A_f}$ represents the pipeline-node incidence matrix, which characterizes the connectivity between pipelines and nodes by using the column indices for nodes and the row indices for pipelines, with the constant-pressure node as the reference node. m denotes the mass flow rate in each pipeline, with units of kg/s. ΔQ represents the net outflow rate at nodes, which remains zero at all times except during pipeline leakage or rupture, with units of kg/s. $\Delta H(m)$ denotes the pressure drop across a single pipeline, analogous to the voltage drop in an electrical circuit branch. S represents the flow resistance, while diag($*$) denotes a diagonal matrix. Z, DH correspond to the pressure provided/consumed by circulation pumps and loads, respectively. $\boldsymbol{B_f}$ is the fundamental loop matrix, which defines the connectivity between pipelines and loops.

Power System: Currently, the formulation of AC power flow equations is well-established and, thus, will not be elaborated in this chapter. The steady-state distribution of power system state variables is typically determined using the AC power flow model. Time-series power flow calculation refers to multiple steady-state power flow calculations performed at specific time intervals throughout the simulation period, from the simulation start to its end. Consequently, its fundamental model is derived from the AC power flow model and is expressed in rectangular coordinates and polar coordinates as follows:

$$\begin{cases} f_{i,t}^{\mathrm{p}}(U,\theta) = P_{i,t}^{\mathrm{bus}}(U,\theta) + P_{i,t}^{\mathrm{d}} - P_{i,t}^{\mathrm{g}} = 0 \\ f_{i,t}^{\mathrm{q}}(U,\theta) = Q_{i,t}^{\mathrm{bus}}(U,\theta) + Q_{i,t}^{\mathrm{d}} - Q_{i,t}^{\mathrm{g}} = 0 \end{cases}, \quad (i \in \mathbb{V}, t \in \mathbb{T}^{\mathrm{e}}) \tag{3.8}$$

$$\begin{cases} P_{i,t}^{\mathrm{bus}}(U,\theta) = U_{i,t}\sum_{j=1}^{N_{\mathrm{n}}'} U_{j,t}\big(G_{ij}\cos\theta_{ij,t} + B_{ij}\sin\theta_{ij,t}\big) \\ Q_{i,t}^{\mathrm{bus}}(U,\theta) = U_{i,t}\sum_{j=1}^{N_{\mathrm{n}}'} U_{j,t}\big(-B_{ij}\cos\theta_{ij,t} + G_{ij}\sin\theta_{ij,t}\big) \end{cases}, \quad (t \in \mathbb{T}^{\mathrm{e}}) \tag{3.9}$$

where P and Q represent the active power and reactive power of the power grid, respectively. The superscripts "bus", "*d*", and "*g*" denote net injection at the node, load injection, and generator injection, respectively. Variables U and θ denote the voltage magnitude and voltage phase angle at grid nodes, respectively. Variables G_{ij} and B_{ij} represent the conductance and susceptance of branch *ij*, respectively. Variable i refers to a node index in the set $\mathbb{V}$. $N_{n'}$ represents the total number of nodes in the power grid. Variable t denotes a time point in the set $\mathbb{T}^e$ for power flow calculation.

CHP Units: The coupling between the power grid and the DHS is primarily achieved through CHP units and electric heating devices. As a key component of integrated energy systems, CHP units can simultaneously generate electricity and heat, playing a crucial role in enhancing energy efficiency and reducing carbon emissions. Based on different operational modes, CHP units can be categorized into back-pressure units and condensing-extraction units. Back-pressure units have gained significant attention in practical applications due to their high energy utilization efficiency and low energy consumption. In contrast, condensing-extraction units offer greater operational flexibility but exhibit a lower degree of heat-electricity coupling, making them more suitable for scenarios with highly fluctuating heat loads.

In this chapter, condensing-extraction CHP units are adopted. A distinguishing feature of these units is their flexible regulation between heat and power outputs, enabling the system to optimally allocate resources based on actual demand, thereby improving operational flexibility. Additionally, condensing-extraction units can maintain high efficiency even under partial load conditions, offering favorable economic and adaptive advantages, particularly in scenarios with significant variations in heat and power demand. Consequently, these units have been widely employed in DHSs, especially where a flexible cogeneration ratio is required. Specifically, the operational characteristics can be mathematically formulated as follows:

$$P_{\text{chp}} + B_{\text{chp}}^{j} h_{\text{chp}} < C_{\text{chp}}^{j} \quad \forall j \in \boldsymbol{J} \tag{3.10}$$

where h and P represent the heat output and electric power output of the unit, respectively. B and C are parameters related to the boundary operating curves. J is a vector representing the boundary operating curves.

3.2.2 Pipeline Leakage Model

When a leakage occurs, the thermal medium escapes from the leakage point, significantly impacting both the hydraulic and thermal conditions of the DHS. To accurately simulate this complex process, this chapter introduces an additional node within the DHS to model the leakage point. The introduction of this node divides the leaking pipeline into two subpipelines, ensuring that the net flow at this node equals the leakage flow, thereby maintaining the system's mass conservation. The mathematical expression for the leakage flow is given by Eq. (3.11).

$$m_i = \Delta Q \tag{3.11}$$

where ΔQ (m^3/h) is determined by the pressure head H(m) of the leaking pipeline and the leakage area A_l (m^2), which can be calculated using Eq. (3.12) [25]:

$$\Delta Q = 3600\sqrt{2gH}A_l \tag{3.12}$$

where g is the gravitational acceleration constant.

It is important to note that pipeline leakage is typically a progressive process, with its impact accumulating and expanding over time. To capture this dynamic characteristic in the simulation model, this chapter employs Eq. (3.13) to describe the gradual increase of the leakage area A_l over time t:

$$A_l = \begin{cases} 0 & 0 < t < t_1 \\ \frac{t-t_1}{t_2-t_1} A_{\max} & t_1 < t < t_2 \\ A_{\max} & t_2 < t < t_3 \end{cases} \tag{3.13}$$

In terms of hydraulics, leakage affects the operation of the DHS system by altering the water pressure distribution. These pressure variations further lead to adjustments in the mass flow rate within the pipeline network, thereby impacting the overall hydraulic balance of the system. The extent of leakage-induced hydraulic changes is primarily determined by two key parameters: the leakage area rise time $\Delta t = t_2 - t_1$ and the maximum leakage area $A_{\max}$. These parameters respectively govern the rate and ultimate magnitude of hydraulic condition variations. A shorter leakage area rise time Δt results in rapid pressure fluctuations, while a larger maximum leakage area $A_{\max}$ significantly exacerbates hydraulic instability in the system.

From a thermal perspective, leakage impacts the system in two main aspects. First, as leakage alters the mass flow rate in the pipeline network, the transmission capacity of the heat carrier is affected, leading to a dynamic adjustment of the temperature distribution within the network. This process can be described by Eqs. (3.1)–(3.4), which primarily capture how changes in mass flow influence heat transfer and distribution. Second, when pipeline leakage occurs, the system must inject lower-temperature make-up water to maintain the operation of constant-pressure nodes. The heating process of this make-up water requires additional thermal power support, further increasing the burden on the system's heat sources. The thermal effects of leakage can be characterized by Eq. (3.14).

$$h_{s,l} = cm_{r,l}\left(T_s^{\mathrm{su}} - T_{s,l}^{\mathrm{re}}\right) + c\Delta Q\left(T_s - T_{\mathrm{sup}}\right) \tag{3.14}$$

The first term on the right-hand side of Eq. (3.14) represents the cascading effect induced by changes in mass flow rate m, while the second term corresponds to the impact of additional make-up water ΔQ.

3.3 Fault Propagation Mechanism Analysis Method

In this section, an integrated simulation method for the CHP-based IES under fault conditions is proposed. This simulation framework aims to accurately capture the dynamic operational characteristics of the system under fault scenarios by employing a staged modeling approach and an iterative solution process among subsystems. The method not only reflects the coupling relationships among different subsystems, such as the DHS, hydraulic system, power system, and coupled equipment, but also accommodates various types of faults, including CHP unit outages and pipeline leakage failures. Through this framework, the impact of faults on the operation of

CHP-based integrated energy systems can be comprehensively analyzed, providing theoretical support for fault diagnosis and operational optimization.

3.3.1 *Fault-State Simulation Framework*

Figure 3.2 illustrates the overall simulation framework for CHP-based IESs under fault conditions. The framework consists of three stages: the initialization stage, the DHS simulation stage, and the CHP equipment and power system calculation stage.

(1) ***Initialization Stage***: In this stage, the fundamental simulation parameters are determined, including system topology, parameters, loads, initial pipeline temperatures, iterative error tolerance, and the assumed initial thermal power output of the heat source.

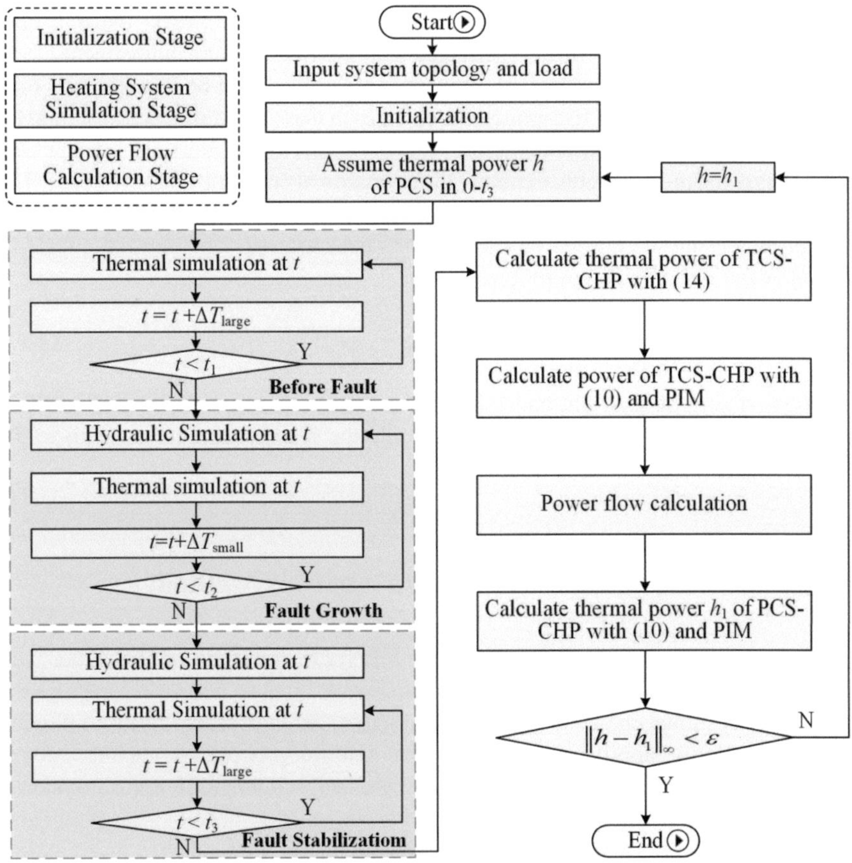

Fig. 3.2 Fault-state HE-IES simulation framework

(2) ***DHS Simulation Stage***: In the DHS simulation, the thermal model can only be solved after the system's hydraulic conditions are determined. Therefore, the simulation process is divided into three phases based on the characteristics of hydraulic variations: the pre-fault period ($0 < t < t_1$), the fault growth period ($t_1 < t < t_2$), and the fault stabilization period ($t_2 < t < t_3$). Before the occurrence of the fault, the hydraulic conditions of the system remain stable, requiring only thermal simulation. This phase is relatively straightforward and primarily serves as a reference baseline for analyzing subsequent dynamic changes after the fault occurs. In the fault growth period, as the leakage flow gradually increases over time, significant variations in the system's hydraulic conditions occur. To accurately simulate this process, an iterative approach is employed, alternating between hydraulic and thermal simulations. In each iteration, hydraulic simulation updates the system's flow rate and pressure distribution, while thermal simulation calculates temperature variations at nodes and within pipelines. Since both hydraulic and thermal conditions exhibit rapid changes during this period, a smaller time step Δt_2 is used to enhance simulation accuracy. In the fault stabilization period, once the leakage flow reaches a steady state, the hydraulic conditions of the system no longer change over time. Consequently, only a single hydraulic simulation is performed at the initial moment ($t = t_2$), while subsequent simulations focus solely on the evolution of thermal conditions. To improve computational efficiency, a larger time step Δt_1 is adopted in this phase.

(3) ***CHP Equipment and Power System Calculation Stage***: In this stage, the thermal power demand for CHP equipment is first determined based on the results of the DHS simulation. The calculation of thermal power is based on the temperature distribution at network nodes and within pipelines, as well as the heat load demand, ensuring that CHP units can meet system heating requirements. Subsequently, the electrical power output of the CHP units is determined using their operational characteristic model Eq. (3.10) and the parameter interaction method (detailed in Sect. 3.3.4). This process establishes the relationship between thermal and electrical power through the thermoelectric coupling characteristics of CHP units. Next, the Newton–Raphson iteration method [29] is applied to perform power flow calculations for the electrical system, determining the electrical power demand of electrically driven CHP units. The power flow analysis considers the power grid topology, load distribution, and operational characteristics of electrical equipment to determine voltage and power distribution at each node, accurately reflecting the operational state of the power system. Finally, the thermal power h_1 of electrically driven CHP units is recalculated using Eq. (3.10) and the parameter interaction method. By comparing h_1 with the initially computed thermal power h, the convergence of the iterative process is assessed. If the difference between h_1 and h falls within the error tolerance, the iteration is considered convergent, and the simulation proceeds to the next stage. Otherwise, h_1 replaces h, and the process returns to the DHS simulation stage for further iterative computation.

3.3.2 Fault-State DHS Simulation Framework

Fig. 3.3a illustrates the DHS simulation algorithm, which consists of the following steps:

Step 1: Determine the initial temperature of the pipeline T_0^{Su} and the computational error tolerance ε. Simultaneously, assume the return water temperature $T_{s,0}^{\mathrm{Re}}$ of the power-controlled heat source.

Step 2: Using the heat source model Eq. (3.1), calculate the supply water temperature $T_{s,0}^{\mathrm{Su}}$ of the power-controlled heat source. Subsequently, apply the thermal pipeline model Eq. (3.3) to compute the temperature distribution in the supply network, based on the heat source supply temperature and the initial temperature of the supply pipeline.

Step 3: After determining the temperature distribution of the supply network, calculate the temperature at load nodes. First, determine the supply temperature at the load node T_l^{Su}, and then use the load node model Eq. (3.4) to compute the return water temperature T_l^{Re}. The temperature calculation at load nodes is based on the heat load demand and the thermal exchange characteristics of the nodes, ensuring that the thermal demand at each load node is met.

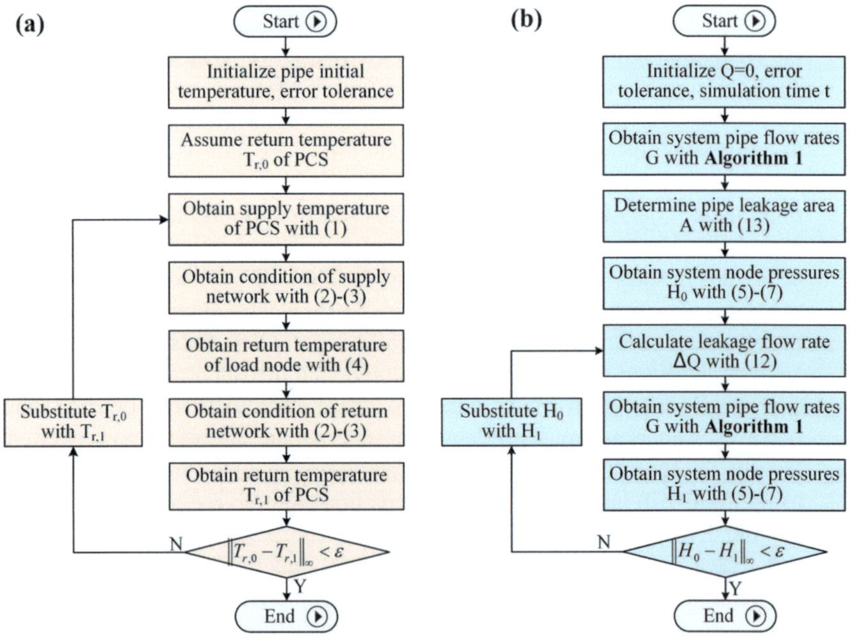

Fig. 3.3 Detailed energy subsystem simulation process: **a** DHS; **b** fault-state hydraulic system

Step 4: Using the thermal pipeline model Eq. (3.3), compute the temperature distribution in the return network by considering the return temperature at the load node T_l^{Re} and the initial temperature of the return pipeline. This step is similar to the supply temperature distribution calculation but must account for heat losses and flow characteristics in the return pipeline. The return network's temperature distribution directly influences the calculation of the heat source's return water temperature.

Step 5: Based on the return network's temperature distribution, determine the return water temperature of the power-controlled heat source $T_{s,1}^{\text{Re}}$. Compare it with the initially assumed return temperature $T_{s,0}^{\text{Re}}$. If the computational error between them is smaller than the predefined tolerance ε, the simulation is considered complete, and the system is regarded as having reached thermal equilibrium. Otherwise, replace $T_{s,0}^{\text{Re}}$ with $T_{s,1}^{\text{Re}}$ and return to Step 2 for further iterative calculations. Through multiple iterations, the system progressively approaches its actual operating state.

Among the above simulation processes, solving the thermal pipeline model Eq. (3.3) is the most complex task. This is because Eq. (3.3) is a PDE, for which an analytical solution is typically difficult to obtain. To address this issue, a numerical method is employed to transform the PDE into algebraic equations for approximate solutions. Specifically, this study adopts the Wendroff differential scheme, which is substituted into Eq. (3.3) to obtain a discrete form of the PDE. The Wendroff scheme is an explicit finite difference method that discretizes the PDE in both time and space, converting the continuous temperature distribution into discrete temperature node values. By introducing time and space steps, this method effectively captures the dynamic heat transfer process of thermal media in pipelines while ensuring high computational accuracy and stability.

By converting the PDE into algebraic equations, solving the thermal pipeline model is reduced to solving a set of algebraic equations. This approach not only enhances computational efficiency but also ensures a certain level of solution accuracy. The implementation of the Wendroff scheme involves the discretization of the temperature field within the pipeline and the appropriate setting of boundary conditions to ensure the physical validity and numerical stability of the solution. For further details on this method, interested readers are referred to [30].

3.3.3 Fault-State Hydraulic Simulation Framework

Figure 3.3b illustrates the simulation process of the hydraulic system under fault conditions. Similar to the DHS simulation algorithm, this process iteratively solves for system flow distribution and node pressures to accurately capture the dynamic variations in hydraulic conditions during faults. The specific steps are as follows:

Step 1: Initialize parameters. Input the simulation error tolerance ε and simulation time t, and assume the initial leakage flow rate $Q = 0$ to establish the necessary foundation for subsequent iterative calculations.

Step 2: Solve the hydraulic equations using the fundamental loop method. The hydraulic Eqs. (3.5)–(3.7) are solved to obtain the system flow distribution G. Based on the current simulation time t and the leakage area growth model (3.13), compute the pipeline leakage area A. The dynamic variation of the leakage area directly affects the hydraulic conditions of the system.

Step 3: Calculate pressure drops and node pressures. Using Eq. (3.6), compute the pressure drop ΔH for each pipeline and determine the node pressure H_0. Then, calculate the leakage flow rate ΔQ using Eq. (3.12).

Step 4: Iteratively update system flow and node pressures. Based on the updated leakage flow ΔQ, apply the fundamental loop method again to recalculate the system flow G and node pressures H_1. If the simulation error between H_0 and H_1 exceeds the tolerance ε, replace H_0 with H_1 and return to Step 3 for further iterations. Otherwise, the simulation is considered complete.

The fundamental loop method used in the above steps is a classical approach for solving flow distribution in hydraulic networks. By iteratively computing pipeline flows and pressure drops, the method gradually converges toward the system's steady-state conditions. The detailed steps of this method are presented in Algorithm 1.

Algorithm 1: Basic Loop Method

Input: system topology and pipeline resistance S
Divide system pipes into trunk pipes and branch pipes.
Generate system topology matrix $\boldsymbol{A}$, $\boldsymbol{A_t}$, $\boldsymbol{A_l}$ and $\boldsymbol{B_f}$
Initialize G_l, $Flag = 1$, $k = 0$, Q and ε.
While Flag
  $G_t = A_t^{-1}(Q - A_l G_l)$; $G = (G_l; G_t)$;
  $M = 2B_f S|G|B_f^T$;
  $\Delta h = B_f(S|G|G + Z - DH)$; $\Delta G_l = -M^{-1}\Delta h$;
  If $\|\boldsymbol{\Delta G_l}\|_\infty < \boldsymbol{\varepsilon}$
    $Flag \leftarrow 0$
  End if
  $G_l \leftarrow G_l + \Delta G$; $k \leftarrow k + 1$;
End while
Output: Pipe Flow rates G
Return T

3.3.4 Cross-System Parameter Interaction Method

In engineering practice, due to the high complexity of scheduling CHP units and power systems, directly implementing control and optimization with short scheduling intervals (e.g., in the order of seconds) is often challenging. Therefore, to balance computational efficiency and engineering feasibility, the scheduling of CHP units and power systems typically adopts a coarser time resolution, usually ranging from 5 to 15 min. In contrast, DHSs exhibit faster dynamic characteristics, requiring high-precision simulations with finer time resolutions, often less than one minute. This discrepancy in time resolution poses a challenge for the integrated modeling of CHP systems. To address this issue, this chapter proposes a cross-system parameter interaction method that enables coordinated interaction between thermal and power systems by transforming and processing data across different time resolutions.

For heat-driven CHP units, their operational characteristics primarily depend on the output of the DHS. The DHS simulation is conducted at a higher time resolution (e.g., on a second scale or less than one minute) to capture the dynamic variations in heat load and heat source output. However, the electrical output of heat-driven CHP units must align with the scheduling time resolution of the power system, which is typically 5–15 min. To achieve this, the electrical power output of heat-driven CHP units is determined based on the average thermal power of the heat source over the scheduling interval. The specific calculation method is provided in Eq. (3.15).

$$h_{\text{chp}} = \frac{\sum_{i=0}^{\tau_{\text{chp}}} h_i}{\tau_{\text{chp}}} \quad \left(\tau_{\text{chp}} = \Delta t_{\text{chp}} / \Delta t_{\text{thermal}}\right) \tag{3.15}$$

For electrically driven CHP units, their operational characteristics primarily depend on the output of the power system. Since power system computations typically use a longer time resolution (e.g., 5–15 min), whereas the DHS requires a higher time resolution to capture transient heat load variations, directly applying the low-resolution results from the power system would lead to insufficient time resolution in thermal power calculations. To address this issue, this study employs a linear interpolation method to fill the numerical gaps in thermal power values within the power system scheduling intervals. The specific calculation method is provided in Eq. (3.16).

$$h_{\text{chp}}(t) = \frac{t_{n+1} - t}{\tau_{\text{chp}}} h(t_n) + \frac{t - t_n}{\tau_{\text{chp}}} h(t_{n+1}) \quad t_n \le t \le t_{n+1}. \tag{3.16}$$

3.4 Case Study

This section evaluates the impact of pipeline leakage on a 31-node combined heating and power IES. All simulations were conducted using MATLAB 2018B on a computer equipped with an Intel i7 processor and 16 GB RAM. By performing a joint simulation of the DHS and power system, the dynamic effects of pipeline leakage on system operation were analyzed.

3.4.1 Experimental Basis

The DHS in this study consists of 20 nodes, including 2 heat source nodes and 4 load nodes. The heat source nodes are Node N1 and Node N6, which exhibit distinct operational characteristics. Node N1 is configured as a temperature-controlled heat source, maintaining a constant supply temperature of 90 °C. In contrast, Node N6 is a power-controlled heat source, delivering a constant thermal power output of 32.67 MW. These two heat source nodes are connected to two CHP units (denoted as CHP#1 and CHP#2), supplying thermal energy to the DHS. In terms of hydraulic conditions, Node N1 is set as a constant-pressure node with a water head pressure of 23 mH_2O, effectively simulating the behavior of different heat sources in real-world DHSs and their influence on system operation.

The power system is adapted from the IEEE 11-node distribution network, comprising 11 nodes, 11 branches, and 5 power plants. The power generation mix includes a 25 MW photovoltaic (PV) power plant, a 30 MW wind farm, two extraction-condensing CHP units, and a coal-fired power plant. The PV and wind power plants represent renewable energy integration, while the coal-fired plant provides stable fossil-fuel-based power output. The electrically driven CHP units reflect the operational characteristics of electrothermal coupling equipment.

Regarding parameter settings for the DHS and power grid, the DHS parameters include heat loss coefficients, pipeline length, diameter, and hydraulic resistance coefficients, while the power system parameters encompass line impedance, node load distribution, and generator operating characteristics. Detailed system parameters can be found in [31], which provides a comprehensive specification of the key operating conditions for both the heating and power networks.

This study conducts a comparative analysis of two operational modes for the combined heating and power IES: grid-connected mode (strongly coupled IES, SC-IES) and islanded mode (weakly coupled IES, WC-IES). The system topologies for these two modes are illustrated in Fig. 3.4, representing different coupling characteristics and energy supply strategies under varying operating conditions.

In the grid-connected mode, CHP#1 and CHP#2 are connected to power system nodes N8 and N9, respectively, and operate as PQ/PV nodes. PQ nodes denote generators that control active and reactive power output. PV nodes represent generators that control active power output and voltage. The system's balancing node is linked to the main power grid, ensuring power supply and demand balance. The main grid provides voltage and frequency support, stabilizing the entire power system.

In the islanded mode, CHP#1 and CHP#2 are connected to power system nodes N1 and N9, forming an independently operated weakly coupled IES (WC-IES). In this mode, the system's balancing node is supported by CHP#2, which is responsible for frequency and voltage regulation, ensuring system stability. Unlike the grid-connected mode, the islanded system must rely entirely on its own generation capacity and the coordinated operation of CHP units to meet both power and heating demands.

In the simulation analysis, pipeline leakage is assumed to occur at the midpoint of pipeline P5, with a maximum leakage area of $6 \times 3600^{-1} m^2$. The leakage area

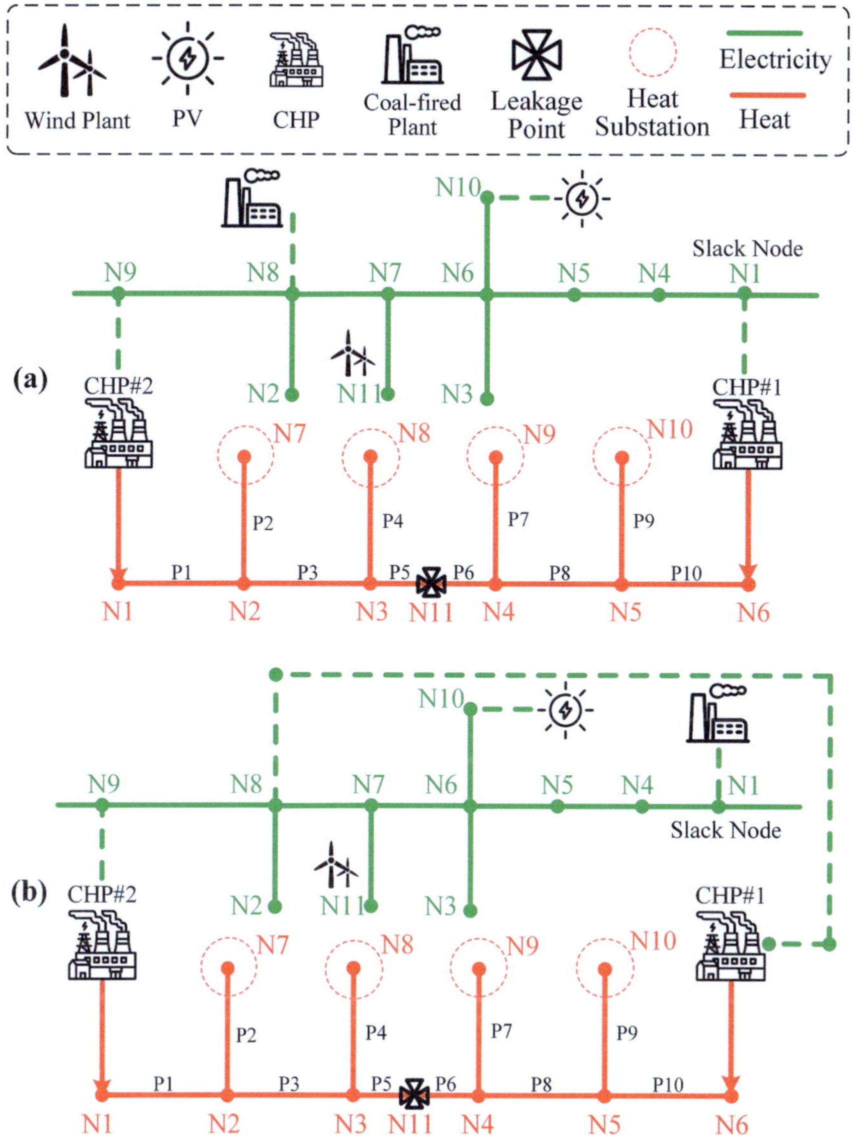

Fig. 3.4 System structure: **a** HE-IES under island mode; **b** HE-IES under grid-connected mode

gradually increases over time, with a growth period of 7200 s, modeling the process from the initial occurrence to full development. To accurately represent the impact of leakage on the system's hydraulic balance and thermal distribution, the leakage point is modeled as an additional node (N11) in the mathematical framework. The pipelines connected to N11 are labeled P5 and P6: P5 connects N11 to N3, and P6 connects N11 to N4.

The simulation process is divided into three phases to comprehensively analyze the dynamic effects of pipeline leakage: pre-fault phase, fault growth phase, and fault stabilization phase. The durations of these three phases are set to 1800, 7200, and 7200 s, respectively, to comprehensively analyze the dynamic process of pipeline leakage from its initial occurrence to full development. In the pre-fault phase, the system operates under normal conditions with no leakage. In the fault growth phase, the leakage area gradually increases, leading to significant changes in the system's hydraulic and thermal distribution. In the fault stabilization phase, the leakage area reaches its maximum, and the system enters a steady operating state.

The simulation of the DHS adopts two time steps to balance computational accuracy and efficiency. The time step Δt_1 is set to 40 s to describe the dynamic behavior of the DHS under normal operating conditions. The time step Δt_2 is set to 10 s to capture rapid changes during the occurrence of the leakage fault. The scheduling interval τ for CHP units and the power system is set to 900 s, reflecting the scheduling cycle of power systems and CHP units in practical engineering applications.

3.4.2 *Fault Cross-System Propagation Mechanism Analysis*

Figure 3.5 illustrates the dynamic response characteristics of leakage point pressure and flow rate as a function of the leakage equivalent area. The results indicate that as the equivalent leakage area A_l increases gradually from 0 to $6 \times 3600^{-1} m^2$, the leakage flow rate significantly rises, ultimately reaching 103.8411 m^3/h, which accounts for 5.53% of the normal flow rate. Meanwhile, the leakage point pressure decreases from its initial value of 16.024 mH_2O to 15.282 mH_2O. The increase in leakage flow rate is primarily driven by the expansion of the equivalent leakage area. As the leakage area grows, the hydraulic resistance at the leakage point decreases, leading to a nonlinear increase in leakage flow. This phenomenon suggests that the hydraulic characteristics of the leakage point significantly impact the overall flow distribution, with larger leakage areas causing more severe disruptions to the system's hydraulic balance. On the other hand, the decrease in leakage point pressure is a result of increased leakage flow causing greater pressure loss in the pipeline. Although the pressure variation is relatively small (a drop from 16.024 mH_2O to 15.282 mH_2O), its impact on the system's hydraulic balance cannot be ignored. The reduction in pressure not only affects the flow distribution in nearby pipelines but may also negatively impact the heating capacity of downstream load nodes.

Direct Impact on the Hydraulic System: According to the detailed simulation results summarized in Table 3.1, pipeline flow variations can be classified into three categories: pipelines with a sharp increase in flow (P1, P3, P5), pipelines with a slight increase in flow (P8, P10), and pipelines with a slight decrease in flow (P2, P4, P6, P7). These changes reflect the significant impact of leakage faults on the hydraulic balance of the DHS.

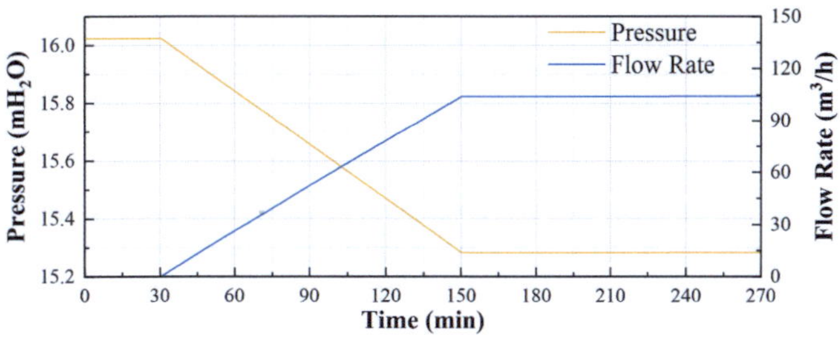

Fig. 3.5 Pipe leakage condition

Table 3.1 Supply network pipe flow rate variation

Pipe	1	2	3	4	5
$\Delta G(\mathrm{m}^3/\mathrm{h})$	81.2	− 5.9	87.1	− 5.9	93.0
Pipe	6	7	8	9	10
$\Delta G(\mathrm{m}^3/\mathrm{h})$	− 10.9	− 8.2	2.7	− 2.2	0.5

For pipelines with a sharp increase in flow (P1, P3, P5), their common characteristic is that they connect the constant-pressure node (N1) to the leakage node (N11). The drop in leakage point pressure directly leads to a significant increase in pressure drop across these pipelines. According to Eq. (3.6), the flow rate in a pipeline is proportional to its pressure drop. Therefore, the increased pressure drop further causes the flow rates of these three pipelines to rise sharply. This phenomenon indicates that pipelines near the leakage point are the most sensitive to the fault, with flow variations significantly larger than those in other pipelines.

On the other hand, pipelines with a slight decrease in flow (P2, P4, P6, P7) are mainly located in the downstream area of the leakage point. Due to the increased pressure drop in P1 and P3, the pressures at nodes N2 and N4 decrease, leading to a slight flow reduction in pipelines P2 and P4, which are connected to these nodes. Additionally, the drop in leakage node (N11) pressure further reduces the pressures at nodes N4 and N5, negatively affecting P7 and P9, which also experience a slight decrease in flow.

For pipelines with a slight increase in flow (P8, P10), their variations are primarily caused by the drop in leakage point pressure. The decrease in pressure at the leakage point causes the pressures at connected nodes N4 and N5 to decrease, resulting in a slight increase in flow rates of P8 and P10. While the magnitude of this change is relatively small, it still reflects the impact of the leakage fault on the flow distribution in downstream pipelines.

Overall, the pressure drop at the leakage node (N11) significantly affects the hydraulic balance of the DHS. The flow rate at the constant-pressure heat source (N1) increases substantially, whereas the flow rates at load nodes generally decrease. This phenomenon suggests that leakage faults have the most significant impact on constant-pressure heat sources, while their effect on load nodes and other heat sources is relatively smaller. Such variations in flow distribution may further influence the heating capacity and operational stability of the DHS, particularly in the vicinity of the leakage point, where hydraulic characteristics change most significantly.

Fault Propagation from the DHS to the Power System: Fig. 3.6 illustrates the operational state of the DHS in the WC-IES, including variations in supply and return temperatures at different nodes. As shown in the figure, nodes N7 and N8 experience minimal impact on supply temperature due to the leakage fault. This is because the thermal medium at these nodes is supplied exclusively by the constant-pressure heat source N1, which maintains a stable supply temperature of 90 °C. However, for nodes N9 and N10, the reduction in flow rate through pipeline P6 leads to a decrease in the heating power supplied by N1, resulting in insufficient thermal supply to meet load demands. To compensate for the heat deficit, the stored thermal energy within the pipeline is released, causing the supply temperature of N6 to drop from 89 to 88.44 °C. This phenomenon indicates that the impact of the leakage fault on the DHS is highly localized. The most significant changes occur in nodes and pipelines adjacent to the leakage point, where variations in thermal conditions are particularly pronounced.

Figure 3.7 illustrates the operational state of the CHP units, including variations in thermal and electrical power output. In Fig. 3.7a, it can be observed that CHP#1's thermal power output increases significantly as the leakage flow rate rises, from 32.94 MW at 30 min to 41.04 MW at 165 min. This change occurs because the leakage alters the thermal load distribution, requiring CHP#1 to provide additional thermal power to maintain the thermal balance of the system. Meanwhile, Fig. 3.7b shows that CHP#1's operating point gradually approaches its constraint curve, leading to a corresponding increase in its electrical power output, which rises from 56 to 70.16 MW. This demonstrates the coupling characteristics of CHP units, where variations in thermal power directly affect electrical power output. In contrast, CHP#2's operating state remains unchanged in the WC-IES. This is because CHP#2 is primarily constrained by power system dispatch in islanded mode, and its thermal power output is not

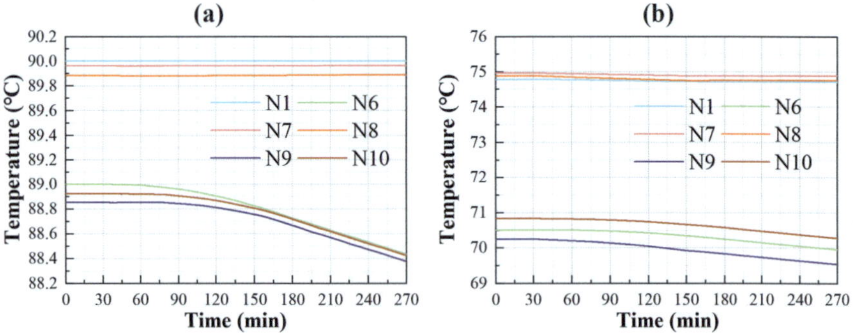

Fig. 3.6 WC-IES thermal condition: **a** supply node temperature; **b** return node temperature

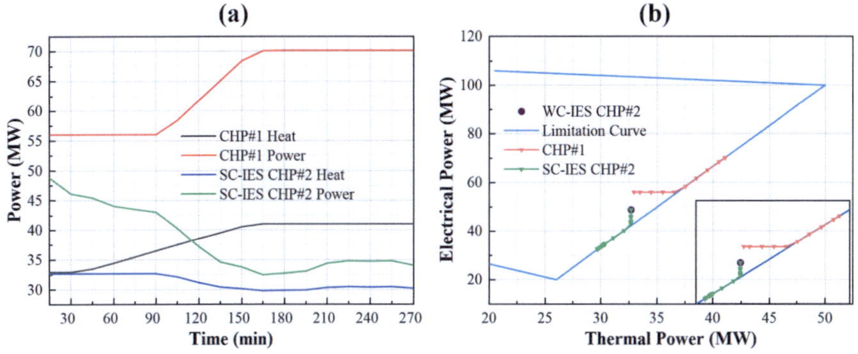

Fig. 3.7 CHP operating condition **a** CHP output; **b** CHP working point

directly affected by the leakage fault. The operational state of CHP#2 in the SC-IES will be analyzed in detail in the next section.

The changes in CHP output induced by pipeline leakage have a significant impact on the operating conditions of the power system. Figure 3.8a presents the voltage level variations at multiple nodes. Among these, N9 experiences the most significant voltage fluctuation, as it is directly connected to CHP#2. As CHP#2's power output increases, the voltage at N9 rises from 1.028 to 1.083. Additionally, voltage levels at nearby nodes (such as N2, N7, N8, and N11) exceed their voltage limits, indicating that the leakage fault adversely affects power system voltage stability through the coupling effect of the CHP units. Figure 3.8b further reveals a significant shift in power distribution within the power system due to the increased CHP power output. The balancing node's power output decreases from 48.78 MW to 34 MW, indicating that CHP units are shouldering a greater share of the power load. This redistribution leads to a significant increase in the load factor of transmission lines near N9, and power flow reversal is observed on lines L2, L3, L5, and L6. This power flow reversal phenomenon suggests that the leakage fault not only affects the DHS's operation but also has a profound impact on power distribution through the coupling effect of CHP units.

Fault Feedback from the Power System to the DHS: In a SC-IES, CHP#2 plays a critical role in supplying power to the slack node in the electrical network. This means that when CHP#1's electrical power output increases, CHP#2's electrical power output must decrease to maintain power system balance. However, due to the coupling relationship between thermal and electrical power output (i.e., the CHP operational limit curve), the reduction in CHP#2's electrical power output also leads to a decrease in its thermal power output. This feedback effect from the power system to the DHS is particularly significant in SC-IESs.

As shown in Fig. 3.7a, during the 0–90 min period, CHP#2's electrical power output gradually decreases from 48.78 to 43.06 MW, while its thermal power output remains constant at 32.68 MW. However, after 90 min, CHP#2's operating point begins to move along its thermoelectric coupling limit curve, leading to a further

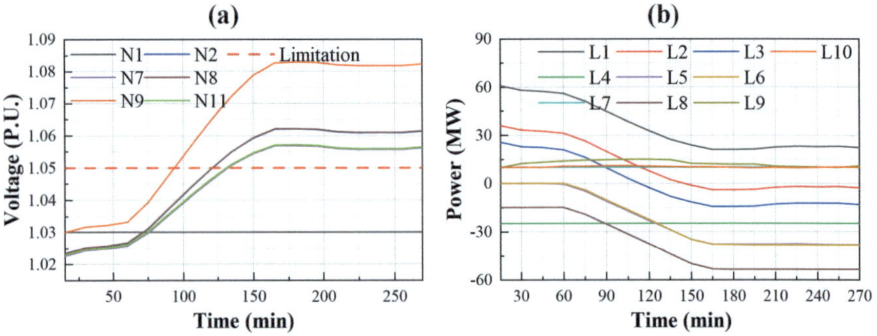

Fig. 3.8 Power system condition **a** power system bus voltage; **b** power system bus power

reduction in both electrical and thermal power output. By 165 min, CHP#2's thermal power output decreases to 30.2 MW, indicating that when CHP#2 operates near its constraint limits, power system dispatch changes directly impact the heating capacity of the DHS. Figure 3.9 illustrates the impact of this fault on the DHS in the SC-IES. Due to the reduction in CHP#2's thermal power output, the supply temperatures at nodes N6, N9, and N10 decrease significantly. For example, the supply temperature at node N6 drops from an initial 89–81.91 °C, which is substantially lower than the 88.44 °C observed in the WC-IES. Similarly, N9 and N10 also experience notable declines in supply temperature, highlighting the pronounced feedback effect of the leakage fault from the power system to the DHS in a SC-IES.

In contrast, in a WC-IES, CHP#2 does not directly participate in balancing the slack node's power demand, resulting in minimal impact on its operating state. Therefore, although the leakage fault leads to an increase in CHP#1's electrical and thermal power output, CHP#2 remains relatively stable, effectively mitigating the fault's impact on the DHS. This difference highlights the distinct fault propagation characteristics between strongly and weakly coupled systems, demonstrating how the level of coupling significantly influences the extent to which a fault in the power system affects the DHS.

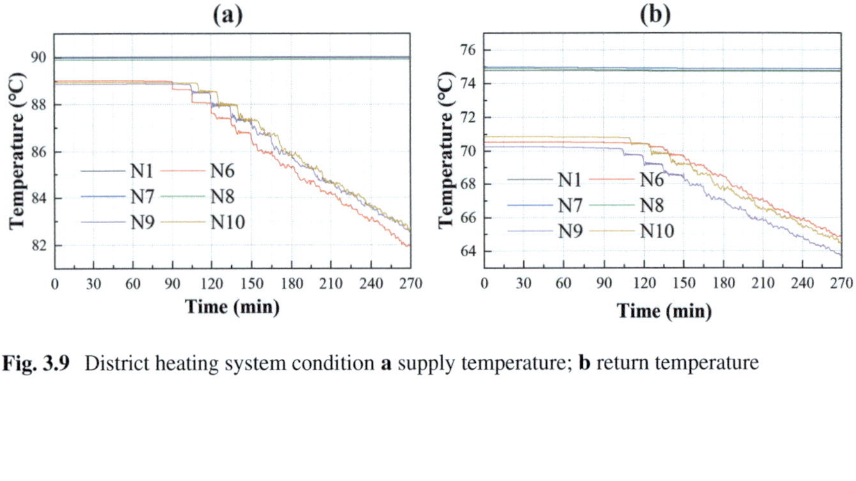

Fig. 3.9 District heating system condition **a** supply temperature; **b** return temperature

3.4.3 Impact on Renewable Energy Integration

Security-Constrained OPF Model: To maximize renewable energy utilization under fault conditions while ensuring the operational security of the power system, this study proposes an optimal power flow (OPF)-based model (3.17) to redispatch the power system. This model optimizes power distribution, coordinating the output of renewable and conventional energy sources to achieve efficient operation while satisfying security constraints. During the simulation of the combined heating and power IES, once the operational state of the power system is determined, a security assessment is conducted to verify whether the system complies with the inequality constraints defined in Eq. (3.17). If the system is identified as operating in an insecure state, it must be redispatched using the proposed OPF model to restore system security and stability.

$$
\begin{aligned}
\text{Max} \quad & E_{\text{Wind}} + E_{\text{PV}} \\
\text{s.t.} \quad & \text{Eq.}(3.9) \\
& \underline{P}_G \le P_G \le \overline{P}_G \\
& \underline{Q}_G \le Q_G \le \overline{Q}_G \\
& \underline{P}_l \le P_l \le \overline{P}_l \\
& \underline{V} \le V \le \overline{V}
\end{aligned}
\tag{3.17}
$$

The objective of the OPF model is to maximize the utilization of renewable energy resources while adhering to a set of operational constraints. These constraints include power flow equations, generator output limits, transmission line power flow limits, and node voltage levels, ensuring that the power system operates within a safe and stable range to prevent voltage violations or transmission line overloads that could lead to system instability. Additionally, the OPF model must account for the operational characteristics of the combined heating and power IES, particularly the thermoelectric coupling properties of CHP units. In a SC-IES, maintaining the quality of heat supply requires that CHP#2's thermal power output must not decrease during operation. This means that when CHP#2 reaches its thermoelectric coupling limit curve, it should stop moving along the constraint curve to prevent further reductions in thermal power output. Consequently, in the strongly coupled system, the lower limit of CHP#2's electrical power output is set to the corresponding electrical power value on the constraint curve, ensuring stable thermal power output. In contrast, in a WC-IES, the constraints on CHP#2's operation are more relaxed, as the power system and DHS are less interdependent. In this case, CHP#2's lower limit on electrical power output is set to the minimum generation output at the slack node, ensuring power system balance and operational security. This approach allows the weakly coupled system to flexibly dispatch generator power output, making full use of the adjustment capability of the slack node and mitigating the impact of the fault on system operation.

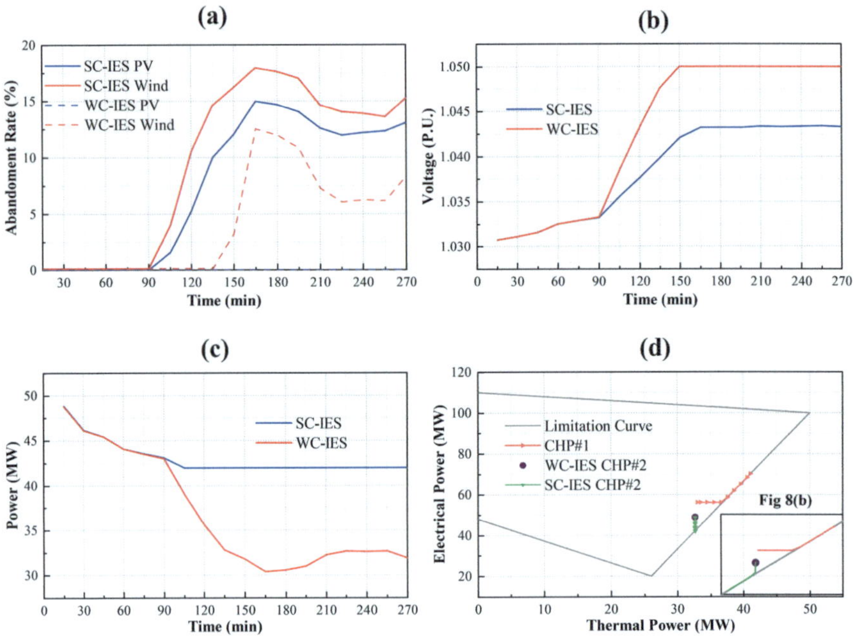

Fig. 3.10 Power system condition after redispatch **a** renewable energy abandonment rate; **b** maximum nodal voltage; **c** power output of slack node; **d** CHP working point

System State After Redispatch: Fig. 3.10 illustrates the power system's operating conditions after redispatch, highlighting key indicators such as renewable energy curtailment rates, maximum node voltage, slack node power output, and CHP operating points. A comparison between strongly coupled and WC-IESs clearly reveals their performance differences under fault conditions.

Figure 3.10a compares the renewable energy curtailment rates between the strongly coupled (solid line) and weakly coupled (dashed line) integrated energy systems. In the strongly coupled system, the CHP operation is constrained by thermal power output limits, leading to higher curtailment rates for wind and PV power, reaching 17.96 and 14.96%, respectively. In contrast, in the weakly coupled system, wind and PV curtailment rates are significantly lower, at 12.52 and 0.25%, respectively. These results indicate that the weakly coupled system allows for more flexible power dispatch strategies, enabling more efficient utilization of renewable energy resources. However, in the strongly coupled system, ensuring power system security and DHS quality necessitates higher renewable energy curtailment rates.

Figure 3.10b shows the maximum node voltage in the power system after redispatch, confirming that the OPF-based strategy effectively regulates system voltage within the secure range. In the strongly coupled system, the maximum node voltage is 1.043. In the weakly coupled system, the maximum node voltage is 1.05. This suggests that the redispatch strategy in the weakly coupled system is primarily constrained by the upper voltage limit, while in the strongly coupled system, the redispatch strategy is more influenced by the lower limit of slack node power output.

Figure 3.10c further validates this finding. In the strongly coupled system, the slack node's electrical power output reaches 42 MW at 75 min and remains constant, indicating that the minimum power output constraint at the slack node plays a key role in the redispatch strategy. In contrast, in the weakly coupled system, the slack node's electrical power output gradually decreases over time, demonstrating that the dispatch strategy is more flexible and can better adapt to the system's dynamic demands.

Figure 3.10d illustrates the operating point variations of CHP units, revealing significant differences compared to Fig. 3.7. In the strongly coupled system, once the CHP operating point reaches the constraint curve, it stops moving further along it. This ensures that CHP#2's thermal power output remains stable, thereby maintaining the quality of the DHS. As a result, the DHS's operating conditions in the redispatched strongly coupled system become identical to those in the weakly coupled system, as shown in Fig. 3.6. This confirms that the OPF model successfully preserves heating quality in the SC-IES while maintaining power system security.

Figure 3.11 illustrates the operating cost variations of the SC-IES after redispatch. In this study, the efficiency of CHP units and generators is set to 0.8 and 0.6, respectively. The natural gas price is 0.553 $/m^3$, and the penalty for curtailing renewable energy is 200 $/MWh. The results indicate that the pipeline leakage fault significantly impacts system economics. CHP#2 experiences a notable increase in operating cost, rising from $6219 to $7775/h. This is because CHP#2 must supply more electrical and thermal power to compensate for the changes in power distribution caused by the fault. In contrast, CHP#1's operating cost decreases from $5698 to $5222/h due to a reduction in its electrical power output, reflecting a load transfer effect between CHP units. Additionally, the pipeline leakage fault leads to an increase in penalties for curtailed renewable energy: The penalty for curtailed photovoltaic power increases from $0 to $815/h. The penalty for curtailed wind power rises from $0 to $1081/h. These results indicate that the SC-IES struggles to fully utilize renewable energy under fault conditions, leading to higher economic costs. Overall, the total system operating cost increases from $16,330 to $19,460/h, representing a 19.2% increase due to the redispatch and additional constraints imposed by the fault.

3.4.4 Sensitivity Analysis

To validate the robustness of the study results, a sensitivity analysis was conducted on the SC-IES after redispatch, focusing on the influence of two key pipeline leakage parameters: maximum leakage area A_{max} and leakage growth time Δt. These experiments provide insights into the variation patterns of renewable energy curtailment rates under different leakage conditions and identify the critical factors affecting system fault tolerance.

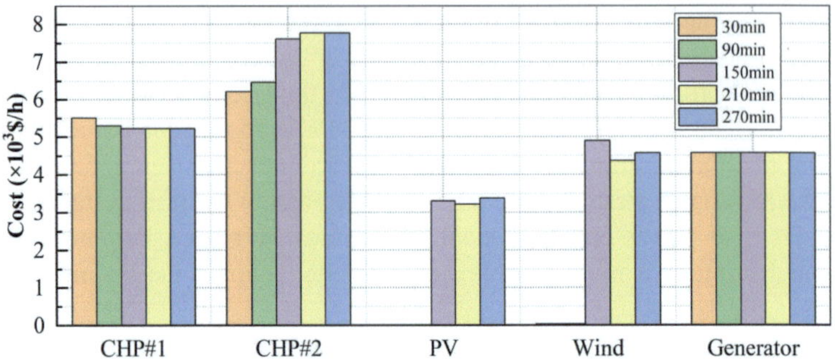

Fig. 3.11 Operational cost of SC-IES

(1) Experiment 1: Impact of Maximum Leakage Area on Renewable Energy Curtailment Rate

Figure 3.12a illustrates the maximum curtailment rates of PV and wind power under different pipeline leakage severities. In this experiment, the leakage growth time Δt is fixed at 7200 s, while the maximum leakage area $A_{\max}$ increases from 0.5×3600^{-1} m^2 to 8×3600^{-1} m^2 in increments of 0.5×3600^{-1} m^2. The experimental results can be categorized into two distinct phases:

Phase S1: When $A_{\max} < 3.5 \times 3600^{-1}$ m^2, renewable energy curtailment remains at zero, indicating that the system exhibits strong fault tolerance in this range. This phase can be further divided into two subphases: For relatively small leakage sizes, CHP#1 does not reach its thermoelectric coupling limit curve, and its electrical power output remains unchanged, ensuring that the power system's operating state is unaffected. As leakage severity increases, CHP#1 reaches its constraint curve, but as long as CHP#2 remains within its operational limits, its electrical power output can be adjusted in coordination with CHP#1, maintaining power balance and preventing

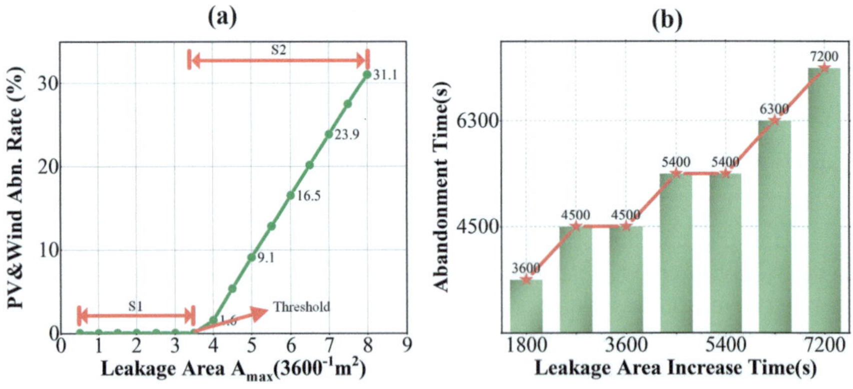

Fig. 3.12 Sensitivity analysis results: **a** sensitivity analysis on $A_{\max}$; **b** sensitivity analysis on Δt

renewable energy curtailment. These findings suggest that the operating margins of CHP#1 and CHP#2 relative to their limit curves are the two key factors determining the fault tolerance of the combined heating and power IES.

Phase S2: When $A_{max} > 3.5 \times 3600^{-1}$ m^2, renewable energy curtailment rises significantly as A_{max} increases. At this stage, CHP#2 reaches its operational limit curve, fixing its electrical power output at 42 MW. Since CHP#2 can no longer adjust its power output, the power balance must be maintained by increasing CHP#1's electrical power output, which requires a corresponding increase in renewable generation. However, CHP#1's operation is constrained by its thermoelectric coupling limit, preventing it from fully accommodating the variability of renewable power. As a result, wind and PV curtailment rates increase progressively. This demonstrates that at higher leakage severities, the system's flexibility is significantly reduced, making it difficult to further enhance renewable energy utilization.

(2) Experiment 2: Impact of Leakage Growth Time on the Onset of Renewable Energy Curtailment

In the second experiment, the impact of leakage growth time Δt on the timing of renewable energy curtailment was investigated. The maximum leakage area A_{max} was fixed at 6×3600^{-1} m^2, while Δt was varied from 900 to 7200 s. Figure 3.12b presents the results, showing a strong correlation between the onset time of renewable energy curtailment and the leakage growth time Δt. Specifically, as Δt increases, the onset time of renewable energy curtailment extends from 1800 to 7200 s. This occurs because a longer leakage growth time results in a slower system response to the fault, meaning that CHP operating points take longer to reach their constraint curves. This delayed effect provides a wider time window for dispatch optimization, allowing the system to dynamically adjust its operating state more effectively, thereby delaying the onset of renewable energy curtailment. These findings suggest that adjusting dispatch strategies in response to the leakage growth rate could enhance system resilience, improving renewable energy integration while maintaining system stability.

3.4.5 *Discussion and Recommendations*

This section summarizes the operational characteristics of the combined heating and power IES under pipeline leakage conditions, examining its impact on the DHS, CHP units, and power system. Additionally, operational recommendations are provided to enhance system reliability and fault tolerance.

(1) Pipeline leakage significantly affects both the hydraulic and thermal properties of the DHS. Leakage causes a pressure drop at load nodes, a significant increase in flow at the constant-pressure heat source, a slight increase in flow at auxiliary heat sources, and a slight decrease in flow at load points. These changes indicate that flow at constant-pressure nodes is most sensitive to leakage. Load

nodes heated by heat-driven CHP units experience minimal impact because their heating temperature remains constant. However, in load nodes heated by electrically driven CHP units, the heating temperature is significantly affected in strongly coupled systems, while in weakly coupled systems, the impact is relatively minor. Overall, end users in the DHS experience minimal impact from pipeline leakage. However, operators should conduct regular safety inspections to detect potential issues early. During inspections: Flow meters and pressure gauges are more effective than temperature sensors for leakage detection. It is recommended to install flow meters around constant-pressure nodes to monitor flow variations. Pressure gauges should be deployed at coupling nodes to improve sensitivity to pressure changes. These measures will enhance the operational security of the DHS.

(2) Pipeline leakage significantly affects the operation of CHP units and the power system. Due to the increase in flow at the constant-pressure node, the thermal power output of heat-driven CHP units rises significantly. Once the CHP operating point reaches its thermoelectric coupling limit curve, its electrical power output must also increase. This may lead to voltage violations at power nodes near heat-driven CHP units and cause significant changes in power flow distribution. Due to power balance constraints, the slack node's power output decreases, further altering the power distribution in the system. The impact of pipeline leakage on the integrated energy system depends not only on leakage severity but also on the operational margin between the CHP units and their limit curves. Operators should closely monitor heating power trends. Under normal conditions, heat demand fluctuates periodically rather than increasing continuously. If continuous heating power increases are observed, adjusting CHP electrical power output accordingly can extend the operational margin between the CHP's operating point and its limit curve. This strategy provides additional fault tolerance for the power system, reduces operational stress, and enhances overall system stability.

(3) The impact of pipeline leakage is significantly greater in SC-IESs than in weakly coupled ones. In weakly coupled systems, the thermal power output of electrically driven CHP units remains unchanged. Flow variations cause only slight imbalances in the DHS and minor fluctuations in heating temperature. In strongly coupled systems, electrically driven CHP units experience reductions in both electrical and thermal power output, leading to severe heat power imbalances in areas heated by power-controlled heat sources. To meet user heating demands, strongly coupled systems must release stored thermal energy from pipelines, causing a significant drop in network temperature. Additionally, to maintain system stability and heating quality, strongly coupled systems must accept higher renewable energy curtailment rates, reducing economic efficiency and overall operational performance. Given these characteristics, the following recommendations are proposed: First, it is recommended that operators consider equipping the system with more energy storage devices to enhance operational flexibility. These storage devices can provide additional regulation capacity in the event of insufficient thermal power, thereby mitigating the impact

of pipeline leakage on system operation. Second, it is advised to increase the distance between the operating point of electrically driven CHP units and their constraint curves to prevent them from reaching their limit states too early. Additionally, implementing heat load reduction measures can effectively prevent thermal power imbalances and improve the system's fault tolerance.

3.5 Conclusions

This chapter investigated the cross-system fault propagation mechanisms in combined heating and power IESs, with a particular focus on pipeline leakage. The core objective of this study was to simulate the operational state of the integrated system under pipeline leakage conditions, providing valuable theoretical insights and practical guidance for ensuring system security. First, a pipeline leakage model was developed, fully considering the hydraulic and thermal effects of leakage on the DHS. Then, an innovative dynamic simulation approach for combined heating and power IESs under fault conditions was proposed. This approach consists of a multi-stage simulation framework and detailed solution algorithms for the DHS, hydraulic system, and coupled equipment. Finally, a case study was conducted to analyze the impacts of pipeline leakage on both strongly and WC-IESs.

The results indicate that in a SC-IES, the impact of pipeline leakage propagates from the DHS to the power system via coupled equipment and subsequently feeds back into the DHS. This strong coupling characteristic makes the system's response to pipeline leakage more complex, where the distance between the operating points of CHP units and their constraint curves becomes a key factor influencing system fault tolerance. When CHP units operate near their constraint curves, the system's regulation capacity decreases significantly, leading to substantial impacts on both the DHS and power system operation. In contrast, a WC-IES, due to its lower coupling degree, exhibits more stable operating points for electrically driven CHP units. As a result, the impact of pipeline leakage remains primarily confined to the DHS, demonstrating higher fault tolerance.

Overall, this study highlights the importance of incorporating fault scenarios into the design and operation of combined heating and power IESs to ensure system security and reliability. The in-depth analysis of pipeline leakage fault propagation mechanisms provides crucial theoretical support and practical references for fault diagnosis, fault-tolerant design, and operational optimization of integrated energy systems. For SC-IESs, the findings suggest that increasing the margin between CHP operating points and constraint curves, optimizing the operation strategies of coupled equipment, and enhancing system energy storage capacity are effective measures to improve system fault tolerance.

References

1. Abomazid, A. M., El-Taweel, N. A., & Farag, H. E. Z. (2022). Optimal energy management of hydrogen energy facility using integrated battery energy storage and solar photovoltaic systems. *IEEE Transactions on Sustainable Energy, 13*(3), 1457–1468.
2. Wang, J., Pinson, P., Chatzivasileiadis, S., et al. (2022). On Machine Learning-Based Techniques for Future Sustainable and Resilient Energy Systems. *IEEE Transactions on Sustainable Energy, 14*(2), 1230–1243.
3. Karimi, H., & Jadid, S. (2023). Multi-layer energy management of smart integrated-energy microgrid systems considering generation and demand-side flexibility. *Applied Energy, 339*, 120984.
4. Lv, C., Liang, R., Jin, W., et al. (2022). Multi-stage resilience scheduling of electricity-gas integrated energy system with multi-level decentralized reserve. *Applied Energy, 317*, 119165.
5. Tian, H., Zhao, H., Liu, C., et al. (2022). A dual-driven linear modeling approach for multiple energy flow calculation in electricity-heat system. *Applied Energy, 314*, 118872.
6. Juanwei, C., Tao, Y., Yue, X., et al. (2019). Fast analytical method for reliability evaluation of electricity-gas integrated energy system considering dispatch strategies. *Applied Energy, 242*, 260–272.
7. Li, G., Bie, Z., Kou, Y., et al. (2016). Reliability evaluation of integrated energy systems based on smart agent communication. *Applied Energy, 167*, 397–406.
8. Liu, L., Wang, D., Hou, K., et al. (2020). Region model and application of regional integrated energy system security analysis. *Applied Energy, 260*, 114268.
9. Fan, Q., Guo, Y., Wu, S., et al. (2019). Two-level diagnosis of heating pipe network leakage based on deep belief network. *IEEE Access, 7*, 182983–182992.
10. Shekofteh, M., Jalili, G. M., & Yazdi, J. (2020). A methodology for leak detection in water distribution networks using graph theory and artificial neural network. *Urban Water Journal, 17*(6), 525–533.
11. Liu, Z., et al. (2023). Risk assessment and alleviation of regional integrated energy system considering cross-system failures. *Applied Energy, 350*, 121714.
12. Qin, X., Sun, H., Shen, X., et al. (2019). A generalized quasi-dynamic model for electric-heat coupling integrated energy system with distributed energy resources. *Applied Energy, 251*, 119165.
13. Yao, S., et al. (2021). Dynamic optimal energy flow in the heat and electricity integrated energy system. *IEEE Transactions on Sustainable Energy, 12*(1), 179–190.
14. Yao, S., et al. (2022). Dynamic energy flow analysis of the heat-electricity integrated energy systems with a novel decomposition-iteration algorithm. *Applied Energy, 322*, 119492.
15. Yu, R., et al. (2022). Non-iterative calculation of quasi-dynamic energy flow in the heat and electricity integrated energy systems. *IEEE Transactions on Power Systems, 38*(5), 4148–4164.
16. Ma, H., Liu, C., Zhao, H., et al. (2023). A novel analytical unified energy flow calculation method for integrated energy systems based on holomorphic embedding. *Applied Energy, 344*, 121163.
17. Lu, S., Xu, Y., Gu, W., et al. (2023). On thermal dynamics embedded static voltage stability margin. *IEEE Transactions on Power Systems, 38*(3), 2982–2985.
18. Zhang, S., et al. (2023). Steady-state security region of integrated energy system considering thermal dynamics. *IEEE Transactions on Power Systems, 39*(2), 4138–4153.
19. Pan, Z., Wu, J., Sun, H., et al. (2019). Quasi-dynamic interactions and security control of integrated electricity and heating systems in normal operations. *CSEE Journal of Power and Energy Systems, 5*(1), 120–129.
20. Zhao, J., et al. (2022). Reliability evaluation of community integrated energy systems based on fault incidence matrix. *Sustainable Cities and Society, 2022*(80), 103769.
21. Zhou, S., O'Neill, Z., & O'Neill, C. (2018). A review of leakage detection methods for district heating networks. *Applied Thermal Engineering, 137*, 567–574.
22. Peng, X., Anyaoha, U., Liu, Z., et al. (2020). Analysis of magnetic-flux leakage (MFL) data for pipeline corrosion assessment. *IEEE Transactions on Magnetics, 56*(6), 1–15.

23. Zhou, S., Liu, C., Zhao, Y., et al. (2022). Leakage diagnosis of heating pipe-network based on BP neural network. *Sustainable Energy, Grids and Networks, 32*, 100869.
24. Zhou, S., Li, H., Gong, P., et al. (2017). Hydraulic modeling of double-source and ring-shaped heating networks. *Applied Thermal Engineering, 119*, 215–221.
25. Zheng, X., et al. (2021). Leak detection of long-distance district heating pipeline, a hydraulic transient model-based approach. *Energy, 237*, 121604.
26. Quinones-Grueiro, M., Milián, M. A., Rivero, M. S., et al. (2021). Robust leak localization in water distribution networks using computational intelligence. *Neurocomputing, 438*, 195–208.
27. Ying, L., & Hong, H. (2020). Application of database systems in leakage fault diagnosis of district eating networks. *Regional Heating, 02*, 32–38.
28. Pan, Z., Guo, Q., & Sun, H. (2016). Interactions of district electricity and heating systems considering time-scale characteristics based on quasi-steady multi-energy flow. *Applied Energy, 167*, 230–243.
29. Guan, A., Zhou, S., Gu, W., et al. (2022). A novel dynamic simulation approach for gas-heat-electric coupled system. *Applied Energy, 315*, 118999.
30. Milano, F. (2008). Continuous Newton's method for power flow analysis. *IEEE Transactions on Power Systems, 24*(1), 50–57.
31. Guan, A. (2023). *Data of case in cross-system fault propagation mechanism in heat-electric integrated energy system*. Available: https://seunic-my.sharepoint.cn/:f:/g/personal/220213085_seu_edu_cn/EuGNPYd0dllDkyKKSIehZ6ABFQJBXCQYz9d_cfbXXFoGgQ?e=8x3QTl

Chapter 4
Resilience Assessment for IES Considering Gas-Thermal Inertia and System Interdependency

Abstract As fundamental attributes of integrated energy systems (IESs), gas-thermal inertia and power-gas-heat interdependency significantly impact system resilience under extreme events. This chapter presents an integrated assessment framework that systematically examines these dual factors' impacts on IES resilience dynamics. The framework synthesizes three critical aspects: (1) operational constraints of the power distribution system (PDS), natural gas system (NGS), and district heating system (DHS); (2) multi-energy system interactions; and (3) gas-thermal inertial dynamics. A dynamic optimization model of energy flows is developed to replicate cascading failure progression and restoration sequences during extreme outages. Based on this, multi-stage resilience curves for the entire IES and its subsystems are derived. A set of interdependency metrics is further proposed based on the geometric features of resilience curves to quantify the vulnerability sensitivity of NGS and DHS arising from their interdependency with PDS. Comparative experimental analysis evaluates the impacts of thermal-gas inertia and system interdependency on IES resilience. Results indicate that the proposed metrics effectively evaluate the spatiotemporal interdependency among the power, gas, and thermal subsystems in terms of resilience, providing actionable support for operators to identify critical links and assess system upgrade strategies to enhance IES resilience.

Keywords Resilience assessment · Integrated energy system · Gas-thermal inertia · Interdependency · Resilience metric

Nomenclature

s	Index of outage scenario
t	Index of time interval
$i/j/k/n$	Index of node
ij/l	Index of power line/gas pipeline/heat pipeline
$\mathcal{N}_{\mathrm{P/G}}$	Set of nodes in PDS/NGS
$\mathcal{N}^{\mathrm{H}}_{\mathrm{Sn/Bn/In}}$	Set of source/load/intersection nodes in DHS

Z. Wu et al., *Integrated Energy System Resilience*,
https://doi.org/10.1007/978-981-95-2363-4_4

$\mathcal{N}^{\mathrm{P}}_{\mathrm{GT/PtG/Com}}$	Set of GT/PtG/compressor nodes in PDS
$\mathcal{N}^{\mathrm{P}}_{\mathrm{CHP/EB/WP}}$	Set of CHP/EB/water pump nodes in PDS
$\mathcal{N}^{\mathrm{G}}_{\mathrm{GT/PtG/Com/Sta}}$	Set of GT/PtG/compressor/gas station nodes in NGS
$\mathcal{N}^{\mathrm{H}}_{\mathrm{CHP/EB/Sn/WP}}$	Set of CHP/EB/source/water pump nodes in DHS
$\mathcal{L}_{\mathrm{P/G/H}}$	Set of power lines/gas pipelines/heat pipelines in PDS/NGS/DHS
$\mathcal{L}^{i+/-}_{\mathrm{H}}$	Set of heat pipelines flow out of/flow into node i
$\mathcal{R}$	Set of source nodes in PDS
$\mathcal{S}$	Set of outage scenarios
$\mathcal{T}$	Set of time interval
$C_{\mathrm{g2}e}/C_{\mathrm{e2g}}$	Gas to electricity/electricity to gas constant
$H^{\mathrm{CHP}}_{i\min}/H^{\mathrm{CHP}}_{i\max}$	Minimum/maximum heat power of CHP, kW
M	Large constant
N_{P}	Number of power nodes
N_{s}	Number of outage scenarios
$P^{\mathrm{D}}_{i,t}/Q^{\mathrm{D}}_{i,t}$	Active/reactive power demand, kW/kVar
$P^{\mathrm{GT}}_{i,\max}/P^{\mathrm{ptG}}_{i,\max}$	Maximum active power of GT/PtG, kW
$P^{\mathrm{CHP}}_{i,\min}/P^{\mathrm{CHP}}_{i,\max}$	Minimum/maximum active power of CHP, kW
P^{WP}_{i}	Rated power consumption of water pump, kW
$Pr_{i,\min}/Pr_{i,\max}$	Lower/upper bounds of gas pressure, Pa
$\mathrm{Pr}^{0}_{i,s,t}/\rho^{0}_{i,s,t}$	Pressure/density of the natural gas source, Pa/[g/m^3]
R_{ij}/X_{ij}	Resistance/reactance of power line
$R^{\mathrm{GT}}_{i\max}/R^{\mathrm{PtG}}_{i\max}$	Maximum ramping rate of GT/PtG, kW/h
φ_{GT}	Power factor of GT
Δt	Duration of time interval
$\xi_{ij,s,t}$	Binary, status of power line (1 if damaged, 0 otherwise)
$L^{\mathrm{G}}_{ij}/A^{\mathrm{G}}_{ij}/D^{\mathrm{G}}_{ij}$	Length/cross-sectional area/diameter of gas pipeline, m/m^2/m
$\overline{\omega}^{\mathrm{G}}_{ij}/\lambda^{\mathrm{G}}_{ij}$	Friction coefficient/ gas velocity of gas pipeline, [p.u.] [m/s]
c/c^{H}_{w}	Sound velocity/heat capacity of water, [m/s]/[kJ/(kg·°C)]
$M^{\mathrm{D}}_{i,t}$	Mass flow rate of gas demand, g/s
Γ_{Com}	Compression ratio of gas compressor
$\lambda^{\mathrm{H}}_{l}/L^{\mathrm{H}}_{l}/A^{\mathrm{H}}_{l}/\rho_{w}$	Heat transfer coefficient/length/cross-sectional area of heat pipeline/water density, [kW/(m °C)]/m/m^2/[kg/m^3]
τ^{amb}_{t}	Ambient temperature of heat pipeline in DHS
τ^{gain}_{t}	Temperature gain of buildings from solar radiation
$\tau^{\mathrm{in}}_{i,\min}/\tau^{\mathrm{in}}_{i,\max}$	Lower/upper bounds of comfortable indoor temperature, °C
$\eta_{\mathrm{GT/PtG/Com}}$	Efficiency of GT/PtG/compressor
$\eta_{\mathrm{EB/WP}}$	Efficiency of EB/water pump
$\eta_{\mathrm{CHP,\,p}}/\eta_{\mathrm{CHP,\,h}}$	Gas to power/gas to heat efficiency of CHP
$\omega^{\mathrm{P}}_{i}/\omega^{\mathrm{G}}_{i}$	Weight of power/gas demand in PDS/NGS
ω^{H}_{i}	Number of buildings of DHS node
$f_{ij,s,t}$	Fictitious power flow through power line
$h^{\mathrm{D}}_{i,s,t}$	Supply status of node heat load
$H^{\mathrm{CHP}}_{i,s,t}/H^{\mathrm{EB}}_{i,s,t}$	Heat power of CHP/EB, kW

$H_{i,t}^{\mathrm{D}}$	Heat demand of buildings in DHS node, kW
$I_{ij,s,t}$	Squared power line current, $(\mathrm{A})^2$
$I_{ij,\min}/I_{ij,\max}$	Minimum/maximum line current flow, $(\mathrm{A})^2$
$S_{ij,\max}^{\mathrm{L}}$	Maximum capacity of power lines
$M_{ij,\min}^{G}/M_{ij,\max}^{G}$	Minimum/maximum gas flow rate in pipeline, g/s
$M_{ij,s,t}^{\mathrm{in}}/M_{ij,s,t}^{\mathrm{out}}$	Inlet/outlet mass flow rate of gas pipeline, g/s
$M_{i,s,t}^{\mathrm{Sta}}/M_{i,s,t}^{\mathrm{CHP}}$	Mass flow rate of gas station/CHP, g/s
$M_{i,\max}^{\mathrm{Sta}}$	Maximum mass flow rate of gas station, g/s
$M_{i,s,t}^{\mathrm{GT}}/M_{i,s,t}^{\mathrm{PtG}}$	Mass flow rate of GT/PtG, g/s
$M_{ij,s,t}^{\mathrm{Com}}$	Mass flow rate through the gas compressor, g/s
$M_{i,s,t}^{\mathrm{Cut}}$	Mass flow rate of curtailed gas demand, g/s
$m_{l,t}^{\mathrm{H}}$	Mass flow rate in heat pipeline, kg/s
M_{l}^{H}	Maximum transmission delay of pipeline in DHS
$H_{l,s,t}^{\mathrm{S,in}}/H_{l,s,t}^{\mathrm{R,in}}$	Inflow heat power of supply/return pipeline, kW
$H_{l,s,t}^{\mathrm{S,out}}/H_{l,s,t}^{\mathrm{R,out}}$	Outflow heat power of supply/return pipeline, kW
$H_{i,s,t}^{\mathrm{D}}$	Heat power of node load, kW
$Pr_{i,s,t}/\rho_{i,s,t}$	Pressure/density of gas node, Pa/[g/m^3]
$Pr_{i,s,t}^{\mathrm{Com}}/Pr_{j,s,t}^{\mathrm{Com}}$	Node pressure at compressor inlet/outlet, Pa
$P_{i,s,t}^{\mathrm{GT}}/P_{i,s,t}^{\mathrm{PtG}}$	Active power of GT/PtG, kW
$P_{i,s,t}^{\mathrm{CHP}}/P_{i,s,t}^{\mathrm{EB}}$	Active power of CHP/EB, kW
$P_{i,s,t}^{\mathrm{WP}}/P_{i,s.t}^{\mathrm{Com}}$	Active power consumption of water pump/compressor, kW
$P_{ij,s,t}^{\mathrm{L}}/Q_{ij,s,t}^{\mathrm{L}}$	Active/reactive power flow on power line, kW/kVar
$P_{i,s,t}^{\mathrm{Cut}}/Q_{i,s,t}^{\mathrm{Cut}}$	Curtailed active/reactive power demand, kW/kVar
$Q_{i,s.t}^{\mathrm{GT}}/Q_{i,s.t}^{\mathrm{CHP}}$	Reactive power of GT/CHP, kVar
$u_{i,s,t}^{\mathrm{GT/PtG/Com}}$	Binary, operating status of GT/PtG/compressor (1 if operated, 0 other- wise)
$u_{i,t}^{\mathrm{CHP/EB/WP}}$	Binary, operating status of CHP/EB/water pump (1 if operated, 0 otherwise)
$V_{i,s,t}$	Squared node voltage, (kV)2
$V_{i,\min}/V_{i,\max}$	Minimum/maximum node voltage
$y_{ij,s,t}$	Binary, auxiliary variable of power line (1 if connected, 0 otherwise)
$\alpha_{ij,s,t}$	Binary, connection status of power line (1 if connected, 0 otherwise)
$\beta_{ij,s,t}$	Binary, fictitious connection status of power line (1 if on, 0 otherwise)
$\eta_{l,t}^{\mathrm{H}}$	Heat loss coefficient of pipeline in DHS
$\tau_{l,s,t}^{\mathrm{S,in}}/\tau_{l,s,t}^{\mathrm{S,out}}$	Inflow/outflow temperature of supply pipeline, °C
$\tau_{l,t}^{\mathrm{R,in}}/\tau_{l,t}^{\mathrm{R,out}}$	Inflow/outflow temperature of return pipeline, °C
$\tau_{i,t}^{\mathrm{in}}$	Indoor temperature of buildings in DHS node, °C

$\tau_{l,t}^{\mathrm{S.out*}}/\tau_{l,t}^{\mathrm{R.out*}}$	Outflow temperature of supply/return pipeline in DHS when ignoring heat losses, °C
$t_{ij,s}^{\mathrm{fault}}/t_{ij,s}^{\mathrm{repair}}$	Fault time/repair time
$\chi_t^{l,k}/\delta_t^{l,k}$	Variables related to pipeline transmission delay in DHS
$a_i^{\mathrm{B}}/b_i^{\mathrm{B}}/c_i^{\mathrm{B}}$	Relevant parameters of the first-order equivalent heat parameter mode for heat load
$c_{\mathrm{v1},i}/c_{\mathrm{v2},i}/c_{\mathrm{m},i}$	Relevant parameters for convex polygon constraints
$Z/R/T/\varepsilon$	Compressibility factor/gas constant/average gas temperature/isentropic exponent
CCM	Convex-concave method
CHP	Combined heat and power
DHS	District heating system
EB	Electric boiler
GT	Gas turbine
IES	Integrated energy system
MILP	Mixed-integer linear programming
NGS	Natural gas system
PDE	Partial differential equation
PDS	Power distribution system
PtG	Power to gas
SOC	Second-order cone
ST	Spinning tree

4.1 Introduction

Resilience assessment and vulnerability identification of the integrated energy systems (IES) serve as the foundation for resilience planning and energy supply restoration research. However, current studies on resilience evaluation of multi-energy coupled IESs still face several critical challenges requiring urgent resolution. First, existing research predominantly focuses on evaluating the holistic resilience attributes of IESs, while insufficient attention has been paid to meticulous modeling of dynamic-static characteristics and deep interdependency among heterogeneous electric-gas-heat subsystems under extreme fault scenarios. This oversight hinders effective evaluation of multi-energy interdependency' impacts on IES resilience. Second, there remains a notable deficiency in establishing micro-level assessment metrics and analytical methodologies for electric-gas-heat interdependency, thereby impeding the identification of system vulnerabilities induced by multi-energy coupling effects. Moreover, the absence of evaluation metrics and analytical methods for electric-gas-heat interdependency from a microscopic perspective impedes the identification of system vulnerabilities caused by multi-energy coupling. Furthermore, gas and thermal inertia, which are critical for enhancing IES operational flexibility during extreme disasters, remain insufficiently addressed in existing IES resilience assessment models.

To address the existing challenges, this chapter proposes a resilience assessment method for the IES that incorporates gas/thermal inertia and multi-energy interdependency. The framework holistically integrates dynamic and static operational characteristics of heterogeneous electrical, gas, and thermal subsystems, including the DistFlow power flow and radial topology constraints in PDS, linepack characteristics and gas flow constraints in NGS, as well as slow dynamic pipeline transmission and building thermal inertia constraints in district heating systems. An extreme contingency-oriented multi-energy flow operation optimization model is thereby established. Compared to existing models, the proposed framework achieves more comprehensive and meticulous modeling of dynamic-static characteristics and deep interdependency across heterogeneous energy subsystems. Subsequently, under a set of extreme fault scenarios, the multi-energy cascading failures and energy restoration processes of the IES are simulated. Based on simulation results, multi-phase resilience curves are generated for both the integrated system and individual subsystems. Furthermore, leveraging geometric features of these resilience curves, a micro-level assessment indicator system with mathematical formulations is proposed from temporal, spatial, systemic, and nodal dimensions. This system enables holistic quantification of multi-energy interdependency under fault conditions and precise identification of vulnerabilities induced by tight multi-energy coupling effects. Finally, through numerical experiments, the specific impacts of multi-energy interdependency and gas/thermal inertia on IES resilience during both multi-energy cascading failure propagation and energy supply restoration phases are systematically evaluated and analyzed.

The chapter is organized as follows: Sect. 4.2 introduces the framework of the proposed IES resilience assessment method. Sect. 4.3 presents the mathematical formulation of the multi-energy flow operation optimization model for IESs under extreme scenarios. Sect. 4.4 details the proposed assessment indicators for quantifying multi-energy interdependency. Sect. 4.5 conducts numerical case studies to validate the methodology. Finally, Sect. 4.6 provides a comprehensive summary of the chapter.

4.2 Resilience Assessment Framework

4.2.1 Multi-energy Interdependency Under Extreme Fault Scenarios

This chapter investigates a distribution-level IES, whose architecture is depicted in Fig. 4.1. Within the IES, the power distribution system couples with the natural gas system through gas turbines, power-to-gas (PtG) facilities, and electric-driven compressors [1, 2], while interconnecting with the district heating system via

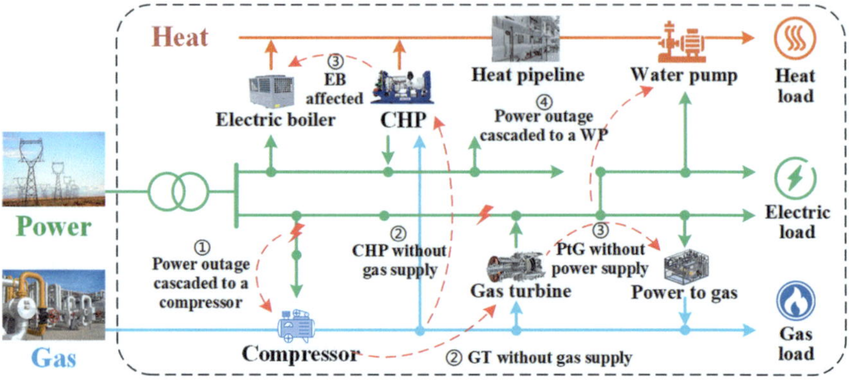

Fig. 4.1 Structure of distribution-level power-gas-heat IES and fault propagation process

combined heat and power (CHP) units, electric boilers, and electric-driven water pumps [2].

To elucidate the multi-energy interdependency within coupled systems under extreme fault scenarios, this section employs the IES illustrated in Fig. 4.1 to demonstrate the cascading propagation of electrical faults to natural gas and heating systems. As is shown in the figure, a fault in the power lines of the distribution system triggers the shutdown of electric-driven compressors, subsequently causing downstream natural gas supply interruptions that disable the normal operation of gas turbines and CHP units. The loss of power supply further halts PtG facilities and electric-driven water pumps, leading to energy supply interruptions for downstream gas and thermal loads [2]. Simultaneously, the malfunction of CHP units exacerbates the operational failure of electric boilers, expanding the scope of the cascading outages. This analysis reveals that the deep coupling among electricity, gas, and heat subsystems in IES significantly amplifies cascading failure risks during disasters, highlighting the potential adverse effects of multi-energy interdependency in extreme events. Conversely, multi-energy interdependency can also play a positive role in post-disaster recovery. Following extreme events, the IES leverages complementary interactions among electricity, gas, and heat subsystems to enable mutual support across heterogeneous energy systems, thereby reducing overall load-shedding. In conclusion, a thorough analysis of multi-energy interdependency is critical for assessing IES resilience attributes and identifying key factors for resilience enhancement.

4.2.2 *Modeling and Characteristics of Gas and Thermal Inertia*

In IES, the slow dynamic characteristics (referred to as inertia) of natural gas and district heating systems provide additional operational flexibility during system failures [3, 4]. This subsection develops mathematical models for gas inertia and thermal inertia, and analyzes their behavioral characteristics under fault scenarios.

Modeling of Gas Inertia: The schematic structure of the natural gas system is illustrated in Fig. 4.2. Gas inertia primarily stems from the slow transmission process of natural gas in pipelines and the compressibility of gas (pipeline storage effect, referred to as linepack). During disasters, the natural gas system may face gas supply interruptions. In the initial phase of a disaster, gas inertia can absorb instantaneous pressure fluctuations caused by abrupt faults, mitigate the impact of rapid gas flow variations on other system components, and reduce cascading failure effects. Furthermore, the linepack of gas networks can temporarily sustain partial gas load supply, alleviating natural gas shortages in disaster-affected areas. Thus, gas inertia significantly influences IES resilience. In contrast to the steady-state gas flow models widely adopted in existing IES assessment methods, this chapter employs a dynamic gas flow model that explicitly captures gas inertia characteristics [5].

The schematic diagram of NGS is shown in Fig. 4.2. The gas dynamic behavior within natural gas pipelines can be described by a set of nonlinear partial differential equations (PDEs). However, directly solving these equations often incurs prohibitive computational costs. To balance computational accuracy with reduced modeling complexity, a linearized quasi-dynamic gas flow model is employed to ensure solution efficiency [5]. Specifically, the hydrodynamic equations for natural gas encompass the Navier–Stokes momentum equation, mass conservation equation, and equation of state, formulated in Eqs. (4.1)–(4.3), respectively. Notably, Eqs. (4.1) and (4.2) are discretized using the Wendroff differencing forms to approximate the PDEs [5, 6].

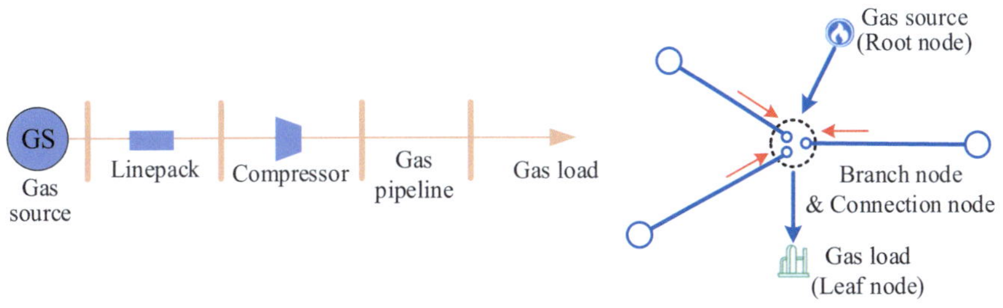

Fig. 4.2 Schematic diagram of the natural gas system

$$\begin{aligned}
&\frac{\lambda_{ij}^{G}\overline{\omega}_{ij}^{G}\Delta t}{4D_{ij}^{G}A_{ij}^{G}}\left(M_{ij,t+1}^{\text{out}}+M_{ij,t+1}^{\text{in}}+M_{ij,t}^{\text{out}}+M_{ij,t}^{\text{in}}\right)\\
&+\frac{1}{A_{ij}^{G}}\left(M_{ij,t+1}^{\text{out}}+M_{ij,t+1}^{\text{in}}-M_{ij,t}^{\text{out}}-M_{ij,t}^{\text{in}}\right)\\
&+\frac{\Delta t}{L_{ij}^{G}}\left(Pr_{j,t+1}-Pr_{i,t+1}+Pr_{j,t}-Pr_{i,t}\right)=0;\quad \forall ij\in\mathcal{L}_{G},t\in\mathcal{T}
\end{aligned}\tag{4.1}$$

$$\begin{aligned}
&\frac{\Delta t}{L_{ij}^{G}A_{ij}^{G}}\left(M_{ij,t+1}^{\text{out}}-M_{ij,t+1}^{\text{in}}+M_{ij,t}^{\text{out}}-M_{ij,t}^{\text{in}}\right)\\
&+\rho_{j,t+1}+\rho_{i,t+1}-\rho_{j,t}-\rho_{i,t}=0;\quad \forall ij\in\mathcal{L}_{G},t\in\mathcal{T}
\end{aligned}\tag{4.2}$$

$$Pr_{i,s,t}=c^{2}\rho_{i,s,t};\forall i\in\mathcal{N}_{G},t\in\mathcal{T}\tag{4.3}$$

Modeling of Thermal Inertia: The schematic diagram of the district heating system is illustrated in Fig. 4.3, comprising a heat source, supply water pipelines, return water pipelines, and thermal loads. The thermal inertia of the heating system primarily originates from the heat storage capacity of the pipelines and buildings. The heat-carrying fluid within the pipelines possesses inherent heat capacity, enabling it to store thermal energy. Consequently, even during heat source interruptions, the fluid retained in the pipelines can continue transferring heat, temporarily sustaining partial thermal load demands. Furthermore, the flow rate and temperature of the heat-carrying fluid can be flexibly regulated within a certain range, thereby enhancing the operational flexibility of heat transfer and release. On the demand side, building components such as walls, floors, and indoor air absorb and store thermal energy. When external heating is disrupted, these components gradually release stored heat, delaying indoor temperature decline and maintaining thermal comfort. This implicit storage capability (i.e., thermal inertia) of both pipelines and buildings improves the operational flexibility of IES under extreme fault scenarios.

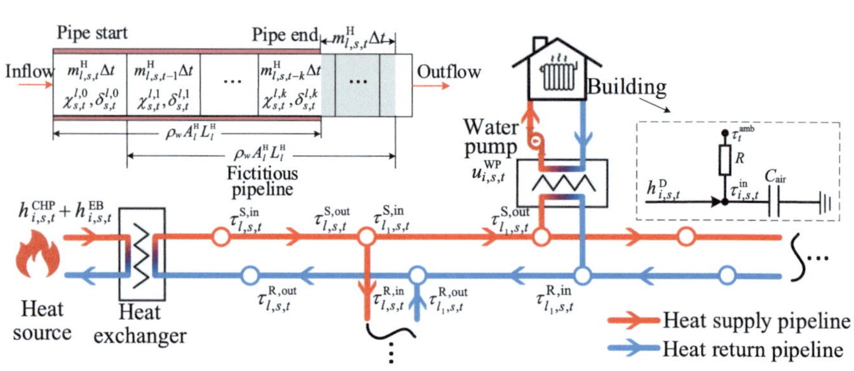

Fig. 4.3 Structure of district heat system

$$c_w^{\mathrm{H}} m_{l,t}^{\mathrm{H}} \tau_{l,t}^{x,\,\mathrm{out}*} = c_w^{\mathrm{H}} \sum_{k=0}^{M_l^{\mathrm{H}}} \left(\chi_t^{l,k} - \delta_t^{l,k}\right) m_{l,t-k}^{\mathrm{H}} \tau_{l,t-k}^{x,\,\mathrm{in}}; \quad \forall x \in \{\mathrm{S}, \mathrm{R}\}, l \in \mathcal{L}_{\mathrm{H}}, t \in \mathcal{T} \tag{4.4}$$

$$\tau_{l,t}^{x,\,\mathrm{out}} = \tau_{l,t}^{x,\,\mathrm{out}*} - \left(1 - \eta_{l,t}^{\mathrm{H}}\right)\left(\tau_{l,t}^{x,\,\mathrm{out}*} - \tau_t^{\mathrm{amb}}\right); \quad \forall x \in \{\mathrm{S}, \mathrm{R}\}, l \in \mathcal{L}_{\mathrm{H}}, t \in \mathcal{T} \tag{4.5}$$

$$\eta_{l,t}^{\mathrm{H}} = \exp\left(-\frac{\lambda_l^{\mathrm{H}} \Delta t}{2 c_w^{\mathrm{H}} \rho_w A_l^{\mathrm{H}}} \left(\sum_{k=0}^{M_l^{\mathrm{H}}} \chi_t^{l,k} + \sum_{k=1}^{M_l^{\mathrm{H}}} \delta_t^{l,k}\right)\right); \quad \forall x \in \{\mathrm{S}, \mathrm{R}\}, l \in \mathcal{L}_{\mathrm{H}}, t \in \mathcal{T} \tag{4.6}$$

The hydraulic constraints of the heating pipelines are formulated in Eqs. (4.7)–(4.10). Equations (4.7)–(4.9) represent the transmission delay constraints of the heat-carrying fluid. The auxiliary variables $\chi_t^{l,k}$ (or $\delta_t^{l,k}$) denote the proportion of the fluid mass block that entered pipeline l (or a virtual pipeline) at time period $t - k$ and remains within the pipeline at time period t. Equation (4.7) defines the feasible range of these auxiliary variables. Equation (4.8) enforces that the inflow rate of the fluid equals the pipeline's capacity. Equation (4.9) specifies that if a fluid mass block entering the pipeline (or virtual pipeline) at time period $t - (k + 1)$ has not fully exited (i.e., $\chi_t^{l,k+1}$ or $\delta_t^{l,k+1}$), then the fluid mass block entering at time period $t - k$ must remain entirely within the pipeline (i.e., $\chi_t^{l,k+1} = 1$ or $\delta_t^{l,k+1} = 1$). Equation (4.10) establishes the flow balance constraint at pipeline junction nodes.

$$0 \le \chi_t^{l,k}, \delta_t^{l,k} \le 1, \quad \delta_t^{l,0} = 1; \quad \forall k = 0, 1, \ldots, M_l^{\mathrm{H}}, l \in \mathcal{L}_{\mathrm{H}}, t \in \mathcal{T} \tag{4.7}$$

$$\sum_{k=0}^{M_l^{\mathrm{H}}} \chi_t^{l,k} m_{l,t-k}^{\mathrm{H}} \Delta t = \rho_w A_l^{\mathrm{H}} L_l^{\mathrm{H}}, \quad \sum_{k=1}^{M_l^{\mathrm{H}}} \delta_t^{l,k} m_{l,t-k}^{\mathrm{H}} \Delta t = \rho_w A_l^{\mathrm{H}} L_l^{\mathrm{H}}; \ \forall l \in \mathcal{L}_{\mathrm{H}}, t \in \mathcal{T} \tag{4.8}$$

$$\left(1 - \chi_t^{l,k}\right) \cdot \chi_t^{l,k+1} = 0, \left(1 - \delta_t^{l,k}\right) \cdot \delta_t^{l,k+1} = 0; \quad \forall k = 0, 1, \cdots, M_l^{\mathrm{H}}, l \in \mathcal{L}_{\mathrm{H}}, t \in \mathcal{T} \tag{4.9}$$

$$\sum_{l \in \mathcal{L}_{\mathrm{H}}^{i-}} m_{l,t}^{\mathrm{H}} \Delta t = \sum_{l \in \mathcal{L}_{\mathrm{H}}^{i+}} m_{l,t}^{\mathrm{H}} \Delta t; \quad \forall i \in \mathcal{N}_{\mathrm{In}}^{\mathrm{H}}, t \in \mathcal{T} \tag{4.10}$$

The thermal inertia constraint for buildings is formulated in Eq. (4.11). When the electrically driven water pump ceases operation due to a power supply interruption (i.e., $u_{i,t}^{\mathrm{WP}} = 0$), the heat injection power to the building becomes zero.

$$\tau_{i,t+1}^{\mathrm{in}} = a_i^{\mathrm{B}} \tau_{i,t}^{\mathrm{in}} + b_i^{\mathrm{B}} H_{i,t}^{\mathrm{D}} u_{i,t}^{\mathrm{WP}} + c_i^{\mathrm{B}} \tau_t^{\mathrm{amb}} + \tau_t^{\mathrm{gain}}; \quad \forall i \in \mathcal{N}_{\mathrm{Bn}}^{\mathrm{H}}, t \in \mathcal{T} \tag{4.11}$$

4.2.3 Resilience Assessment Framework for IES

The proposed resilience assessment framework for IES is illustrated in Fig. 4.4, comprising three key steps: (1) generation of stochastic failure scenarios under disaster-induced uncertainties, (2) simulation of multi-energy cascading failures and energy supply restoration processes in IES, and (3) resilience evaluation of IES based on electric-gas-heat interdependency metrics. This chapter focuses on investigating the propagation of cascading failures from electrical faults to natural gas and heating systems. It is assumed that power lines are vulnerable to extreme disaster events, while underground gas and heating pipelines remain unaffected by direct impacts [7].

Generation of Failure Scenario Sets Under Extreme Disasters: To characterize the uncertainty of disaster causation in extreme hazard events, we first employ the methodology from Ref. [8] to generate a stochastic failure scenario set$\mathcal{S}$, establishing the foundational scenarios for IES cascading failure and energy restoration model. Each scenario $\forall\ s \in \mathcal{S}$ comprehensively describes the fault status $\xi_{ij,s,t}$ (where 1/0 denotes fault/normal operation) for every power line $\forall_{ij} \in \mathcal{L}_{\mathrm{P}}$ throughout the disaster timeline. The operational state of each line is decided by two critical parameters: the failure initiation time $t_{ij,s}^{\mathrm{fault}}$ determined by the spatiotemporal intensity of the evolving hazard and the line's structural vulnerability, and the recovery time $t_{ij,s}^{\mathrm{repair}}$ dependent on crew repair efficiency. For illustrative clarity, consider an 8-h disaster event with hourly resolution. A power line ij remains fully operational if undamaged ($t_{ij,s}^{\mathrm{fault}} = 0$, $t_{ij,s}^{\mathrm{repair}} = 0$), exhibiting a continuous normal status sequence $\xi_{ij,s} = [0,0,0,0,0,0,0,0]$. Conversely, a line experiencing damage initiation at $t_{ij,s}^{\mathrm{fault}} = 3$ and restoration at $t_{ij,s}^{\mathrm{repair}} = 7$ would demonstrate a failure progression $\xi_{ij,s} = [0, 0, 1, 1, 1, 1, 1, 0]$, where the unit-time transitions between operational states explicitly capture the infrastructure's outage window from failure onset to repair completion.

Multi-energy Cascading Failure and Energy Restoration Process Simulation: By comprehensively considering the dynamic-static characteristics of heterogeneous subsystems (electricity, gas, and heat) and operational constraints of coupling devices, this study develops a multi-energy flow operation optimization model for the IES under fault conditions. Through simulating multi-energy cascading failures and

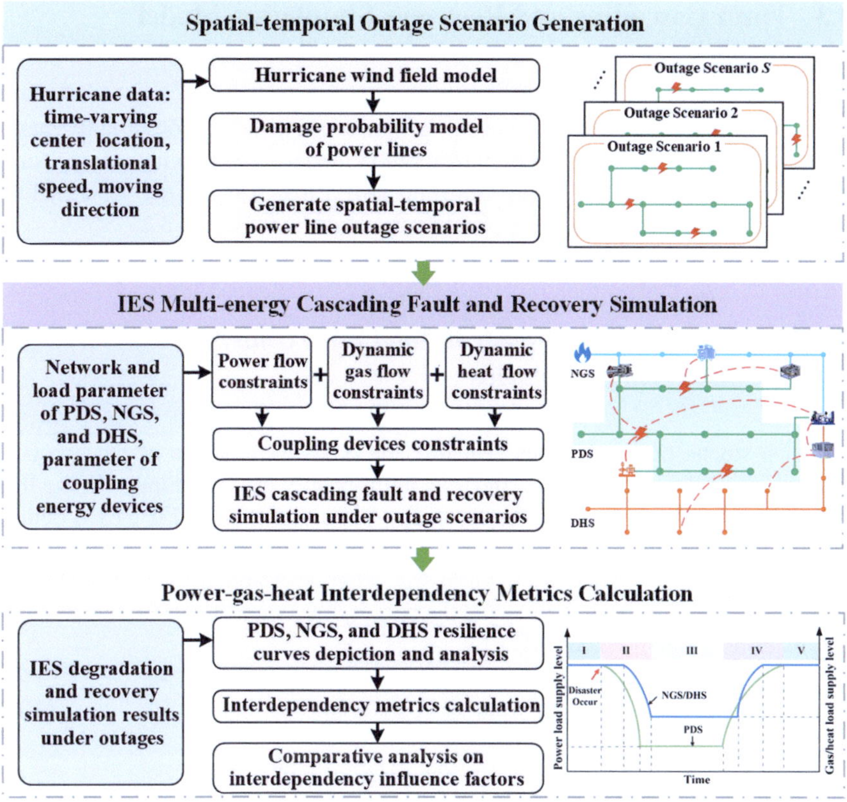

Fig. 4.4 Resilience assessment framework for power-gas-heat IES

energy supply restoration processes under various fault scenarios, this work subsequently generates multi-stage resilience curves that quantify the temporal evolution of load supply levels for both the entire IES and its individual subsystems. The proposed methodology enables systematic evaluation of system-wide and subsystem-specific service recovery capabilities throughout fault progression phases, supported by comprehensive simulation results.

Construction of Electric-Gas-Heat Interdependency Metrics and Resilience Evaluation for IES: Leveraging multi-dimensional geometric features of resilience curves, this study establishes a multi-dimensional comprehensive evaluation index system with mathematical formulations to characterize multi-energy interdependency (electric-gas-heat) under contingency scenarios. Through numerical experiments, the impacts of gas/thermal inertia and multi-energy interdependency on IES resilience are systematically investigated across cascading failure propagation and energy restoration phases, enabling quantitative identification of system vulnerabilities induced by cross-domain energy couplings.

4.3 Fault Cascading and Recovery Simulation Model for IES

This section presents the mathematical formulations of the proposed multi-energy cascading failure and energy supply restoration model for IES, with specific focus on two core components: the quantification of IES load loss in Sect. 4.3.1 and the operational constraints of the multi-energy flow system in Sect. 4.3.2.

4.3.1 Quantification of System Load Loss Under Contingency Scenarios

For each stochastic contingency scenario, the objective function of the proposed model is formulated to minimize the total multi-energy load curtailment in the IES, as mathematically expressed in Eqs. (4.12)–(4.15). Specifically, Eq. (4.13) defines the weighted load curtailment in the power distribution system, while Eq. (4.14) quantifies its counterpart in the natural gas network. Thermal load curtailment is characterized by the deviation beyond the permissible comfort temperature range of buildings, as explicitly modeled in Eq. (4.15).

$$\min : f_{\mathrm{P}} + f_{\mathrm{G}} + f_{\mathrm{H}} \tag{4.12}$$

$$f_{\mathrm{P}} = \sum_{t \in \mathcal{T}} \sum_{i \in \mathcal{N}_{\mathrm{P}}} P_{i,s,t}^{\mathrm{Cut}} \omega_i^{\mathrm{P}} \Delta t; \tag{4.13}$$

$$f_{\mathrm{G}} = \sum_{t \in \mathcal{T}} \sum_{i \in \mathcal{N}_{\mathrm{G}}} M_{i,s,t}^{\mathrm{Cut}} \omega_i^{\mathrm{G}} \Delta t; \tag{4.14}$$

$$f_{\mathrm{H}} = \sum_{t \in \mathcal{T}} \sum_{i \in \mathcal{N}_{\mathrm{Bn}}^{\mathrm{H}}} \left(\max\left\{0, \tau_{i,s,t}^{\mathrm{in}} - \tau_{i,\max}^{\mathrm{in}}, \tau_{i,\min}^{\mathrm{in}} - \tau_{i,s,t}^{\mathrm{in}} \right\} \right). \tag{4.15}$$

4.3.2 System Operation Constraints Considering Gas and Thermal Inertia

Power Distribution System Constraints: The power flow in PDS is modeled via a second-order cone relaxation-based *DistFlow* formulation [2], as explicitly defined in Eqs. (4.16)–(4.19). Specifically, Eqs. (4.16) and (4.17) represent the active and reactive power balance constraints at each electrical node, respectively. Equation (4.18) represents the voltage drop constraint along distribution lines, where the big-M method [9] is employed to relax voltage deviations for de-energized lines, with binary variable $\alpha_{ij,s,t} \in \{0,1\}$ indicating the operational status of line ij(1: energized, 0: de-energized). The second-order cone constraint in Eq. (4.19) serves as a convex relaxation of the original non-convex circular relation $P^2 + Q^2 = VI$, whose exactness under typical operating conditions has been rigorously validated in prior works [2, 10].

$$\sum_{k:k\to i} \left(P^{\mathrm{L}}_{ki,s,t} - I_{ki,s,t}R_{ki}\right) - \sum_{j:i\to j} P^{\mathrm{L}}_{ij,s,t} + P^{\mathrm{GT}}_{i,s,t} + P^{\mathrm{CHP}}_{i,s,t} - P^{\mathrm{PtG}}_{i,s,t}$$
$$- P^{\mathrm{Com}}_{i,s,t} - P^{\mathrm{EB}}_{i,s,t} - P^{\mathrm{WP}}_{i,s,t} = P^{\mathrm{D}}_{i,t} - P^{\mathrm{Cut}}_{i,s,t}; \forall i \in \mathcal{N}_{\mathrm{P}}, s \in \mathcal{S}, t \in \mathcal{T} \tag{4.16}$$

$$\sum_{k:k\to i} \left(Q^{\mathrm{L}}_{ki,s,t} - I_{ki,s,t}X_{ki}\right) - \sum_{ji\to j} Q^{\mathrm{L}}_{ji,s,t} + Q^{\mathrm{GT}}_{i,s,t} + Q^{\mathrm{cTH}}_{i,s,t}$$
$$= Q^{\mathrm{D}}_{i,t} - Q^{\mathrm{Cut}}_{i,s,t}; \quad \forall i \in \mathcal{N}_{\mathrm{P}}, s \in \mathcal{S}, t \in \mathcal{T} \tag{4.17}$$

$$\left|V_{i,s,t} - V_{j,s,t} - 2P^{\mathrm{L}}_{ij,s,t}R_{ij} - 2Q^{\mathrm{L}}_{ij,s,t}X_{ij} - \left(R^2_{ij} + X^2_{ij}\right)I_{ij,s,t}\right|$$
$$\le \left(1 - \alpha_{ij,s,t}\right)M; \ \forall ij \in \mathcal{L}_{\mathrm{P}}, s \in \mathcal{S}, t \in \mathcal{T} \tag{4.18}$$

$$\left\| 2P^{\mathrm{L}}_{ij,s,t} \quad 2Q^{\mathrm{L}}_{ij,s,t} \quad I_{ij,s,t} - V_{i,s,t} \right\|_2 \le I_{ij,s,t} + V_{i,s,t};$$
$$\forall ij \in \mathcal{L}_{\mathrm{P}}, s \in \mathcal{S}, t \in \mathcal{T} \tag{4.19}$$

The power curtailment constraints for electrical loads are governed by Eqs. (4.20)–(4.21), where Eq. (4.20) bounds the active and reactive load-shedding magnitudes within their respective demand limits, while Eq. (4.21) represents the quadratic coupling relationship between active power curtailment and reactive power curtailment.

$$[0,0] \le \left[P^{\mathrm{Cut}}_{i,s,t}, Q^{\mathrm{Cut}}_{i,s,t}\right] \le \left[P^{\mathrm{D}}_{i,t}, Q^{\mathrm{D}}_{i,t}\right]; \quad \forall i \in \mathcal{N}_{\mathrm{P}}, s \in \mathcal{S}, t \in \mathcal{T} \tag{4.20}$$

$$P^{\mathrm{Cut}}_{i,s,t}Q^{\mathrm{D}}_{i,t} - Q^{\mathrm{Cut}}_{i,s,t}P^{\mathrm{D}}_{i,t} = 0; \forall i \in \mathcal{N}_{\mathrm{p}}, s \in \mathcal{S}, t \in \mathcal{T} \tag{4.21}$$

The operational constraints for gas turbines (GTs), including active power bounds, ramping rate limits, and reactive power capacity, are formulated in Eqs. (4.22)–(4.24). Analogous constraint structures with identical variable bounding mechanisms apply to PtG units, EBs, and CHP plants, which are therefore not restated here to avoid redundancy.

$$0 \le P^{\mathrm{GT}}_{i,s,t} \le P^{\mathrm{GT}}_{i,\max}u^{\mathrm{GT}}_{i,s,t}; \forall i \in \mathcal{N}^{\mathrm{P}}_{\mathrm{GT}}, s \in \mathcal{S}, t \in \mathcal{T} \tag{4.22}$$

$$\left|P^{\mathrm{GT}}_{i,s,t} - P^{\mathrm{GT}}_{i,s,t-1}\right| \le R^{\mathrm{GT}}_{i,\max}\Delta t + \left(1 - u^{\mathrm{GT}}_{i,s,t}\right)M; \forall i \in \mathcal{N}^{\mathrm{P}}_{\mathrm{GT}}, s \in \mathcal{S}, t \in \mathcal{T} \tag{4.23}$$

$$\left|Q_{i,s,t}^{\mathrm{GT}}\right| \leq P_{i,s,t}^{\mathrm{GT}} \tan\left(\cos^{-1} \varphi_{\mathrm{GT}}\right); \forall i \in \mathcal{N}_{\mathrm{GT}}^{\mathrm{P}}, s \in \mathcal{S}, t \in \mathcal{T} \tag{4.24}$$

The constraints on node voltages, line currents, as well as the upper and lower bounds for active and reactive power flows in distribution lines are given by Eqs. (4.25)–(4.27), respectively.

$$V_{i,\min}^2 \leq V_{i,s,t} \leq V_{i,\max}^2; \forall i \in \mathcal{N}_{\mathrm{P}}, s \in \mathcal{S}, t \in \mathcal{T} \tag{4.25}$$

$$I_{ij,\min}^2 \leq I_{ij,s,t} \leq I_{ij,\max}^2; \forall ij \in \mathcal{L}_{\mathrm{P}}, s \in \mathcal{S}, t \in \mathcal{T} \tag{4.26}$$

$$-S_{ij,\max}^{\mathrm{L}} \cdot \alpha_{ij,s,t} \leq P_{ij,s,t}^{\mathrm{L}}, Q_{ij,s,t}^{\mathrm{L}} \leq S_{ij,\max}^{\mathrm{L}} \cdot \alpha_{ij,s,t}; \forall ij \in \mathcal{L}_{\mathrm{P}}, s \in \mathcal{S}, t \in \mathcal{T} \tag{4.27}$$

To ensure radial operation of distribution network topology under extreme fault scenarios, the spanning tree (ST) constraints (Eqs. 4.28–4.29) and virtual power flow constraints (Eqs. 4.30–4.34) are implemented to model the radial network structure [11]. Here, variables y and β are employed to construct a virtual spanning tree for the power network topology under normal operating conditions. Equations (4.28)–(4.29) denote that in the virtual spanning tree, the source nodes and load nodes are the root nodes and child nodes, respectively. Equations (4.30)–(4.32) describe the virtual power flow constraints within the spanning tree framework. Equation (4.33) specifies that the actual network topology must constitute a subgraph of the virtual spanning tree, enabling the distribution system to reconfigure into multiple microgrids during faults, thereby enhancing operational flexibility. Equation (4.34) represents the operational prohibition of damaged power lines.

$$\sum_{j:j\to i} y_{ji,s,t} = 0; \quad \forall i \in \mathcal{R}, \quad s \in \mathcal{S}, \quad t \in \mathcal{T} \tag{4.28}$$

$$\sum_{j:j\to i} y_{ji,s,t} = 1; \quad \forall i \in \mathcal{N}_{\mathrm{P}} \backslash \mathcal{R}, \quad s \in \mathcal{S}, \quad t \in \mathcal{T} \tag{4.29}$$

$$\sum_{k:k\to i} f_{ki,s,t} - \sum_{j:i\to j} f_{ij,s,t} = 1; \quad \forall i \in \mathcal{N}_{\mathrm{p}} \backslash \mathcal{R}, \quad s \in \mathcal{S}, \quad t \in \mathcal{T} \tag{4.30}$$

$$\beta_{ij,s,t} = y_{ij,s,t} + y_{ji,s,t}; \quad \forall ij \in \mathcal{L}_{\mathrm{P}}, \quad s \in \mathcal{S}, t \in \mathcal{T} \tag{4.31}$$

$$\left|f_{ij,s,t}\right| \leq \beta_{ij,s,t} N_{\mathrm{P}}; \quad \forall ij \in \mathcal{L}_{\mathrm{P}}, s \in \mathcal{S}, t \in \mathcal{T} \tag{4.32}$$

$$\alpha_{ij,s,t} \leq \beta_{ij,s,t}; \quad \forall ij \in \mathcal{L}_{\mathrm{P}}, \quad s \in \mathcal{S}, \quad t \in \mathcal{T} \tag{4.33}$$

$$\alpha_{ij,s,t} \leq 1 - \xi_{ij,s,t}; \quad \forall ij \in \mathcal{L}_{\mathrm{P}}, \quad s \in \mathcal{S}, t \in \mathcal{T} \tag{4.34}$$

Natural Gas System Constraints: The operational constraints of the natural gas system encompass not only the gas hydrodynamic constraints (Eqs. 4.1–4.3) but also the following components. Equation (4.35) represents the nodal mass flow balance at pipeline junctions. To ensure system stability, the gas pressure and density at city gate stations are maintained as constant values [12], as specified in Eq. (4.36). The operational limitations of gas compressors are decided by Eqs. (4.37)–(4.38). Equation (4.37) establishes the pressure ratio constraint between compressor inlet and outlet, while Eq. (4.38) imposes upper and lower bounds on the mass flow through compressors.

$$\sum_{k:k\to i}\frac{M_{ki,s,t}^{\text{out}}}{A_{ki}}-\sum_{j:i\to j}\frac{M_{ij,s,t}^{\text{in}}}{A_{ij}}+M_{i,s,t}^{\text{Sta}}+M_{i,s,t}^{\text{PG}}-M_{i,s,t}^{\text{GT}}-M_{i,s,t}^{\text{CHP}}$$
$$=M_{i,t}^{\text{D}}-M_{i,s,t}^{\text{Cut}};\quad \forall i\in\mathcal{N}_{\text{G}},s\in\mathcal{S},t\in\mathcal{T} \tag{4.35}$$

$$\begin{cases} Pr_{i,s,t}=Pr_{i,s,t}^{0}; \\ \rho_{i,s,t}=\rho_{i,s,t}^{0}; \end{cases} \quad \forall i\in\mathcal{N}_{\text{sta}}^{\text{G}},s\in\mathcal{S},t\in\mathcal{T} \tag{4.36}$$

$$Pr_{i,s,t}^{\text{Com}}\le Pr_{j,s,t}^{\text{Com}}\le\Gamma_{\text{Com}}Pr_{i,s,t}^{\text{Com}};\ \forall ij\in\mathcal{N}_{\text{Com}}^{\text{G}},s\in\mathcal{S},t\in\mathcal{T} \tag{4.37}$$

$$0\le M_{ij,s,t}^{\text{Com}}\le M_{ij,\max}^{\text{Com}}u_{ij,s,t}^{\text{Com}};\ \forall ij\in\mathcal{N}_{\text{Com}}^{\text{G}},s\in\mathcal{S},t\in\mathcal{T} \tag{4.38}$$

The operational constraints are mathematically formulated as follows: Eq. (4.39) establishes the upper and lower bounds for nodal gas pressure. Equation (4.40) defines the permissible range of nodal gas mass flow rates. The operational limits on gas throughput at city gate stations are decided by Eq. (4.41). Equation (4.42) constrains the gas intake boundaries of CHP units, while Eq. (4.43) specifies the allowable range for gas load curtailment quantities, completing the comprehensive constraint framework for natural gas network operation.

$$Pr_{i,\min}\le Pr_{i,s,t}\le Pr_{i,\max};\ \forall i\in\mathcal{N}_{\text{G}},s\in\mathcal{S},t\in\mathcal{T} \tag{4.39}$$

$$M_{ij,\min}^{\text{G}}\le M_{ij,s,t}^{\text{in}},M_{ij,s,t}^{\text{out}}\le M_{ij,\max}^{\text{G}};\ \forall ij\in\mathcal{L}_{\text{G}},s\in\mathcal{S},t\in\mathcal{T} \tag{4.40}$$

$$0\le M_{i,s,t}^{\text{Sta}}\le M_{i,\max}^{\text{Sta}};\ \forall i\in\mathcal{N}_{\text{Sta}}^{\text{G}},\quad s\in\mathcal{S},t\in\mathcal{T} \tag{4.41}$$

$$0\le M_{i,s,t}^{\text{CHP}}\le M_{i,\max}^{\text{CHP}};\ \forall i\in\mathcal{N}_{\text{CHP}}^{\text{G}},s\in\mathcal{S},t\in\mathcal{T} \tag{4.42}$$

$$0\le M_{i,s,t}^{\text{Cut}}\le M_{i,t}^{\text{D}};\ \forall i\in\mathcal{N}_{\text{G}},s\in\mathcal{S},t\in\mathcal{T} \tag{4.43}$$

District Heat System Constraints: The operational constraints regulating the district heating system are comprehensively formulated as follows: In addition to the thermal pipeline transmission constraints (Eqs. 4.4–4.10) and building thermal inertia constraints (Eq. 4.11), the framework additionally encompasses the following critical components. Equation (4.44) represents the thermal power balance at heat source nodes, while Eq. (4.45) establishes the thermodynamic coupling between nodal heating loads and temperature distributions in supply/return water pipelines. The thermal power conservation at pipeline inlet/outlet nodes is rigorously maintained through Eq. (4.46). Equation (4.47) ensures hydraulic consistency by requiring identical supply water temperatures at pipe inlets sharing the same source node, whereas Eq. (4.48) guarantees thermal energy equilibrium at pipeline junction nodes through nodal power balance constraints, thereby completing the thermohydraulic modeling architecture.

$$\begin{aligned} & H_{i,s,t}^{\mathrm{CHP}} + H_{i,s,t}^{\mathrm{EB}} = c_w^{\mathrm{H}} m_{l,s,t}^{\mathrm{H}} \left(\tau_{l,s,t}^{\mathrm{S,\,in}} - \tau_{l,s,t}^{\mathrm{R,\,out}} \right); \\ & \forall i \in \mathcal{N}_{\mathrm{Sn}}^{\mathrm{H}}, l \in \mathcal{L}_{\mathrm{H}}^{i+}, s \in \mathcal{S}, t \in \mathcal{T} \end{aligned} \tag{4.44}$$

$$\begin{aligned} & H_{i,s,t}^{\mathrm{D}} = c_w^{\mathrm{H}} m_{l,s,t}^{\mathrm{H}} \left(\tau_{l,s,t}^{\mathrm{S,\,out}} - \tau_{l,s,t}^{\mathrm{R,\,in}} \right); \\ & \forall i \in \mathcal{N}_{\mathrm{Bn}}^{\mathrm{H}}, l \in \mathcal{L}_{\mathrm{H}}^{i-}, s \in \mathcal{S}, t \in \mathcal{T} \end{aligned} \tag{4.45}$$

$$\begin{aligned} & H_{l,s,t}^{x,\,\mathrm{in}} = c_w^{\mathrm{H}} m_{l,s,t}^{\mathrm{H}} \tau_{l,s,t}^{x,\,\mathrm{in}}, \quad H_{l,s,t}^{x,\,\mathrm{out}} = c_w^{\mathrm{H}} m_{l,s,t}^{\mathrm{H}} \tau_{l,s,t}^{x,\,\mathrm{out}}; \\ & \forall x \in \{\mathrm{S}, \mathrm{R}\}, l \in \mathcal{L}_{\mathrm{H}}, s \in \mathcal{S}, t \in \mathcal{T} \end{aligned} \tag{4.46}$$

$$\begin{aligned} & \tau_{l_1,s,t}^{x,\,\mathrm{in}} = \tau_{l_2,s,t}^{x,\,\mathrm{in}} \Big|_{\forall l_1, l_2 \in \mathcal{L}_{\mathrm{H}}^{i+}}; \\ & \forall x \in \{\mathrm{S}, \mathrm{R}\}, i \in \mathcal{N}_{\mathrm{In}}^{\mathrm{H}}, s \in \mathcal{S}, t \in \mathcal{T} \end{aligned} \tag{4.47}$$

$$\begin{aligned} & \sum_{l \in \mathcal{L}_{\mathrm{H}}^{i-}} H_{l,s,t}^{x,\,\mathrm{in}} \Delta t = \sum_{l \in \mathcal{L}_{\mathrm{H}}^{i+}} H_{l,s,t}^{x,\,\mathrm{out}} \Delta t; \\ & \forall x \in \{\mathrm{S}, \mathrm{R}\}, i \in \mathcal{N}_{\mathrm{In}}^{\mathrm{H}}, s \in \mathcal{S}, t \in \mathcal{T} \end{aligned} \tag{4.48}$$

Equation (4.49) imposes convex polyhedral constraints [6] on the CHP units to restrict their feasible operating region in terms of electrical and thermal power generation. This formulation mathematically describes the thermodynamic coupling between electricity and heat outputs, ensuring the operational limits of cogeneration systems are rigorously maintained within the multi-dimensional solution space.

$$\begin{cases} P_{n,s,t}^{\mathrm{CHP}} \le P_{n,\max}^{\mathrm{CHP}} + c_{\mathrm{v1},i} H_{i,s,t}^{\mathrm{CHP}} \\ P_{n,s,t}^{\mathrm{CHP}} \ge P_{n,\min}^{\mathrm{CHP}} + c_{\mathrm{v2},i} H_{i,s,t}^{\mathrm{CHP}} \\ P_{n,s,t}^{\mathrm{CHP}} \ge \psi_{\mathrm{CHP}} + c_{\mathrm{m},i} H_{i,s,t}^{\mathrm{CHP}} \\ H_{i,\min}^{\mathrm{CHP}} \le H_{i,s,t}^{\mathrm{CHP}} \le H_{i,\max}^{\mathrm{CHP}} \end{cases} \quad \forall n \in \mathcal{N}_{\mathrm{CHP}}^{\mathrm{P}}, i \in \mathcal{N}_{\mathrm{CHP}}^{\mathrm{G}}, s \in \mathcal{S}, t \in \mathcal{T} \tag{4.49}$$

Interdependency Constraints of PDS, NGS, and DHS: The coupling relationships between the power distribution system and natural gas system are formulated in Eqs. (4.50)–(4.52). Specifically, the energy conversion characteristics of gas turbines and PtG facilities are mathematically described by Eqs. (4.50) and (4.51), respectively. Equation (4.52) quantifies the electrical power consumption of gas compressors. The thermal power interdependency between the power distribution system and district heating system is established through three coupling components: CHP units, EBs, and electric-driven water pumps, as explicitly modeled in Eqs. (4.53)–(4.55). The natural gas consumption of CHP units exhibits a positive correlation with both their electrical power output and thermal energy generation, as formulated in Eq. (4.53). The relationship between the input electric power and output thermal power of an electric boiler is expressed by Eq. (4.54). Equation (4.55) characterizes the electrical power demand of circulation pumps in the heating network.

$$P_{i,s,t}^{\mathrm{GT}} = C_{\mathrm{g2e}} M_{n,s,t}^{\mathrm{GT}} \eta_{\mathrm{GT}}; \quad \forall i \in \mathcal{N}_{\mathrm{GT}}^{\mathrm{P}}, n \in \mathcal{N}_{\mathrm{GT}}^{\mathrm{G}}, s \in \mathcal{S}, t \in \mathcal{T} \tag{4.50}$$

$$M_{n,s,t}^{\mathrm{PtG}} = C_{\mathrm{e2g}} P_{i,s,t}^{\mathrm{PtG}} \eta_{\mathrm{PtG}}; \quad \forall i \in \mathcal{N}_{\mathrm{PtG}}^{\mathrm{P}}, n \in \mathcal{N}_{\mathrm{PtG}}^{\mathrm{G}}, s \in \mathcal{S}, t \in \mathcal{T} \tag{4.51}$$

$$\begin{aligned} P_{n,s,t}^{\mathrm{Com}} &= \frac{ZRT\varepsilon}{(\varepsilon-1)\eta_{\mathrm{Com}}}\left[(\Gamma_{\mathrm{Com}})^{\frac{\varepsilon-1}{\varepsilon}} - 1\right] M_{ij,s,t}^{\mathrm{Com}}; \\ &\forall n \in \mathcal{N}_{\mathrm{Com}}^{\mathrm{P}}, ij \in \mathcal{N}_{\mathrm{Com}}^{\mathrm{G}}, s \in \mathcal{S}, t \in \mathcal{T} \end{aligned} \tag{4.52}$$

$$\begin{aligned} &M_{n,s,t}^{\mathrm{CHP}} C_{\mathrm{g2e}} = P_{i,s,t}^{\mathrm{CHP}} / \eta_{\mathrm{CHP,p}} + H_{j,s,t}^{\mathrm{CHP}} / \eta_{\mathrm{CHP,h}}; \\ &\forall n \in \mathcal{N}_{\mathrm{CHP}}^{\mathrm{G}}, i \in \mathcal{N}_{\mathrm{CHP}}^{\mathrm{P}}, j \in \mathcal{N}_{\mathrm{CHP}}^{\mathrm{H}}, s \in \mathcal{S}, t \in \mathcal{T} \end{aligned} \tag{4.53}$$

$$H_{i,s,t}^{\mathrm{EB}} = P_{n,s,t}^{\mathrm{EB}} \eta_{\mathrm{EB}}; \quad \forall i \in \mathcal{N}_{\mathrm{EB}}^{\mathrm{H}}, n \in \mathcal{N}_{\mathrm{EB}}^{\mathrm{P}}, s \in \mathcal{S}, t \in \mathcal{T} \tag{4.54}$$

$$P_{i,s,t}^{\mathrm{WP}} = P_i^{\mathrm{WP}} u_{n,s,t}^{\mathrm{WP}}; \quad \forall i \in \mathcal{N}_{\mathrm{WP}}^{\mathrm{P}}, n \in \mathcal{N}_{\mathrm{WP}}^{\mathrm{H}}, s \in \mathcal{S}, t \in \mathcal{T} \tag{4.55}$$

Linearization of the Model: The developed multi-energy flow operation optimization model for IES incorporates multiple nonlinear constraints, including those specified in Eqs. (4.4)–(4.6), (4.8)–(4.9), (4.11), (4.15), and (4.44)–(4.46), which pose significant computational challenges for solution derivation. To address this computational complexity, this subsection proposes a systematic linearization methodology involving a series of piecewise linearization techniques and constraint relaxations to transform the original nonlinear formulation into a computationally tractable mixed-integer linear programming (MILP) model.

$$A \geq B, \quad A \geq C; \tag{4.56}$$

For the nonlinear heat loss constraints (4.5)–(4.6), a linear approximation is implemented based on the relatively low heat transfer coefficients characteristic of district heating pipelines, where the term $1 - \eta_{l,t}^{\mathrm{H}} \approx \lambda_l^{\mathrm{H}} L_l^{\mathrm{H}} / \left(c_w^{\mathrm{H}} m_{l,t}^{\mathrm{H}}\right)$ is adopted for operational simplification [13]. Subsequently, through systematic integration with the hydraulic-thermal coupling relationships established in Eqs. (4.8) and (4.46) within the proposed methodological framework, the linearized expressions derived from the original constraints can be reformulated as the computationally efficient Eq. (4.57).

$$H_{l,s,t}^{x,\,\mathrm{out}} = H_{l,s,t}^{x,\,\mathrm{out}\,*} - \lambda_l^{\mathrm{H}} L_l^{\mathrm{H}} \left(\tau_{l,s,t}^{x,\,\mathrm{out}*} - \tau_t^{\mathrm{amb}} \right); \quad \forall x \in \{S, R\}, l \in \mathcal{L}_{\mathrm{H}}, s \in \mathcal{S}, t \in \mathcal{T} \tag{4.57}$$

Equation (4.11) contains bilinear terms involving the product of continuous variables and binary variables, mathematically formulated as $z = xy$, where x represents a binary (0–1) variable and y denotes a continuous variable. This nonlinearity is addressed by applying the Big-M linearization method [14], as formally expressed in Eq. (4.58). To ensure the precision of this linearization, careful parameterization of the Big-M coefficient is critical. Suboptimal selection of the Big-M value may lead to computational inefficiency: an undersized coefficient could yield infeasible or suboptimal solutions, while an oversized coefficient might compromise numerical stability due to ill-conditioned constraint matrices. In this chapter, a systematic parameter selection strategy is implemented where the Big-M coefficient is set as 10 times the upper bound of variable y. This calibrated parameterization guarantees both the compactness of the reformulated model and the exact equivalence between the original nonlinear formulation and its linear counterpart through rigorous mathematical derivation.

$$0 \le z \le x\bar{y}, \quad y - (1 - x)\bar{y} \le z \le y; \tag{4.58}$$

The bilinear terms in Eqs. (4.4)–(4.6), (4.8)–(4.9), and (4.44)–(4.46), which involve products of continuous variables with mathematical structure $z = xy$ (where both x and y are continuous variables), are addressed through a dual-stage reformulation. First, auxiliary variables z_1 and z_2 are introduced to decompose the bilinear terms, enabling their reconfiguration into two quadratic constraints as formalized in Eq. (4.59). Subsequently, leveraging convex relaxation theory [13], the quadratic constraints in Eq. (4.59) are transformed into second-order cone (SOC) constraints through geometric convexification, as rigorously expressed in Eq. (4.60). This two-phase methodology preserves convexity while maintaining mathematical equivalence under bounded variable domains.

$$4z_1 = (x + y)^2, \quad 4z_2 = (x - y)^2, \quad z = z_1 - z_2; \tag{4.59}$$

$$\left\| \begin{matrix} x + y \\ 1 - z_1 \end{matrix} \right\|_2 \le 1 + z_1, \quad \left\| \begin{matrix} x - y \\ 1 - z_2 \end{matrix} \right\|_2 \le 1 + z_2. \tag{4.60}$$

The convexified and linearized model can be effectively solved using the convex-concave method (CCM) [13], which guarantees the exactness of SOC relaxation while enabling the acquisition of high-quality local optimal solutions with provable convergence properties [10].

4.4 Evaluation Metrics for Electric-Gas-Thermal Interdependency Under Extreme Fault Scenarios

Based on the simulation results of multi-energy cascading failures and supply restoration in IES under stochastic fault scenario sets, the multi-stage resilience curves for electrical, gas, and thermal subsystems as well as the IES can be plotted. This subsection presents the proposed evaluation metrics for electric-gas-thermal interdependency.

4.4.1 Interdependency Assessment Framework

The multi-stage resilience curves depict the time-varying characteristics of load supply levels in IES during disaster scenarios. Figure 4.5 illustrates schematic diagrams of these resilience curves for the PDS, NGS, and DHS. Specifically, the evolution process of load supply levels in the PDS during disasters, and its cascading failure impacts on the NGS and DHS, can be divided into five distinct phases: pre-disaster phase, fault propagation phase, post-fault phase, supply restoration phase, and post-restoration phase.

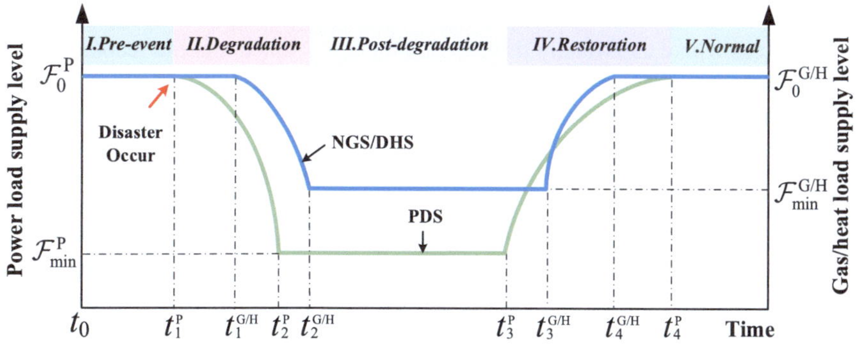

Fig. 4.5 Power, gas, and heat resilience curves during extreme events

$$\mathcal{F}_t^{\mathrm{P}} = 1 - \frac{1}{N_s}\sum_{s\in\mathcal{S}}\left(\sum\nolimits_{i\in\mathcal{N}_{\mathrm{P}}}(P_{i,s,t}^{\mathrm{Cut}}\omega_i^{\mathrm{P}}) \Big/ \sum\nolimits_{i\in\mathcal{N}_{\mathrm{P}}}(P_{i,t}^{\mathrm{D}}\omega_i^{\mathrm{P}})\right) \tag{4.61}$$

$$\mathcal{F}_t^{\mathrm{G}} = 1 - \frac{1}{N_s}\sum_{s\in\mathcal{S}}\left(\sum\nolimits_{i\in\mathcal{N}_{\mathrm{G}}}(M_{i,s,t}^{\mathrm{Cut}}\omega_i^{\mathrm{G}}) \Big/ \sum\nolimits_{i\in\mathcal{N}_{\mathrm{G}}}(M_{i,t}^{\mathrm{G}}\omega_i^{\mathrm{G}})\right) \tag{4.62}$$

$$\mathcal{F}_t^{\mathrm{H}} = \frac{1}{N_s}\sum_{s\in\mathcal{S}}\left(\sum\nolimits_{i\in\mathcal{N}_{\mathrm{Bn}}}(h_{i,s,t}^{\mathrm{D}}\omega_i^{\mathrm{H}}) \Big/ \sum\nolimits_{i\in\mathcal{N}_{\mathrm{Bn}}}\omega_i^{\mathrm{H}}\right) \tag{4.63}$$

where N_s denotes the number of fault scenarios, ω_i^{H} represents the number of buildings contained in thermal load node i, and $h_{i,s,t}^{\mathrm{D}}$ indicates the supply status of nodal thermal load. The binary variable $h_{i,s,t}^{\mathrm{D}}$ is assigned a value of 1 if the indoor temperature remains within the comfortable range; otherwise, it is set to 0, with its mathematical formulation explicitly defined in Eq. (4.64).

$$h_{i,s,t}^{\mathrm{D}} = \begin{cases} 1, & \tau_{i,\min}^{\mathrm{in}} \le \tau_{i,s,t}^{\mathrm{in}} \le \tau_{i,\max}^{\mathrm{in}} \\ 0, & \text{others} \end{cases}. \tag{4.64}$$

Considering the slow dynamic characteristics of natural gas and district heating systems, the cascading failure propagation processes induced by power distribution system disturbances exhibit similar patterns. For illustrative simplification, the electricity-gas interdependency is herein exemplified. As depicted in Fig. 4.4, at time instant t_1^{P}, an extreme disaster causes power line outages, leading to a gradual decline in electrical load supply level from the initial value $\mathcal{F}_0^{\mathrm{P}}$, reaching its lowest point at t_2^{P}. Owing to the delayed response of gas-fired coupled equipment to initial power contingencies and the buffering effect of gas linepack, gas load curtailment commences at a later time t_1^{G}. Concurrently, the inherent gas inertia further postpones the attainment of minimum gas supply level $\mathcal{F}_{\min}^{\mathrm{G}}$ until t_2^{G}. During the energy restoration phase, similar situations may arise where the PDS and NGS differ in their initiation and completion times, determined by the progress of power supply restoration in areas where electrically driven equipment is located and the reconstruction rate of natural gas linepack.

To holistically quantify the susceptibility of natural gas and district heating systems to power supply interruptions under extreme disaster scenarios, this chapter develops a multi-dimensional electric-gas-thermal interdependency assessment framework, as systematically organized in Table 4.1. The framework comprises twelve metrics categorized along temporal-spatial axes: the first eight temporal metrics characterize multi-energy interdependency during cascading failure propagation and supply restoration processes, while the remaining four spatial metrics quantify interdependency magnitudes at both system-level (bulk energy transmission) and nodal-level (critical energy hub) granularities.

Table 4.1 Resilience metrics for power-gas-heat interdependency

Stage	Resilience metric	Symbol
II	NGS/DHS sensitivity to PDS degradation	$\mathcal{D}_G/\mathcal{D}_H$
	Extension time for NGS/DHS degradation	$\Delta t_D^G/\Delta t_D^H$
IV	NGS/DHS sensitivity to PDS restoration	$\mathcal{R}_G/\mathcal{R}_H$
	Extension time for NGS/DHS restoration	$\Delta t_R^G/\Delta t_R^H$
IV	System interdependency of NGS/DHS to PDS power outage	$\mathcal{I}_G/\mathcal{I}_H$
	Nodal interdependency of NGS/DHS to PDS power outage	$\mathcal{I}_{G,i}/\mathcal{I}_{H,i}$

4.4.2 Quantification of the Interdependency Metrics

This subsection presents the mathematical formulations of the proposed interdependency evaluation metrics in Table 4.1.

NGS/DHS sensitivity to PDS degradation: The sensitivity ratios $\mathcal{D}_G$ and $\mathcal{D}_H$ quantify the relative degradation rates of gas/thermal supply levels to power supply degradation in the cascading failure propagation phase, with their mathematical formulations explicitly defined in Eq. (4.65). A higher metric value indicates stronger dynamic coupling between the NGS/DHS and PDS during multi-energy cascading failures.

$$\mathcal{D}_\sigma = \frac{\mathcal{F}_0^\sigma - \mathcal{F}_{\min}^\sigma}{\mathcal{F}_0^P - \mathcal{F}_{\min}^P} \cdot \frac{t_2^P - t_1^P}{t_2^\sigma - t_1^\sigma}, \quad \sigma \in \{G, H\} \tag{4.65}$$

NGS/DHS sensitivity to PDS restoration: The restoration sensitivity ratios $\mathcal{R}_G$ and $\mathcal{R}_H$ quantify the relative recovery rates of natural gas/thermal supply compared to power supply restoration during the post-fault recovery phase, as mathematically formulated in Eq. (4.66). A higher value of $\mathcal{R}_G$ or $\mathcal{R}_H$ indicates stronger operational dependency of the NGS/DHS restoration progress on the PDS's recovery trajectory.

$$\mathcal{R}_\sigma = \frac{\mathcal{F}_0^\sigma - \mathcal{F}_{\min}^\sigma}{\mathcal{F}_0^P - \mathcal{F}_{\min}^P} \cdot \frac{t_4^P - t_3^P}{t_4^\sigma - t_3^\sigma}, \quad \sigma \in \{G, H\} \tag{4.66}$$

Extension time for NGS/DHS degradation and restoration: The temporal decoupling metric $\Delta t_D^G/\Delta t_D^H$ quantifies the duration that NGS/DHS maintain full-load operation following power faults, with its mathematical formulation rigorously defined in Eq. (4.67). The magnitude of this metric is decided by the spatial location of electrical faults and the inertial persistence inherent to NGS/DHS dynamics. The restoration time differentials $\Delta t_D^G/\Delta t_D^H$ characterize the temporal lag between NGS/DHS recovery completion and PDS restoration baselines, as analytically expressed in Eq. (4.68). These metrics are influenced by three interdependent mechanisms: gas transmission network linepack replenishment kinetics, thermal energy propagation delays in district heating pipelines, and building thermal inertia-induced temperature restoration dynamics.

$$\Delta t_{\mathrm{D}}^{\sigma} = t_1^{\sigma} - t_1^{\mathrm{P}}, \quad \sigma \in \{\mathrm{G}, \mathrm{H}\} \tag{4.67}$$

$$\Delta t_{\mathrm{R}}^{\sigma} = t_4^{\mathrm{P}} - t_4^{\sigma}, \quad \sigma \in \{\mathrm{G}, \mathrm{H}\} \tag{4.68}$$

System interdependency of NGS/DHS to PDS power outage: The interdependency indices $\mathcal{I}_{\mathrm{G}}$ and $\mathcal{I}_{\mathrm{H}}$ evaluate the holistic coupling strength between DHS/NGS and PDS throughout extreme event scenarios, with their mathematical formulations rigorously derived in Eq. (4.69). These metrics employ the mathematical construct of cross-correlation [15, 16] to quantify the similarity of multi-stage resilience curves across electric, gas, and thermal subsystems. A higher $\mathcal{I}_{\mathrm{G},i}$ or $\mathcal{I}_{\mathrm{H},i}$ value demonstrates stronger dynamic interdependency among heterogeneous electric-gas-thermal subsystems during the entire disaster timeline, from fault propagation to full recovery.

$$\mathcal{I}_{\sigma} = \max_{\tau} \left\{ \sum_{t=-\infty}^{+\infty} \left(1 - \mathcal{F}_t^{\mathrm{P}}\right) \cdot \left(1 - \mathcal{F}_{t+\tau}^{\sigma}\right) \right\}, \quad \sigma \in \{\mathrm{G}, \mathrm{H}\} \tag{4.69}$$

Nodal interdependency of NGS/DHS to PDS power outage: The interdependency indices $\mathcal{I}_{\mathrm{G}}$ and $\mathcal{I}_{\mathrm{H}}$ assess the system-level coupling dynamics among electric, gas, and thermal subsystems. Owing to the heterogeneous impacts of power contingencies on gas/heating nodes, the spatial heterogeneity of gas-thermal interdependency with the power distribution system manifests as localized nodal vulnerabilities. To address this, Eq. (4.70) extends the framework of Eq. (4.69) by introducing nodal sensitivity metrics $\mathcal{I}_{\mathrm{G},i}$ and $\mathcal{I}_{\mathrm{H},i}$, which quantify the susceptibility of individual gas/thermal nodes to power disturbances. These nodal indices empower IES operators to pinpoint critical gas/thermal nodes exhibiting strong operational reliance on power supply integrity, thereby enabling precision hardening strategies for identified weak points.

$$\mathcal{I}_{\sigma,i} = \max_{\tau} \left\{ \sum_{t=-\infty}^{+\infty} \left(1 - \mathcal{F}_t^{\mathrm{P}}\right) \cdot \left(1 - \mathcal{F}_{i,t+\tau}^{\sigma}\right) \right\}, \quad \sigma \in \{\mathrm{G}, \mathrm{H}\} \tag{4.70}$$

where $\mathcal{F}_{i,t}^{\mathrm{P}}$, $\mathcal{F}_{i,t}^{\mathrm{G}}$ and $\mathcal{F}_{i,t}^{\mathrm{H}}$ denote the load supply levels at time instant t for electric power node i, natural gas node j, and district heating node k, respectively, with their mathematical formulations explicitly defined in Eqs. (4.71)–(4.73).

$$\mathcal{F}_{i,t}^{\mathrm{P}} = 1 - \frac{1}{N_s} \sum_{s \in \mathcal{S}} \left(P_{i,s,t}^{\mathrm{Cut}} / P_{i,t}^{\mathrm{D}}\right) \tag{4.71}$$

$$\mathcal{F}_{i,t}^{\mathrm{G}} = 1 - \frac{1}{N_s} \sum_{s \in \mathcal{S}} \left(M_{i,s,t}^{\mathrm{Cut}} / M_{i,t}^{\mathrm{D}}\right) \tag{4.72}$$

$$\mathcal{F}_{i,t}^{\mathrm{H}} = \frac{1}{N_s} \sum_{s \in \mathcal{S}} h_{i,s,t}^{\mathrm{D}} \tag{4.73}$$

The proposed electric-gas-heat interdependency assessment index system serves the following functions: Firstly, the interdependency metrics provide quantitative analytical tools for systematically evaluating the coupling tightness among electricity, gas, and heat subsystems from a resilience perspective, enabling comprehensive quantification of multi-energy interdependency under fault conditions across multiple dimensions. Secondly, the time-dependent interdependency indicators facilitate the evaluation of resilience enhancement strategies' effectiveness during different phases of extreme disasters, offering critical assessment criteria for developing practical resilience reinforcement strategies in IES. Thirdly, the spatial-dimension interdependency metrics deliver crucial information for IES operators to identify system vulnerabilities caused by strong multi-energy coupling relationships, thereby supporting the formulation of more targeted resilience improvement solutions.

4.5 Case Studies

To validate the effectiveness of the proposed resilience assessment method, this section conducts case studies on the IES E33-G21-H27 system illustrated in Fig. 4.6 and the IES E123-G40-H51 system described in Sect. 4.5.5. In the IES E33-G21-H27 system, the PDS couples with the NGS through two gas turbines, one PtG device, and two compressors, while coupling with the DHS via one CHP unit, one EB, and four electric-driven pumps. Electrical and gas loads are categorized into primary, secondary, and tertiary levels with weighting coefficients of 10, 5, and 1, respectively. To balance computational efficiency and accuracy in the multi-energy flow model, the stochastic fault scenario set contains 20 scenarios, each scenario having an equal probability of 1/20 [17]. The power line restoration time is assumed to follow a uniform distribution U(2,6). Detailed parameters of the case studies are provided in Ref. [18]. All simulations are implemented on a laptop with a 3.60 GHz CPU and 32 GB RAM using MATLAB programming and solved via the Gurobi solver.

4.5.1 Multi-energy Cascading Failures and Energy Supply Restoration Results

This subsection first verifies the correctness of the proposed IES cascading failure and energy supply restoration model under fault conditions. Taking fault scenario 10 as an example, the operational results of IES under this scenario are analyzed.

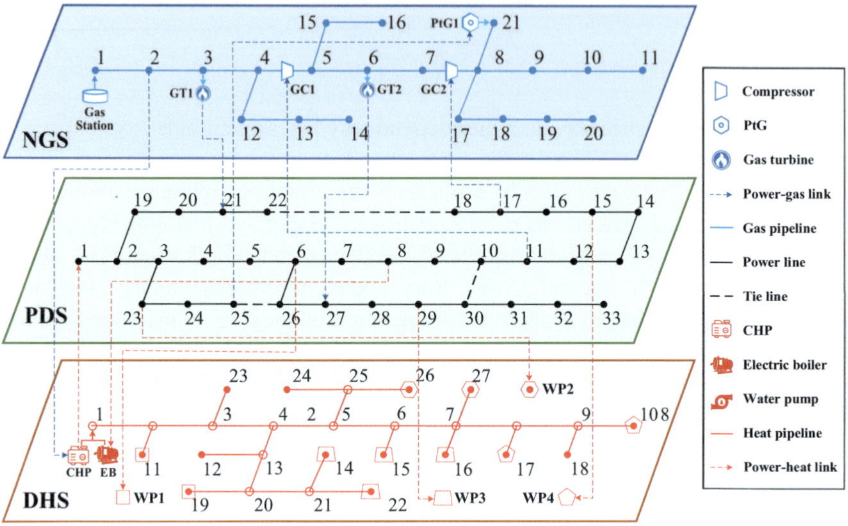

Fig. 4.6 IES E33-G21-H27 test system

Figure 4.7 illustrates the topology reconfiguration of the PDS during the fault and restoration phases. As shown in Fig. 4.7a, at $t = 10$, the system is in the fault phase, with eight power lines experiencing failures. The PDS is divided into six subregions, including two islanded microgrids formed by gas turbines. At this time, 65.5% of the electrical loads are disconnected, and compressor 1, compressor 2, pump 3, and pump 4 cease operation due to power loss. As shown in Fig. 4.7b, at $t = 21$, the system enters the energy restoration phase, where all power lines except lines 13–14, 20–21, and 28–29 have been repaired. At this stage, the electrical loads of the PDS are fully restored, and all equipment resumes normal operation. Throughout the fault process, the network topology of the PDS maintains radial structure operation.

A power outage in the distribution system triggered cascading failures in the NGS and DHS. As shown in Fig. 4.8a, b, during the cascading failure phase at $t = 9$, compressor 1 and compressor 2 ceased operation due to power loss, resulting in

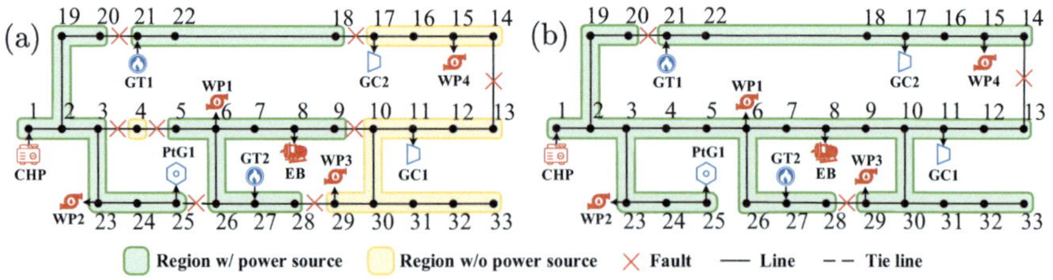

Fig. 4.7 PDS network reconfiguration at **a** $t = 10$, power load supply level: 34.45%; **b** $t = 21$, power load supply level: 100%

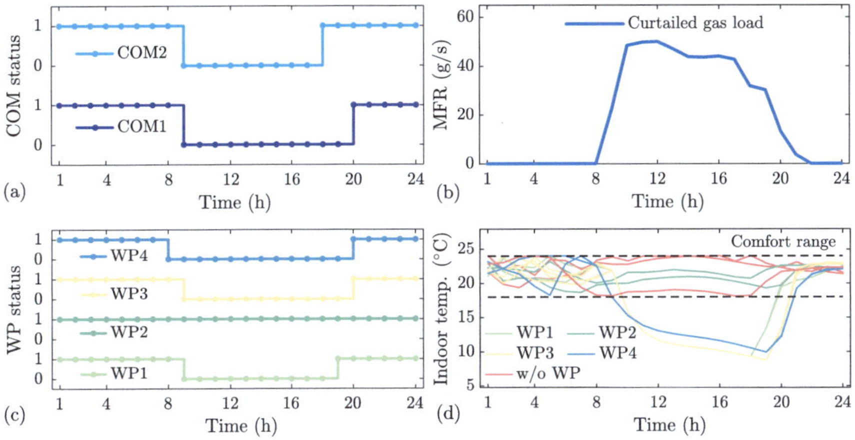

Fig. 4.8 Results of NGS and DHS: **a** status of compressors; **b** mass flow rate of curtailed gas load; **c** status of WPs; **d** indoor temperature of buildings

the disconnection of downstream natural gas loads. During the energy restoration phase from $t = 18$ to $t = 20$, the two compressors regained power supply and resumed operation successively, restoring load supply in the natural gas system. Figure 4.8c, d illustrates the operational status of pumps in the DHS and changes in indoor temperatures of buildings. During the outage, electrically driven pumps 1, 3, and 4 failed to operate due to power loss, leaving corresponding buildings without thermal supply and causing indoor temperatures to fall outside the comfort range. Owing to the thermal inertia of the buildings, the indoor temperature gradually decreased to approximately 10 °C.

Figure 4.9 presents the multi-stage resilience curves of the PDS, NGS, DHS, and the IES. It is evident that during the fault propagation phase, the failures in the NGS and DHS occur later than those in the PDS. This delay arises due to the following reasons: the initial outage area lacks multi-energy coupling components, thus not directly affecting the normal operation of the NGS and DHS; during the initial fault propagation phase, gas inertia mitigates the fault impact while utilizing linepack to support partial natural gas load supply; and the thermal inertia of heating pipelines and buildings slows the decline in indoor temperatures. Additionally, during the energy restoration phase, the restoration times for the natural gas and heating systems are $t = 21$, later than that of the power distribution system at $t = 20$.

4.5.2 *Impacts of Gas and Thermal Inertia on IES Resilience*

To investigate the impacts of gas and thermal inertia on the resilience of the IES, this subsection establishes four case groups (Case 1–Case 4) for comparative analysis, with specific case configurations detailed in Table 4.2. Here, "×" denotes the use

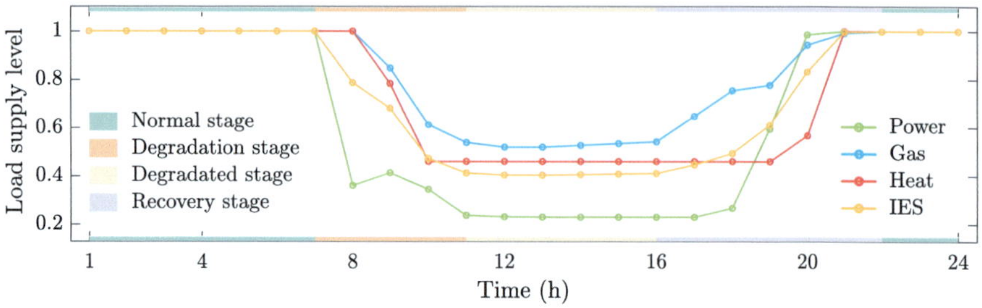

Fig. 4.9 System resilience curves for each subsystem in IES

of a static power flow model, while "√" indicates the adoption of a quasi-dynamic power flow model. Figure 4.10 compares the multi-stage resilience curves of the IES under the four case groups.

The results indicate that during the multi-phase extreme disaster process, the load supply levels of the IES in Case 1 and Case 2 (employing the natural gas static model) are significantly lower than those in Case 3 and Case 4, demonstrating the critical supporting role of gas inertia in cascading failures and energy restoration processes. Further comparison between Case 1 and Case 2 (or Case 3 and Case 4) reveals that the load supply level of Case 1 during the energy restoration phase is slightly lower than that of Case 2, confirming that the thermal inertia of the heating system contributes to enhancing the energy restoration performance of IESs.

For further analysis of the specific impacts of gas and thermal inertia on IES operation during disasters, Fig. 4.11 compares the differentiated operational states of selected IES equipment in Cases 1–4 under fault scenario 10. The results demonstrate that: (1) GT 2 maintains operation during the initial phase of upstream gas supply

Table 4.2 Settings of gas-thermal inertia in Cases 1–4

	Case1	Case2	Case3	Case4
Gas inertia	×	×	√	√
Thermal inertia	×	√	×	√

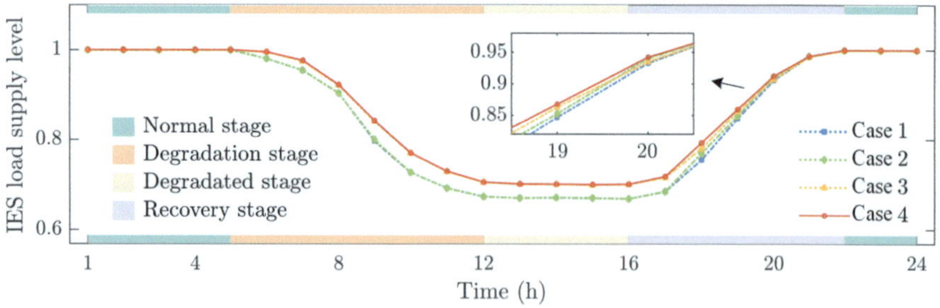

Fig. 4.10 IES resilience curves in Cases 1–4

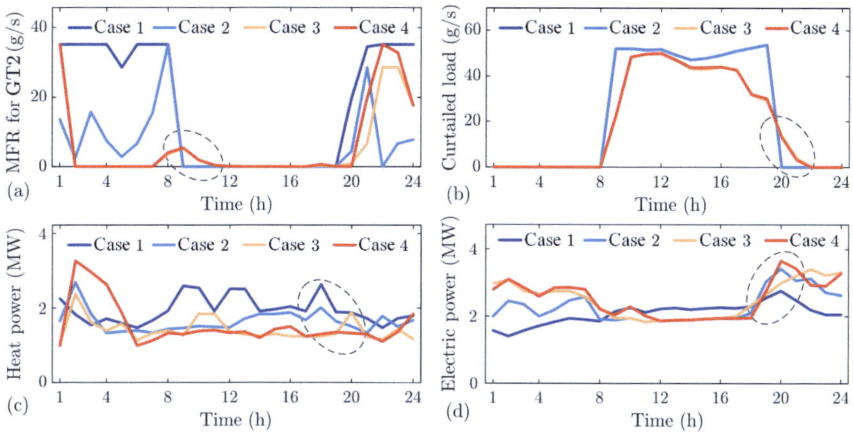

Fig. 4.11 Comparison of IES operating status: **a** Mass flow rate (MFR) of GT2; **b** MFR of curtailed gas load in NGS; **c** Heat power of CHP; **d** Electric power of CHP in Cases 1-4

loss ($t = 9$–10 in Fig. 4.11a) due to the support of natural gas linepack, ensuring power supply to critical loads in the PDS, which illustrates the specific role of gas inertia in enhancing IES resilience during cascading failures; (2) During the energy restoration phase, the reconstruction process of natural gas linepack may delay the recovery of the NGS, as shown in the $t = 20$–21 period in Fig. 4.11b; (3) As depicted in Fig. 4.11c–d, owing to the thermal storage effects and transmission delays in heating pipelines, the thermal power output of CHP units in Case 2 and Case 4 (accounting for thermal inertia) is lower than that in Case 1 and Case 3 during fault recovery, while their electrical power output increases, thereby accelerating the power supply restoration process. The above analysis elucidates the specific mechanisms through which gas and thermal inertia influence IES resilience during disasters.

4.5.3 Assessment of Electric-Gas-Heat Interdependency and Analysis of Influencing Factors

This subsection validates the effectiveness of the proposed electric-gas-heat interdependency metrics and systematically analyzes key factors influencing multi-energy interdependency under fault conditions. Considering the impacts of gas and thermal inertia on IES resilience, further evaluation of Cases 1–4 is conducted based on the proposed interdependency metrics, with results summarized in Table 4.3.

For the electricity-gas interdependency aspect, the metrics $\mathcal{D}_{\mathrm{G}}$ and $\mathcal{R}_{\mathrm{G}}$ in Case 3 and Case 4 (accounting for gas inertia) are significantly lower than those in Case 1 and Case 2, indicating that the presence of gas inertia reduces the electricity-gas sensitivity during both cascading failure and energy restoration phases. Furthermore, the metric $\mathcal{I}_{\mathrm{G}}$ in Case 3 and Case 4 is smaller compared to Case 1 and Case 2,

Table 4.3 Comparison of interdependency metrics in Cases 1–4

	$\mathcal{D}_G$	$\mathcal{D}_H$	$\mathcal{R}_G$	$\mathcal{R}_H$	$\mathcal{I}_G$	$\mathcal{I}_H$	Δt_D^G	Δt_D^H	Δt_R^G	Δt_R^H
Case1	0.814	0.960	0.692	1.000	0.882	1.210	0.65	1.05	0.25	0.05
Case2	0.798	0.956	0.670	1.003	0.868	1.187	0.50	1.25	0.40	– 0.25
Case3	0.552	1.003	0.435	1.004	0.502	1.202	0.95	1.00	– 0.25	0.00
Case4	0.550	1.003	0.435	1.004	0.497	1.153	0.70	1.25	– 0.25	– 0.20

demonstrating that gas inertia mitigates the electricity-gas sensitivity throughout the entire disaster process.

For the electricity-heat interdependency aspect, the results indicate that thermal inertia has a minor impact on metrics Δt_D^G and $\mathcal{R}_H$. The metric $\mathcal{I}_H$ in Case 2 and Case 4 (accounting for thermal inertia) shows a slight decrease compared to Case 1 and Case 3, suggesting that thermal inertia marginally reduces electricity-heat sensitivity throughout the disaster process. Analysis of fault propagation time through the interdependency metrics Δt_D^G and Δt_D^H across Cases 1–4 reveals that the presence of gas and thermal inertia delays the initial fault occurrence times in the NGS and DHS. Regarding energy restoration time, the metrics Δt_D^G and Δt_D^H exhibit negative values, indicating that the restoration times of the NGS and DHS lag behind the power distribution system. This demonstrates that the influence of gas and thermal inertia on IES resilience is not always positive. These findings further validate the specific impacts of gas and thermal inertia on the IES from the perspective of multi-energy interdependency.

Next, three critical factors influencing electric-gas-heat interdependency under fault conditions are selected for analysis.

Impact of Backup Generators: In practical engineering, critical energy equipment is typically equipped with backup generators to address emergencies. To analyze the specific impacts of backup power supplies for electrically driven equipment (natural gas compressors and pressurized water pumps) on multi-energy interdependency, this subsection establishes three case groups (Case 5.1-Case 5.3) for comparison with the benchmark case (Case 4). The operational durations of backup power supplies in Case 5.1, Case 5.2, and Case 5.3 are set to 1, 2, and 3 h, respectively. Comparative results of resilience curves and interdependency metrics under the four case groups are shown in Figs. 4.12 and 4.13.

Figure 4.12 reveals that the deployment of backup power supplies effectively mitigates the degradation rate of IES energy supply levels during the fault propagation phase and reduces the total load loss. Further analysis of Fig. 4.13 indicates that compared to the benchmark Case 4 (without backup power), the metrics $\mathcal{D}_G$ and $\mathcal{D}_H$ in Cases 5.1–5.3 are relatively smaller. This demonstrates reduced electricity-gas and electricity-heat sensitivities during the cascading failure phase in cases 5.1–5.3, lowering the risk of cascading failures induced by multi-energy coupling and thereby enhancing IES resilience. As the operational duration of backup generators increases,

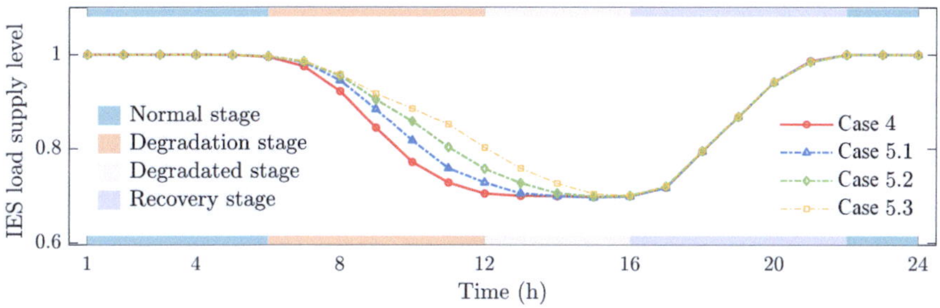

Fig. 4.12 IES resilience curves for Case 4 and Cases 5.1–5.3

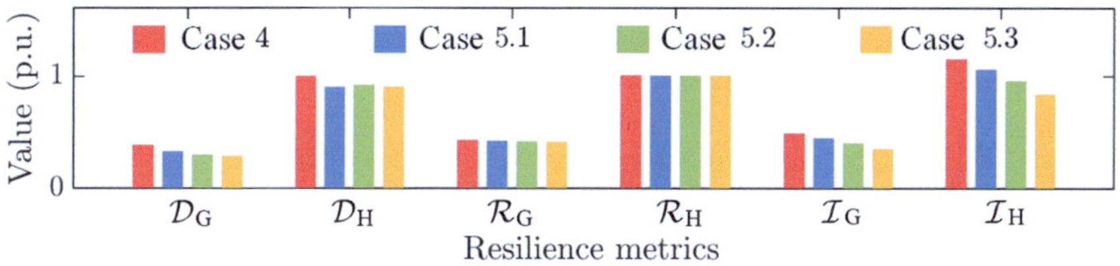

Fig. 4.13 Impact of the backup generators on interdependency metrics

the metrics $\mathcal{I}_G$ and $\mathcal{I}_H$ exhibit a decreasing trend, indicating that deploying higher-capacity backup power supplies achieves better resilience improvement effects in this case study.

Capacity and location of coupling devices: To analyze the specific impact of this factor on multi-energy interdependency, this subsection establishes three case groups (Case 6.1-Case 6.3) for comparison with the benchmark case (Case 4). In Case 6.1, the capacities of GTs and PtG devices are reduced to half of the benchmark case; Case 6.2 configures water pumps 1 and 2 to be powered by independent electrical supplies; Case 6.3 combines both modifications, setting GTs and PtG devices to half capacity while providing independent power supplies to water pumps 1 and 2. Comparative results of resilience curves and interdependency metrics under the four case groups are illustrated in Figs. 4.14 and 4.15.

Comparative analysis of Case 6.1 with Case 4 (or Case 6.3 with Case 6.2) reveals that the energy supply level in Case 6.1 (or Case 6.3) significantly declines compared to the latter, indicating that GTs and PtG devices play a critical role in the mutual support between electricity and gas systems under extreme fault scenarios. As shown in Fig. 4.15, the metric $\mathcal{I}_G$ in Case 6.1 is higher than in Case 4, demonstrating enhanced electricity-heat interdependency during disasters. The reduced capacities of GTs and PtG devices weaken the complementary effects of electricity-gas energy flows under fault conditions, thereby lowering the resilience level of the IES. Furthermore, in Case 6.2 and Case 6.3, the reduced number of water pumps dependent on the PDS for electricity supply diminishes electricity-heat interdependency, which aligns with the comparative results of metrics $\mathcal{D}_H$, $\mathcal{R}_H$, and $\mathcal{I}_H$ in Fig. 4.14.

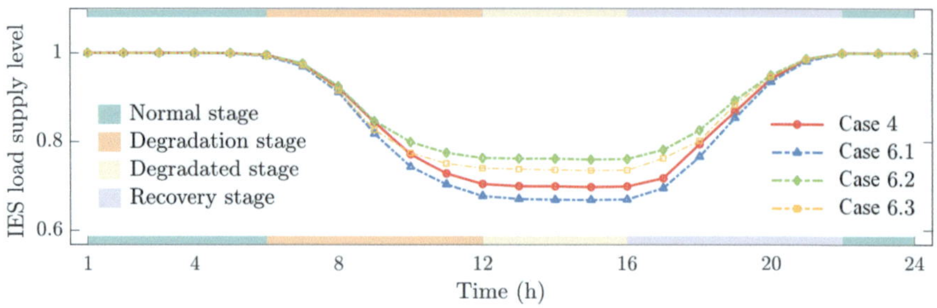

Fig. 4.14 IES resilience curves for Case 4 and Cases 6.1–6.3

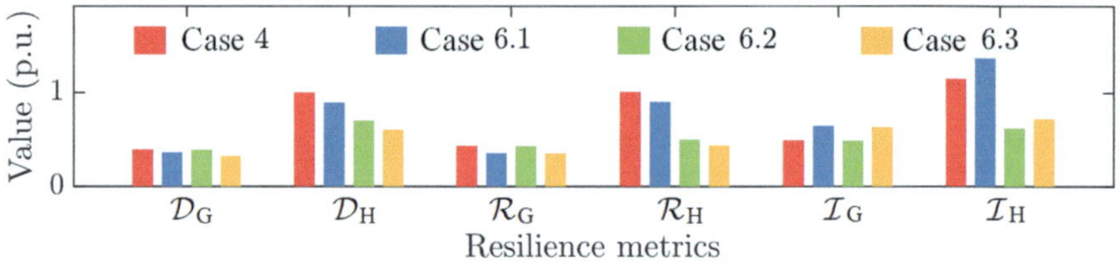

Fig. 4.15 Impact of the coupling devices on interdependency metrics

Number of reconfigurable power lines: This factor directly affects the network reconfiguration capability and load transfer flexibility of the PDS. To evaluate the impact of the number of reconfigurable lines on multi-energy interdependency, this subsection establishes three case groups (Case 7.1-Case 7.3) with 15, 10, and 5 reconfigurable lines, respectively, for comparison with the benchmark case, Case 4 (20 reconfigurable lines). Comparative results of IES resilience curves and interdependency metrics under the four case groups are shown in Figs. 4.16 and 4.17.

Figure 4.16 demonstrates that as the number of reconfigurable power lines decreases, the resilience level of the IES exhibits an accelerated decline trend, validating the critical influence of power network reconfiguration flexibility on IES resilience. Figure 4.17 reveals that with fewer reconfigurable power lines, the values

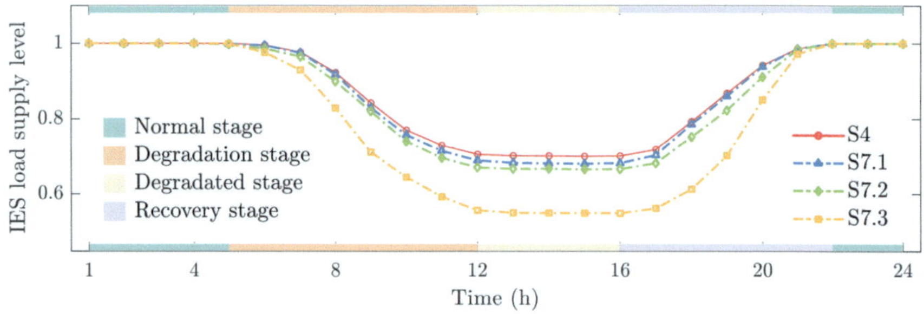

Fig. 4.16 IES resilience curves for Case 4 and Cases 7.1–7.3

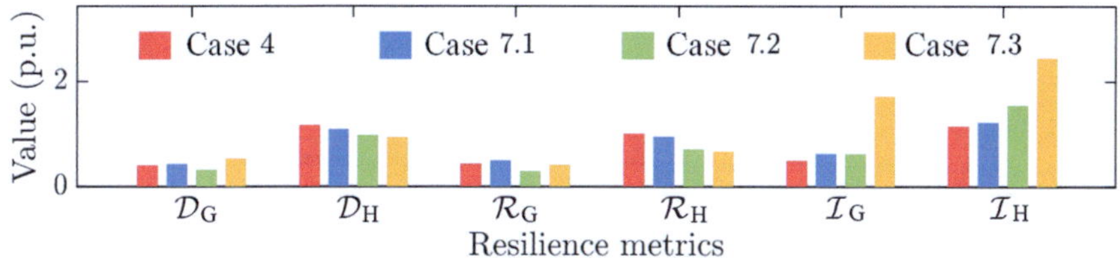

Fig. 4.17 Impact of the number of reconfigurable power lines on interdependency metrics

of metrics $\mathcal{I}_G$ and $\mathcal{I}_H$ gradually increase, indicating enhanced interdependency among electric-gas-heat systems during disasters. Reduced power network reconfiguration capability elevates fault propagation risks across multi-energy flow systems, consequently diminishing the resilience level of the IES. Therefore, the reconfiguration flexibility of the power network plays a critical role in maintaining the energy supply levels of the IES during disasters and mitigating fault propagation risks.

In summary, the established interdependency assessment index system enables multi-dimensional quantification of electric-gas-heat interdependency under extreme disasters, providing an effective evaluation and analysis framework for assessing the specific impact mechanisms of resilience enhancement measures on IES as well as mitigating operational risks caused by multi-energy coupling.

4.5.4 Vulnerability Identification Based on Node Interdependency Metrics

The node-level interdependency metrics $\mathcal{I}_G$ and $\mathcal{I}_{H,n}$ enable the quantification of sensitivity to multi-energy cascading failures at different natural gas and heating nodes during disasters, thereby aiding in the identification of system vulnerabilities. To validate the effectiveness of the proposed node interdependency metrics and compare the influence of multi-energy coupling device locations on node interdependency, this subsection establishes two case groups (Case 8.1-Case 8.2) for comparison with the benchmark case, Case 4. Specific configurations for Cases 8.1–8.2 are detailed in Table 4.4, and the evaluation results of interdependency metrics across the three case groups are illustrated in Fig. 4.18.

As shown in Fig. 4.18, significant differences in node interdependency metrics across the three case groups indicate that the locations of multi-energy coupling

Table 4.4 Energy device coupling locations in PDS

	GT	Compressor	PtG	Electric boiler	Water pump
Case 8.1	17, 28	7, 21	16	5	2, 20, 31, 12
Case 8.2	4, 20	9, 11	8	22	27, 28, 17, 12

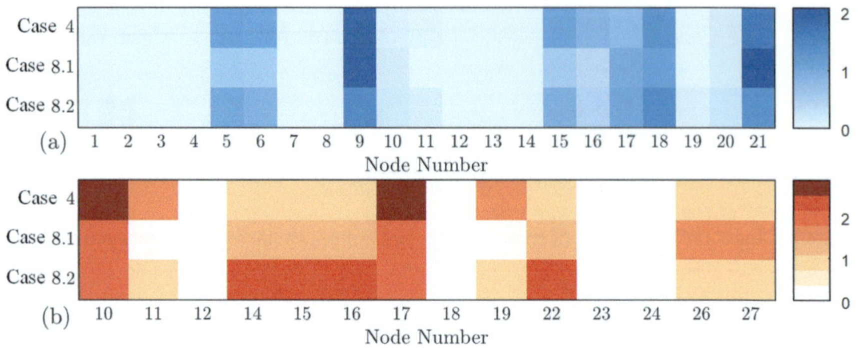

Fig. 4.18 Interdependency of nodal load curtailment in **a** NGS; **b** DHS to PDS outage in Case 4 and Cases 8.1–8.2

devices directly influence the spatial distribution of system cascading failure risks during disasters. In the NGS, nodes 1–4 and 12–14, which have maintained stable natural gas supply, exhibit node interdependency metrics of zero in all cases. In the DHS, water pumps at nodes 12, 18, 23, and 24 are powered by independent electrical supplies, making their operation independent of the PDS; thus, their node interdependency metrics also remain zero. Darker-colored nodes, such as natural gas nodes 9 and 21 in Case 8.1 and heating nodes 10 and 17 in Case 4, demonstrate higher vulnerability during fault conditions. Based on node interdependency metrics, IES operators can identify system vulnerabilities caused by multi-energy coupling and deploy reinforcement measures at critical nodes prior to extreme events to enhance the resilience level of the IES.

4.5.5 Scalability Testing

To validate the scalability of the constructed evaluation method, this subsection conducts scalability testing on the IES E123-G40-H51 system. The system topology is illustrated in Fig. 4.19. The system includes multi-energy coupling devices such as three GTs, one PtG device, three compressors, three GTs, one CHP unit, one EB, and six electric-driven water pumps. Following the four case configurations in Sect. 4.5.2, the expanded system was tested and compared under Cases 1–4 to validate the impacts of gas and thermal inertia on IES resilience in larger-scale systems. Simulation results are presented in Fig. 4.20 and Table. 4.5.

The test results of the scalable system align with those in Sect. 4.5.2. Figure 4.20 and Table. 4.5 demonstrate the critical supporting role of gas and thermal inertia in mitigating multi-energy cascading failures and facilitating energy restoration processes in the IES. These inertial properties partially weaken the electric-gas-heat multi-energy interdependency under extreme disaster scenarios, contributing

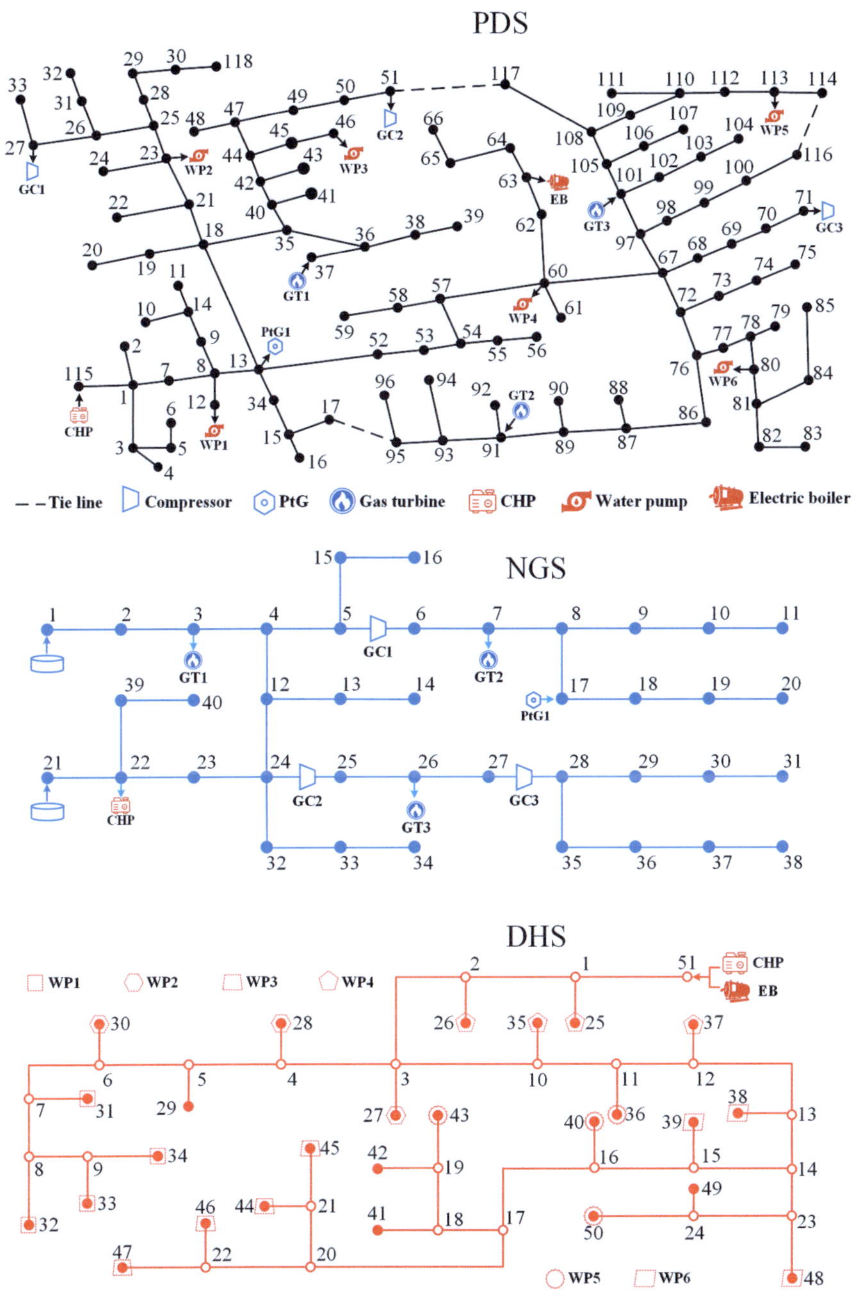

Fig. 4.19 IES E123-G40-H51 test case

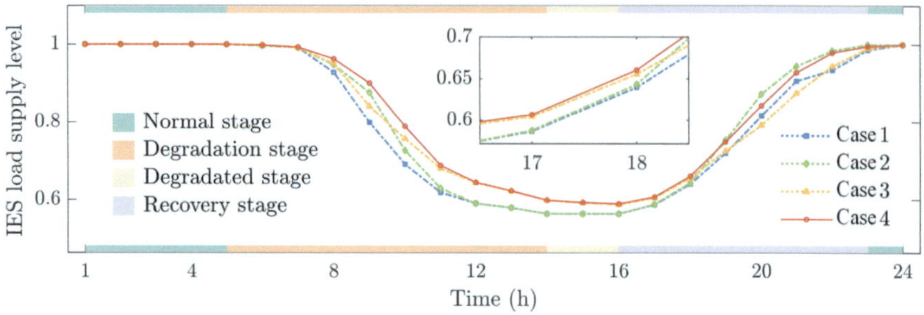

Fig. 4.20 IES resilience curves for IES E123-G40-H51 system in Cases 1–4

Table 4.5 Comparison of interdependency metrics in Cases 1–4

	$\mathcal{D}_G$	$\mathcal{D}_H$	$\mathcal{R}_G$	$\mathcal{R}_H$	$\mathcal{I}_G$	$\mathcal{I}_H$	Δt G D	Δt H D	Δt G R	Δt H R
Case 1	0.451	1.057	0.365	0.868	1.055	2.534	1.10	1.45	0.75	– 2.00
Case 2	0.450	1.088	0.354	0.815	1.045	2.405	1.10	2.30	0.70	– 1.35
Case 3	0.349	1.092	0.275	0.909	0.687	2.315	1.70	1.65	– 0.90	– 1.90
Case 4	0.345	1.094	0.268	1.007	0.670	2.191	1.70	2.30	– 0.60	– 1.30

to reduced total load loss throughout the disaster. Notably, the differences in multi-stage resilience curves between Cases 1–4 become more pronounced in the scalable system. This amplification is likely attributed to the increased lengths of natural gas and heating pipelines in the expanded system, which enhance gas and thermal inertia characteristics, thereby exerting a more significant influence on system resilience.

4.6 Conclusions

This chapter systematically evaluates the specific impacts of gas/thermal inertia and electric-gas-heat interdependency on IES resilience under extreme disasters. First, a multi-energy flow simulation model incorporating dynamic and static characteristics of electrical, gas, and thermal subsystems was established to simulate multi-energy cascading failures and energy restoration processes under stochastic fault scenarios. Subsequently, multi-stage resilience curves for heterogeneous subsystems and the IES were characterized, with an interdependency assessment index system developed across temporal, spatial, systemic, and nodal dimensions for extreme disaster conditions. Finally, systematic evaluations were conducted on the specific impacts of gas/thermal inertia on IES resilience, including in-depth analyses of how factors such as backup power supplies, coupling devices, and reconfigurable power lines influence multi-energy interdependency and their underlying mechanisms affecting system resilience. By assessing IES resilience attributes from the perspective of multi-energy interdependency, this chapter provides a diagnostic framework for identifying system vulnerabilities induced by strong multi-energy flow coupling, while offering critical vulnerability information for subsequent resilience planning and restoration studies. Future work will focus on refined modeling of gas/thermal inertia to enhance IES resilience assessment precision, along with incorporating additional flexibility resources into the evaluation framework.

References

1. Zhang, H., Wang, P., Yao, S., et al. (2020). Resilience assessment of interdependent energy systems under hurricanes. *IEEE Transactions on Power Systems, 35*(5), 3682–3694.
2. Li, J., Xu, Y., Wang, Y., Li, M., He, J., Liu, C.-C., & Schneider, K. P. (2021). Resilience-motivated distribution system restoration considering electricity-water-gas interdependency. *IEEE Transactions on Smart Grid, 12*(6), 4799–4812.
3. Sun, W., Wang, Q., Ye, Y., & Tang, Y. (2022). Unified modelling of gas and thermal inertia for integrated energy system and its application to multitype reserve procurement. *Applied Energy, 305*, 117963.
4. Lu, S., Gu, W., Meng, K., Yao, S., Liu, B., & Dong, Z. Y. (2020). Thermal inertial aggregation model for integrated energy systems. *IEEE Transactions on Power Systems, 35*(3), 2374–2387.
5. Fang, J., Zeng, Q., Ai, X., Chen, Z., & Wen, J. (2018). Dynamic optimal energy flow in the integrated natural gas and electrical power systems. *IEEE Transactions on Sustainable Energy, 9*(1), 188–198.
6. Tao, R., Zhao, D., Xu, C., Wang, H., & Xia, X. (2023). Resilience enhancement of integrated electricity-gas-heat urban energy system with data centres considering waste heat reuse. *IEEE Transactions on Smart Grid, 14*(1), 183–198.
7. Zhou, Y. Z., Wei, Z. O., Shahidehpour, M., & Chen, S. (2021). Distributionally robust resilient operation of integrated energy systems using moment and Wasserstein metric for contingencies. *IEEE Transactions on Power Systems, 36*(4), 3574–3584.
8. Rajaei, F., Latify, M. A., & Ebrahimi, A. (2023). Incorporating DSO's situational awareness in resilience-oriented distribution system planning. *IEEE Transactions on Smart Grid, 14*(3), 1985–1997.
9. Liu, P. X., Wu, Z., Gu, W., Lu, Y. P., Yang, X., Sun, K., & Sun, Q. R. (2022). Security-constrained AC-DC hybrid distribution system expansion planning with high penetration of renewable energy. *International Journal of Electrical Power & Energy Systems, 142*, 108285.
10. Sun, Q., Wu, Z., Gu, W., Zhang, X.-P., Lu, Y., Liu, P., Lu, S., & Qiu, H. (2024). Resilience assessment for integrated energy system considering gas-thermal inertia and system interdependency. *IEEE Transactions on Smart Grid, 15*(2), 1509–1524.
11. Lei, S., Chen, C., Song, Y., et al. (2020). Radiality constraints for resilient reconfiguration of distribution systems: Formulation and application to microgrid formation. *IEEE Transactions on Smart Grid, 11*(5), 3944–3956.
12. Yan, M. Y., He, Y. B., Shahidehpour, M., Ai, X. M., Li, Z. Y., & Wen, J. Y. (2019). Coordinated regional-district operation of integrated energy systems for resilience enhancement in natural disasters. *IEEE Transactions on Smart Grid, 10*(5), 4881–4892.
13. Lu, S., Gu, W., Zhang, C., Meng, K., & Dong, Z. Y. (2020). Hydraulic-thermal cooperative optimization of integrated energy systems: A convex optimization approach. *IEEE Transactions on Smart Grid, 11*(6), 4818–4832.
14. Li, Z. M., Xu, Y., Wang, P., & Xiao, G. X. (2023). Coordinated preparation and recovery of a post-disaster multi-energy distribution system considering thermal inertia and diverse uncertainties. *Applied Energy, 336*, 120736.
15. Rodriguez-Garcia, L., Parvania, M., Hosseini, M. M., & Mosier, T. (2022). Resilience analytics for interdependent power and water distribution systems. *IEEE Transactions on Power Systems, 37*(6), 4244–4257.

16. Lyon, D. (2010). The discrete Fourier transform, part 6: Cross-correlation. *Journal of Object Technology, 9*(2), 17–22.
17. Yang, X., Zhou, Z., Zhang, Y., Liu, J., Wen, J., Wu, Q., & Cheng, S. (2023). Resilience-oriented co-deployment of remote-controlled switches and soft open points in distribution networks. *IEEE Transactions on Power Systems, 38*(2), 1350–1365.
18. Sun, Q. Test data [Online]. Retrieved from: https://docs.google.com/spreadsheets/d/1iFlx1LFQMNdwqK1tF4Ede5uErTeEYKod/edit

Chapter 5
Accelerated Resilience Assessment for IES Considering Non-convex Gas Flow Dynamics (Part I): Handling Non-convex Constraints

Abstract At the distribution network level, integrated electricity-gas systems (IEGSs) exhibit significant non-convexity due to the non-negligible impedance of power grids and the complexity of dynamic gas transmission. Moreover, considering the stochastic nature of disaster-inducing mechanisms of extreme events, resilience assessment requires load-shedding calculations under multiple fault scenarios, significantly increasing evaluation time. To accelerate the resilience assessment process, this chapter and the subsequent one propose a fast resilience assessment method for distribution-level IEGS that accounts for the non-convexity of dynamic gas flow. This chapter focuses on the optimal energy flow problem in distribution-level IEGS and introduces an efficient non-convex optimization algorithm. First, based on the physical characteristics of distribution-level IEGS, a non-convex optimal energy flow model is formulated, incorporating alternating current (AC) power flow equations for the electricity network and quasi-dynamic energy flow equations for the gas network. Second, an equivalent transformation is applied to standardize nonlinear and non-convex constraints of different mathematical forms into a standard bilinear equality constraint format. Next, within the framework of the spatial branch-and-bound (SB&B) algorithm, the concept, generation method, and proof process of the bilinear branching feasible region are proposed to guide branching decisions and accelerate algorithm convergence. Then, the SB&B algorithm is embedded using the *callback* mechanism of the CPLEX solver. Finally, by comparing the computational performance of 12 solution algorithms for mixed-integer bilinear optimization problems on two IEGS benchmark test cases of different scales, the accuracy and efficiency of the proposed non-convex optimization algorithm are validated.

Keywords Integrated electricity-gas system · Optimal electricity-gas flow · Non convex optimization · Second order cone constraints

Nomenclature

t	Index for the scheduling time period

Z. Wu et al., *Integrated Energy System Resilience*,
https://doi.org/10.1007/978-981-95-2363-4_5

i	Index for electricity grid nodes
m	Index for gas grid nodes
ij	Index for power line index
mn	Index for natural gas pipeline index
Ω_t	Set of optimization periods
Ω_s	Set of substation nodes
Ω_g	Set of distributed generation nodes
Ω_w	Set of gas gate station nodes
Ω_b	Set of power grid nodes
Ω_n	Set of gas network nodes
α_i^+	Set of lines with node i as the starting node
α_i^-	Set of lines with node i as the ending node
Ω_i^s	Sets of substations
Ω_i^g	Sets of distributed generation units
Ω_i^u	Sets of gas power plant nodes
Ω_l	Set of lines
Ω_u	Set of gas power plant nodes
$\|\Omega_b\|$	Total number of nodes in the system
$\|\Omega_s\|$	Total number of substation nodes
Ω_m^a	Sets of gas storage tanks connected to node m
Ω_m^w	Sets of gas gate stations connected to node m
Ω_m^u	Sets of gas-fired power plants connected to node m
γ_m	Set of gas pipelines connected to node m
p	Set of gas pipelines
Ω_a	Set of gas storage tank nodes
C_i^s	Generation cost coefficient of distributed generation units
C_g^i	Generation cost coefficient of distributed power generation
C_m^w	Cost coefficient for purchasing gas from the upper-level gas grid
C_i^e	Outage costs for power network load nodes i
C_m^f	Outage costs for gas network load nodes m
$p_{i,t}^s$	Power at substation i
$p_{i,t}^g$	Output of distributed generator i
$f_{m,t}^w$	Gas purchase quantity at gas station m
$u_{i,t}^p$	Active load-shedding amounts at grid load node i
$u_{i,t}^q$	Reactive load-shedding amounts at grid load node i
$u_{m,t}^f$	Load-shedding amount at the gas load node m
$D_{i,t}^p$	Active loads at node i
$D_{i,t}^q$	Reactive loads at node i
R_{ij}	Resistance of line ij
X_{ij}	Reactance of line ij
M	A sufficient large number
$P_{ij}^{\max}$	Maximum active power transmission on line ij
$Q_{ij}^{\max}$	Maximum reactive power transmission on line ij
$V_i^{\max}$	Maximum voltage magnitude limits at node i

$V_i^{\min}$	Minimum voltage magnitude limits at node i
$I_{ij}^{\max}$	Maximum current limit for line ij
$P_i^{s,\max}$	Maximum active power transmission from the substation at node i
$Q_i^{s,\max}$	Maximum reactive power transmission from the substation at node i
$P_i^{g,\max}$	Maximum active power output from distributed generation at node i
$Q_i^{g,\max}$	Maximum reactive power output from distributed generation at node i
$P_i^{u,\max}$	Maximum active power output from gas power plant at node i
$Q_i^{u,\max}$	Maximum reactive power output from gas power plant at node i
χ	Proportion of maximum load-shedding capacity
$p_{ji,t}$	Active power flowing through line ij
$q_{ji,t}$	Reactive power flowing through line ij
$p_{i,t}^s$	Active power injected into the system by the substation at node i
$q_{i,t}^s$	Reactive power injected into the system by the substation at node i
$p_{i,t}^g$	Output active power of the distributed generation units
$q_{i,t}^g$	Output reactive power of the distributed generation units
$p_{i,t}^u$	Output active power of the gas power plant
$q_{i,t}^u$	Output reactive power of the gas power plant
$v_{ji,t}$	Imbalance of voltage drop on line ij
y_{ij}	Status of the line (1 for closed, 0 for open)
$l_{ji,t}$	Square of the line current
$v_{i,t}$	Square of the node voltage
z_{ij}	Virtual power of the line ij
z_i^s	Virtual power of the substation
$D_{m,t}^g$	Gas load at node m
W_{mn}	Weymouth equation coefficient for pipeline mn
$F_m^{w,\text{ramp}}$	Regulation capacity of the gas gate station
T_m	Gas-fired power plant's gas generation ratio
$\Pi_m^{\min}$	Minimum pressure at node m
$\Pi_m^{\max}$	Maximum pressure at node m
$F_{mn}^{\max}$	Maximum gas flow capacity of pipeline mn
$F_m^{a,\min}$	Minimum supply gas flow of the gas storage tank at node m
$F_m^{a,\max}$	Maximum supply gas flow of the gas storage tank at node m
$F_m^{u,\min}$	Minimum supply gas flow of the gas-fired power plant at node m
$F_m^{u,\max}$	Maximum supply gas flow of the gas-fired power plant at node m
$F_m^{w,\min}$	Minimum supply gas flow of the gas gate station at node m
$F_m^{w,\max}$	Maximum supply gas flow of the gas gate station at node m
$f_{n,t}^a$	Gas flow rates at the gas storage tank
$f_{n,t}^w$	Gas flow rates at the gas gate station
$f_{n,t}^u$	Gas flow rates at the gas-fired power plant nodes
$f_{mn,t}$	Gas flow through pipeline mn
$p_{i,t}^u$	Output of the gas-fired power plant
$U_{m,t}^f$	Gas user load shedding at node m
$\pi_{m,t}$	Gas pressure at node m
$s_{ij,t}$	Apparent power of the transmission line

λ_{mn}	Gas flow direction in the pipeline mn
IEGS	Integrated electricity-gas system
SB&B	Spatial branch and bound
OEGF	Optimal electricity-gas flow
OPF	Optimal power flow
AC	Alternating current
DC	Direct current
SOC	Second-order cone
OGF	Optimal gas flow
FDM	Finite difference method
LP	Linear programming
QP	Quadratic programming
SOCP	Second-order cone programming
B&B	Branch and bound
OA	Outer approximation
MINLP	Mixed-integer nonlinear programming
NLP	Nonlinear programming
PWL	Piecewise linearization
MILP	Mixed-integer linear programming
MISOCP	Mixed-integer second-order cone programming
CCP	Convex-concave procedure
NP-hard	Nondeterministic polynomial hard
MIBLP	Mixed-integer bilinear programming
DGs	Distributed generation units

5.1 Introduction

At the urban level, distribution-level integrated electricity-gas systems (IEGSs) are located at the terminal end of the urban energy system, connecting upstream to the transmission power and gas networks while directly supplying various types of end-user loads downstream. Among these, building loads represent a crucial component of the urban energy system. The optimal electricity-gas flow (OEGF) model for distribution-level IEGS is inherently a complex non-convex optimization problem. Furthermore, at the distribution level, the non-negligible resistance of the power distribution network and the finer granularity of gas network scheduling further exacerbate the model's non-convexity, increasing its computational complexity.

At the urban spatial scale, urban electricity and gas networks serve as both the terminal stage of the provincial energy transmission system and the initial stage of the regional energy supply system, playing a crucial role in bridging these two hierarchical levels. Therefore, optimizing urban-level IEGS is of great significance for improving energy system operation, enhancing the economic efficiency and security of urban energy supply.

5.1.1 Optimization Model

Optimal Power Flow Model for Power System: The research on optimal power flow (OPF) can be traced back to 1958 [1]. Since the 1960s, as security concerns arising from overloaded transmission lines have intensified, gained increasing attention on power system operation optimization. Since its initial formulation in 1962, the OPF model has gradually become a central issue in power system operation optimization [2]. The steady-state power flow model of electrical systems is a set of nonlinear equations formulated for AC networks, incorporating branch active power P, branch reactive power Q, voltage magnitude v, and voltage phase angle θ. The objective of OPF is to determine an optimal set of solutions based on predefined performance metrics. Consequently, the original OPF formulation is typically a nonlinear optimization problem, commonly referred to as the AC-OPF model. Given the distinct physical properties of transmission and distribution networks, various adaptations of the AC-OPF model have emerged. In high-voltage transmission networks, where line resistance is significantly smaller than reactance and both voltage magnitudes and phase angle differences between terminals are nearly zero, the AC-OPF model can be simplified into the direct current (DC) model, known as DC-OPF. Due to its simplicity and computational efficiency, the DC-OPF model has been widely applied in economic dispatch [3, 4] and electricity markets [5, 6]. However, while the linear nature of DC-OPF offers notable advantages, it neglects branch reactive power and voltage magnitudes, leading to a loss of accuracy. In response, linearized power flow models based on AC-OPF have gained growing academic interest. For instance, in [7], v^2 and $v^2\theta$ were treated as independent variables to derive a linear approximation of the AC-OPF model. In Ref. [8], a decoupled linear OPF model was proposed by retaining voltage magnitude v in branch power flow equations while decoupling active and reactive power. Meanwhile, Ref. [9] employed a first-order Taylor expansion around a known operating point, using v^2 as an independent variable to construct a more accurate linear OPF model. Compared to transmission networks, distribution networks have significantly higher resistance, which increases the nonlinearity of the OPF model. Among the various formulations, the branch flow model (also known as the DistFlow model) [10] and its second-order cone (SOC) relaxation [11–13] have been most widely applied in distribution networks. Additionally, depending on accuracy requirements and application scenarios, distribution network OPF models have multiple variants, including linearized branch flow models [14] and semidefinite relaxation models [15].

***Optimal Gas Flow Model for Gas System*:** The modeling of gas system is inherently more complex than that of power system. Unlike power flow equations, which are formulated as algebraic equations, gas flow in pipeline networks is governed by a set of differential equations, including the continuity, momentum, energy, and state equations [16]. The solution to this system describes the spatiotemporal distribution of pipeline pressure and flow, which forms the core of the optimal gas flow (OGF) problem. Based on the degree of simplification applied to these fluid dynamics equations, the OGF model can be classified into three main categories: steady-state models, quasi-dynamic models, and dynamic models. The steady-state model is primarily used to describe applications with relatively stable gas flow. It assumes negligible temporal variations in pipeline pressure and flow, omitting the time derivative terms from the governing equations. By performing spatial integration on the remaining differential equations, a set of nonlinear algebraic equations, known as the Weymouth equations, can be derived [17]. Since the steady-state model assumes stable gas flow, it is mainly applied to the operational optimization of regional gas distribution networks, where gas flow stabilizes rapidly [18], as well as in planning problems where the accuracy of gas flow is not critical [19]. However, gas flow in pipelines generally does not reach a steady state in a short time, imposing significant limitations on the steady-state model. In contrast, the quasi-dynamic model builds on the steady-state model by further considering the linepack characteristics of the gas pipeline, which accounts for the dynamic changes in pipeline pressure and flow over time to a certain extent [20]. As a result, it provides more accurate computational results. Currently, quasi-dynamic models that consider linepack characteristics have been widely applied in the operational optimization of gas transmission networks [21–23]. However, in applications related to safety and stability analysis, the accuracy of the quasi-dynamic model may still be insufficient to meet practical requirements. In recent years, scholars have started to attempt discretizing the mentioned partial differential equations using the finite difference method (FDM) [24], transforming them into a set of nonlinear algebraic equations. This class of models based on FDM is referred to as the dynamic OGF model [25]. Compared to the quasi-dynamic model, the dynamic model offers higher computational accuracy, making it more suitable for short-term operational optimization of gas networks [26].

Electricity-Gas Coordinated Optimization Model: The electricity and gas networks in urban areas can be coupled through gas-fired generators and power-to-gas devices, making the IEGS more optimized and offering additional synergy compared to independent subsystems. This synergy is significant for enhancing the system's operational flexibility [27]. Typically, the construction of an electricity-gas coordinated optimization model is based on the integration of the OPF and OGF models, with the introduction of mathematical models for coupling devices to describe the energy flow conversion process between the electrical and gas networks. Compared to energy hub models that do not consider network constraints [28, 29], the optimal electricity-gas flow (OEGF) model, centered around OPF and OGF, can more accurately reflect the distribution of energy flows in IEGS, which is essential for urban-scale energy management. Building on this foundation, Ref. [30] extended the N-1 security constraint from the power grid to the gas grid, developing an OEGF model that considers integrated energy N-1 security constraints. Ref. [31] further incorporated the uncertainty of renewable energy output into the operational model, proposing a tri-level robust optimization model with a min–max-min structure. Ref. [32] addressed the energy and information interaction issues in multi-area IEGS, proposing a distributed cooperative optimization method that effectively ensures privacy protection. Ref. [33] proposed a comprehensive electrical coordination plan from the perspective of economic operation, covering the process from day-ahead scheduling to actual operation. Ref. [34] developed a post-disaster restoration for IEGS after natural disasters or extreme weather disruptions, focusing on improving system resilience. Finally, Ref. [35] incorporated smart buildings as flexible resources in the OEGF model, constructing a joint optimization model.

Electricity-Gas Coordinated Optimization Planning Model: In the actual operation of IEGS, many issues, such as pipeline overpressure, flow blockages, and curtailment of wind and solar power, are often caused by unreasonable early-stage planning. These issues are typically difficult to resolve solely through optimization of the operational process. Therefore, reasonable planning for IEGS has become an urgent problem to address. In fact, whether for independent electricity grids [36], independent gas networks [37], or highly coupled IEGS [38], the core of the planning problem is to determine the optimal capacity configuration for related equipment under conditions that ensure supply security and reliability, while maximizing economic, environmental, and social benefits. Planning models generally include two levels: configuration and operation. At the configuration level, the planning model must determine the capacity and location of various equipment, such as gas-fired power plants [39], new transmission lines/pipelines [40, 41], power/gas sources [42], and refueling stations [43]. At the operation level, the OEGF model mentioned earlier is typically used to simulate the operation under the current planning scheme and evaluate its feasibility and economic viability. Considering the high uncertainty of factors such as load, climate, prices, and policies over the full life cycle of equipment, the planning model cannot simultaneously simulate all possible operating scenarios. Therefore, the mainstream approach in academia is to use stochastic optimization techniques, applying Monte Carlo or Latin Hypercube Sampling [44] to historical data or randomly generated scenarios, resulting in representative typical day data [45], which are then included as probabilistic scenarios in the planning model for solving. Robust optimization is another approach to address uncertainty. The goal of robust optimization is to ensure the optimality of the system under the worst-case scenarios. As a result, the planning solutions obtained tend to be conservative and are typically not used for medium-to-long-term planning. Instead, they are used to guide the reinforcement of critical equipment in IEGS [46, 47] to prevent significant damage from extreme weather or natural disasters.

5.1.2 Solution Algorithms

From a mathematical modeling perspective, both optimization operation and optimization planning can be abstracted as mathematical programming problems that include objective functions and constraint conditions. Currently, for specific standard forms of mathematical optimization problems, such as linear programming (LP), quadratic programming (QP), second-order cone programming (SOCP), and their mixed-integer forms, academia and industry have developed a mature and comprehensive set of solution mechanisms. For continuous optimization problems that do not contain integer variables, common optimization algorithms include the Simplex method [48] and Interior Point methods [49]. The Simplex method is primarily used for solving LP problems, while interior point methods—mainly in the "primal–dual" and "barrier" forms—can solve a variety of convex optimization problems, including LP, QP, and SOCP [50]. For mixed-integer programming, well-established algorithms include branch and bound (B&B) [51], cutting-plane methods [52], Generalized Benders Decomposition [53], and outer approximation (OA) [54], among others. Since the term "mathematical programming" was first introduced in the early twentieth century, it has undergone continuous improvements and refinements, resulting in a complete and mature theoretical system and algorithmic framework. With the development of computer software and hardware technologies, commercial and open-source solvers that integrate various mathematical programming algorithms, such as Gurobi [55], CPLEX [56], IPOPT [57], and SCIP [58], have been widely adopted in both academia and industry. However, real-world engineering problems often exhibit complex non-convexity and nonlinearity, and the abstracted mathematical models are primarily formulated as mixed-integer nonlinear programming (MINLP) problems. For example, in IEGS, the AC-OPF in power networks is a classic nonlinear optimization problem, and the steady-state, quasi-dynamic, and dynamic OGF models in gas networks also involve numerous nonlinear equations, making them unsolvable by direct methods. To address this issue, scholars both domestically and internationally have proposed a series of solutions from various perspectives, which are summarized as follows.

Approximation and Relaxation: Currently, approximation and relaxation are the most commonly used methods for addressing the nonlinear programming (NLP) components in MINLP problems. In terms of approximation, the academic community primarily employs the concept of piecewise linearization (PWL), where a piecewise function is used to approximate the original nonlinear function, transforming the MINLP problem into a mixed-integer linear programming (MILP) problem. For example, in Ref. [59], the PWL method is applied to linearly approximate the Weymouth equation for gas pipelines; Ref. [60] reconstructs a gas compressor model using a linearized approach based on PWL. In addition to linear functions, Ref. [61] explores the use of piecewise McCormick envelopes to approximate quadratic functions in the OEGF model. While piecewise approximation methods are simple and easy to implement, the large number of piecewise variables can lead to a significant model expansion, thus reducing solution efficiency. Therefore, they are not suitable for large-scale systems or multi-stage optimization problems. Relaxation is another approach to solving model non-convexity. The core idea of relaxation is to expand the solution space by replacing the original non-convex set with a larger convex set. Common relaxation methods include SOC relaxation and semidefinite relaxation. For instance, Ref. [62] performs convex relaxation on the non-convex branch flow equation, Weymouth equation, and gas compressor model, establishing a mixed-integer second-order cone programming (MISOCP) model for the OEGF problem. Ref. [63] presents a model reconstruction method based on SOCP to address the non-convexity in the mixed processes of multiple types of gas; Ref. [64] proposes a tighter

OEGF model relaxation method based on semidefinite programming compared to SOC relaxation. Since relaxation, to some extent, enlarges the solution space, it can lead to reduced solving efficiency. As a result, existing research has explored several improvement measures. For example, Ref. [65] further reduces the solution space by constructing convex hulls based on convex relaxation, improving solution efficiency. Ref. [66] improves solution quality by adding the relaxation error as a penalty term to the objective function, guiding the solver to search in areas closer to the boundary of the solution space.

Local Optimization: Although approximations and relaxations can transform the MINLP problem into MILP or MISOCP problems, the non-convexity of the original problem means that these transformed problems often do not guarantee the feasibility of the solution. Taking the SOC relaxation method for the OEGF model as an example, a common approach in academia is to relax the non-convex equality constraints in the model into inequality constraints and then solve the relaxed problem. If the resulting solution satisfies the original non-convex equality constraints, then the solution to the relaxed problem is also the solution to the original problem. In fact, for the OPF model in distribution networks, the conditions to ensure the tightness of the relaxation are often very stringent, making it impractical for real-world applications [67, 68]. For the OGF model, no research has yet provided sufficient and necessary conditions to guarantee the tightness of the relaxation. Therefore, finding a high-quality local optimal solution while ensuring feasibility remains a major research focus in academia, with heuristic algorithms being a representative approach. Heuristic algorithms generally start with one or more initial solutions and, through a series of heuristic rules, progressively explore the solution space to achieve higher-quality solutions [69]. In addition to common algorithms like genetic algorithm, particle swarm optimization, and simulated annealing, a series of heuristic algorithms, such as neighborhood search [70], have also been integrated into solvers like Gurobi and CPLEX to assist their built-in B&B algorithms in finding integer solutions more quickly. Heuristic algorithms typically iterate starting from one or more feasible solutions. However, in practical engineering, finding even a feasible solution can be exceptionally difficult. This has led to increasing attention on the convex-concave procedure (CCP) in recent years [71]. CCP is an iterative optimization framework where the core idea is to decompose non-convex constraints into convex and concave functions. The convex functions can be directly solved by the solver, while the concave functions require a first-order Taylor expansion at the convex function's solution to construct a linear penalty term, which is applied in the next iteration until the resulting solution satisfies the non-convex constraints. As an efficient local optimization algorithm, CCP has been widely applied to optimization problems in IEGS in recent years [31, 72–74]. Since the CCP focuses only on non-convex constraints, each subproblem in the iteration might still be a MILP or MISOCP problem containing integer variables, which complicates the solution process. To address this, Ref. [75] proposed a fully LP-based iterative framework based on CCP, which further enhances the solution efficiency.

Global Optimization: The drawbacks of local optimization algorithms are also evident for non-convex, nonlinear IEGS optimization problems. On the one hand, local optimization generally results in a local optimal solution rather than a global optimal solution, and it is difficult to quantify the gap in quality between the obtained solution and the global optimum. On the other hand, local optimization is highly dependent on the quality of the initial solution. When the initial solution is poor, leading the search process into an infeasible local region, local optimization algorithms often fail to escape this region, resulting in a failure to solve the problem. Therefore, global optimization algorithms have garnered increasing attention in recent years. The concept behind global optimization is to partition the non-convex solution space into multiple convex subspaces, solving each subproblem to determine the upper and lower bounds of the optimal solution, thereby finding the global optimum. The partitioning of the solution space is at the core of global optimization, with spatial branching-and-bound (SB&B) being the most commonly used strategy. SB&B can be seen as an extension of the integer branching strategy from the traditional B&B algorithm to continuous spaces [76]. In fact, the optimality criterion for SB&B is similar to that of B&B; for a minimization problem $f(x) : x \in X$, if there exists a feasible solution x^* such that for all $x \in X, f(x^*) \leq f(x) + \varepsilon$, then it is considered that $f(x)$ has been solved to "ε-optimal" [77]. Thanks to its comprehensive theoretical foundation, this algorithm has been integrated into solvers like Gurobi and SCIP. However, since non-convex optimization is classified as a non-deterministic polynomial hard (NP-hard) problem, the computational efficiency of the SB&B algorithm is generally low. As a result, scholars both domestically and internationally have proposed a series of improvements tailored to various problems. For example, Ref. [78] introduces a novel intersection cutting plane within the SB&B algorithm to improve the solution efficiency of mixed-integer bilinear programming (MIBLP) problems; similarly, Ref. [79] employs adaptive multi-segment McCormick envelope generation to replace the original binary branching strategy in SB&B, proposing an iterative algorithm with faster lower bound convergence; Ref. [80] introduces a cutting-plane generation strategy based on SOCP and boundary-tightening strategies to accelerate the SB&B algorithm for the AC-OPF problem; Ref. [81] incorporates the SOC relaxation error from the OEGF model as auxiliary information into the spatial branching strategy, effectively improving the efficiency of SB&B in solving the OEGF model; Ref. [82] designs a branching strategy based on three-dimensional spatial reference points for SOC equality constraints, significantly enhancing SB&B's efficiency when solving these specialized non-convex constraints. It is worth noting that, while global optimization algorithms theoretically can approach or even achieve the global optimum, in practical applications, a balance between accuracy and efficiency is typically achieved by appropriately setting thresholds according to specific needs.

This chapter proposes a non-convex optimization algorithm for the OPF model of urban distribution-level integrated electricity-gas systems. The subsequent sections are organized as follows: Sect. 5.2 constructs the OPF model for distribution-level IEGS, Sect. 5.3 reformulates the proposed model and provides its matrix expression, Sect. 5.4 introduces a non-convex optimization algorithm designed to address

SOC equality constraint, Sect. 5.5 presents case studies and analysis, and Sect. 5.6 concludes the summary of the chapter.

5.2 Optimal Power Flow Model for Distribution-Level IEGS

At the distribution network level, the construction of the OPF model is based on the following assumptions: (a) The topology of the distribution network can be altered by changing the status of its segment and interconnection switches, i.e., considering network reconfiguration [36]; (b) due to the relatively small geographical scale of the gas network and the limited capacity of the pipeline storage, the dynamic process of gas flow in pipelines is neglected, and a steady-state model is used for approximation [83]. In this section, Sect. 5.2.1 provides the objective function of the proposed model, while Sect. 5.2.2 presents the corresponding constraints.

5.2.1 *Objective Function*

The optimal power flow model for distribution-level IEGS is formulated within the context of day-ahead or real-time scheduling. The objective function is shown in Eq. (5.1), where the goal is to minimize the total operating cost of the system.

$$\begin{aligned} \min \quad & \sum_{t\in\Omega_t}\sum_{i\in\Omega_s} C_i^s p_{i,t}^s + \sum_{t\in\Omega_t}\sum_{i\in\Omega_g} C_i^g p_{i,t}^g + \sum_{t\in\Omega_t}\sum_{m\in\Omega_w} C_m^w f_{m,t}^w \\ & + \sum_{t\in\Omega_t}\sum_{i\in\Omega_b} C_i^e \left(u_{i,t}^p + u_{i,t}^q\right) + \sum_{t\in\Omega_t}\sum_{m\in\Omega_n} C_m^f u_{m,t}^f \end{aligned} \tag{5.1}$$

where t denotes the scheduling time period index, and i, m are node indices. Ω_t represents the set of optimization periods, Ω_s represents the set of substation nodes, Ω_g represents the set of distributed generation nodes, Ω_w represents the set of gas gate station nodes, Ω_b represents the set of power system nodes, and Ω_n represents the set of gas network nodes. The coefficients represent various cost factors: the cost coefficient for purchasing electricity from the upper-level power grid is denoted, C_i^s represents the generation cost coefficient of distributed generation units, C_g^i is the generation cost coefficient of distributed power generation, C_m^w is the cost coefficient for purchasing gas from the upper-level gas system, and C_i^e and C_m^f represent the outage costs for power and gas network load nodes i and m, respectively. The variable $p_{i,t}^s$ represents the power flow at substation i during time period t, $p_{i,t}^g$ represents the output of distributed generator i during time period t, $f_{m,t}^w$ represents the gas purchase quantity at gas station m during time period t, $U_{i,t}^p$ and $U_{m,t}^q$ represent the active and reactive load-shedding amounts at the grid load node i during time period t, respectively, and $U^f_{m,t}$ represents the load-shedding amount at the gas load node m during time period t. The objective function (5.1) mainly includes the cost of energy purchase from the upper-level IEGS, the generation cost of distributed generators, and the load-shedding costs of both power and gas systems.

5.2.2 Constraints

Power Flow Constraints: On the power system side of the IEGS, based on the branch flow model for the distribution network [10] and considering the network reconfiguration characteristics, the corresponding power flow constraints are constructed as follows:

$$\sum_{ji\in\alpha_i^+}\left(p_{ji,t}-l_{ji,t}R_{ij}\right)-\sum_{ik\in\alpha_i^-}p_{ik,t}+\sum_{j\in\Omega_i^s}p^s_{j,t}+\sum_{j\in\Omega_i^g}p^g_{j,t}+\sum_{j\in\Omega_i^u}p^u_{j,t}$$
$$=D^p_{i,t}-u^p_{i,t},\quad \forall i\in\Omega_b,\forall t\in\Omega_t \tag{5.2}$$

$$\sum_{ji\in\alpha_i^+}\left(q_{ji,t}-l_{ji,t}X_{ij}\right)-\sum_{ik\in\alpha_i^-}q_{ik,t}+\sum_{j\in\Omega_i^s}q^s_{j,t}+\sum_{j\in\Omega_i^g}q^g_{j,t}+\sum_{j\in\Omega_i^u}q^u_{j,t}$$
$$=D^q_{i,t}-u^q_{i,t},\quad \forall i\in\Omega_b,\forall t\in\Omega_t \tag{5.3}$$

$$v_{ij,t}=v_{i,t}-v_{j,t}-2(R_{ij}p_{ij,t}+X_{ij}q_{ij,t})+(R_{ij}^2+X_{ij}^2)l_{ij,t},\quad \forall ij\in\Omega_l,\forall t\in\Omega_t \tag{5.4}$$

$$-M(1-y_{ij})\le v_{ij}\le M(1-y_{ij}),\quad \forall ij\in\Omega_l,\forall t\in\Omega_t \tag{5.5}$$

$$p^2_{ij,t}+q^2_{ij,t}=v_{i,t}l_{ij,t},\quad \forall ij\in\Omega_l,\forall t\in\Omega_t \tag{5.6}$$

$$-P^{\max}_{ij}y_{ij}\le p_{ij,t}\le P^{\max}_{ij}y_{ij},\quad \forall ij\in\Omega_l,\forall t\in\Omega_t \tag{5.7}$$

$$-P^{\max}_{ij}y_{ij}\le q_{ij,t}\le P^{\max}_{ij}y_{ij},\quad \forall ij\in\Omega_l,\forall t\in\Omega_t \tag{5.8}$$

$$(V_i^{\min})^2\le v_{i,t}\le (V_i^{\max})^2,\quad \forall i\in\Omega_b,\forall t\in\Omega_t \tag{5.9}$$

$$0\le l_{ij,t}\le (I_{ij}^{\max})^2y_{ij},\quad \forall ij\in\Omega_l,\forall t\in\Omega_t \tag{5.10}$$

$$-P_i^{s,\max}\le p^s_{i,t}\le P_i^{s,\max},\ -Q_i^{s,\max}\le q^s_{i,t}\le Q_i^{s,\max},\quad \forall i\in\Omega_s,\forall t\in\Omega_t \tag{5.11}$$

$$0\le p^g_{i,t}\le P_i^{g,\max},\ -Q_i^{g,\max}\le q^g_{i,t}\le Q_i^{g,\max},\quad \forall i\in\Omega_g,\forall t\in\Omega_t \tag{5.12}$$

$$0 \le p_{i,t}^u \le P_i^{u,\max},\ -Q_i^{u,\max} \le q_{i,t}^u \le Q_i^{u,\max}, \quad \forall i \in \Omega_u, \forall t \in \Omega_t \tag{5.13}$$

$$0 \le u_{i,t}^p \le \chi D_{i,t}^p,\ 0 \le u_{i,t}^q \le \chi D_{i,t}^q \quad \forall i \in \Omega_b, \forall t \in \Omega_t \tag{5.14}$$

$$\sum_{ji \in \alpha_i^+} z_{ji} - \sum_{ik \in \alpha_i^-} z_{ik} + \sum_{j \in \Omega_i^s} z_j^s = 1, \quad \forall i \in \Omega_b \tag{5.15}$$

$$\sum_{ij \in \Omega_l} y_{ij} = |\Omega_b| - |\Omega_s| \tag{5.16}$$

where α_i^+ and α_i^- represent the set of lines with node i as the starting or ending node, respectively. Ω_i^s, Ω_i^g, and Ω_i^u represent the sets of substations, distributed generation units (DGs), and gas power plant nodes connected to node i, respectively. Ω_l represents the set of lines, Ω_s represents the set of substation nodes, Ω_g represents the set of distributed generation nodes, and Ω_u represents the set of gas power plant nodes. In terms of coefficients, $D_{i,t}^p$ and $D_{i,t}^q$ represent the active and reactive loads at node i, respectively. R_{ij} and X_{ij} represent the resistance and reactance of line ij, respectively. M is a sufficiently large positive real number. $P_{ij}^{\max}$ and $Q_{ij}^{\max}$ represent the upper limits of active and reactive power transmission on line ij, respectively. $V_i^{\max}$ and $V_i^{\min}$ represent the voltage magnitude limits at node i, respectively. $I_{ij}^{\max}$ represents the current limit for line ij. $P_i^{s,\max}$ and $Q_i^{s,\max}$ represent the upper limits of active and reactive power transmission from the substation at node i, respectively. $P_i^{g,\max}$, $Q_i^{g,\max}$, $P_i^{u,\max}$, and $Q_i^{u,\max}$ represent the upper limits of active and reactive power output from distributed generation and gas power plant units, respectively. χ represents the proportion of maximum load-shedding capacity. In terms of variables, $p_{ji,t}$ and $q_{ji,t}$ represent the active and reactive power flow through line ij, respectively. $p_{i,t}^s$ and $q_{i,t}^s$ represent the active and reactive power injected into the system by the substation at node i, respectively. $p_{i,t}^g$ and $q_{i,t}^g$ represent the output power of the distributed generation units, respectively. $p_{i,t}^u$ and $q_{i,t}^u$ represent the output power of the gas power plant, respectively. $v_{ji,t}$ represents the imbalance of voltage drop on line ij. y_{ij} represents the status of the line (1 for closed, 0 for open). $l_{ji,t}$ represents the square of the line current (treated as an independent variable). $v_{i,t}$ represents the square of the node voltage (treated as an independent variable). z_{ij} represents the virtual power of the line, and z_i^s represents the virtual power of the substation. Additionally, the operator $|\cdot|$ denotes the number of elements in the set. $|\Omega_b|$ represents the total number of nodes in the system, with $|\Omega_s|$ representing the number of substation nodes.

Equations (5.2)–(5.3) represent the active and reactive power balance at the grid nodes, while Eqs. (5.4)–(5.5) correspond to the voltage drop across the lines. Equation (5.6) defines the branch flow equation for the distribution network, and Eqs. (5.7)–(5.8) give the upper and lower limits of active and reactive power on the lines. Equation (5.9) sets the upper and lower bounds for the node voltage magnitudes, and constraint (5.10) specifies the upper limit for the line current. Equations (5.11)–(5.13) impose the upper and lower limits on the output of the substations, distributed generation units, and gas power plants, respectively. Equation (5.14) defines the upper limit of load shedding for the users. Equations (5.15)–(5.16) limit the radial topology of the distribution network, where the former ensures network connectivity by introducing the concept of virtual power, and the latter guarantees the acyclic nature of the distribution network by limiting the number of active lines.

Gas Flow Constraints: On the gas system side of the IEGS, based on the steady-state process model of natural gas transmission [83], the corresponding energy flow constraints are constructed as follows:

$$\sum_{n\in\Omega_m^a} f_{n,t}^a + \sum_{n\in\Omega_m^w} f_{n,t}^w - \sum_{n\in\Omega_m^u} f_{n,t}^u - \sum_{mn\in\gamma_m} f_{mn,t} = D_{m,t}^g - u_{m,t}^f, \quad \forall m \in \Omega_n, \forall t \in \Omega_t \tag{5.17}$$

$$f_{mn,t} = \text{sgn}(\pi_{m,t}, \pi_{n,t}) \sqrt{W_{mn} \left|\pi_{m,t}^2 - \pi_{n,t}^2\right|}, \quad \forall mn \in \Omega_p, \forall t \in \Omega_t \tag{5.18}$$

$$\text{sgn}(\pi_{\mathrm{m,t}}, \pi_{\mathrm{n,t}}) = \begin{cases} 1, & \text{if } \pi_{m,t} \geq \pi_{n,t} \\ -1, & \text{if } \pi_{m,t} < \pi_{n,t} \end{cases} \tag{5.19}$$

$$-F_m^{w,\text{ramp}} \leq f_{m,t}^w - f_{m,t-1}^w \leq F_m^{w,\text{ramp}}, \quad \forall m \in \Omega_w, \forall t \in \Omega_t \tag{5.20}$$

$$p_{i,t}^u = T_m f_{m,t}^u, \quad \forall i, m \in \Omega_u, \forall t \in \Omega_t \tag{5.21}$$

$$0 \leq u_{m,t}^f \leq \chi D_{m,t}^g, \quad \forall m \in \Omega_n, \forall t \in \Omega_t \tag{5.22}$$

$$\Pi_m^{\min} \leq \pi_{m,t} \leq \Pi_m^{\max}, \quad \forall m \in \Omega_n, \forall t \in \Omega_t \tag{5.23}$$

$$-F_{mn}^{\max} \leq f_{mn,t} \leq F_{mn}^{\max}, \quad \forall mn \in \Omega_p, \forall t \in \Omega_t \tag{5.24}$$

$$F_m^{a,min} \leq f_{m,t}^a \leq F_m^{a,\max}, \quad \forall m \in \Omega_a, \forall t \in \Omega_t \tag{5.25}$$

$$F_m^{u,min} \leq f_{m,t}^u \leq F_m^{u,\max}, \quad \forall m \in \Omega_u, \forall t \in \Omega_t \tag{5.26}$$

$$F_m^{w,min} \leq f_{m,t}^w \leq F_m^{w,\max}, \quad \forall m \in \Omega_w, \forall t \in \Omega_t \tag{5.27}$$

where Ω_m^a, Ω_m^w, and Ω_m^u represent the sets of gas storage tanks, gas gate stations, and gas-fired power plants connected to node m, respectively; γ_m represents the set of gas pipelines connected to node m; Ω_p represents the set of gas pipelines; Ω_w represents the set of gas gate station nodes; Ω_u represents the set of gas-fired power plant nodes; Ω_a represents the set of gas storage tank nodes. $D^{g}_{m,t}$ represents the gas load at node m; W_{mn} represents the Weymouth equation coefficient for pipeline mn; $F^{w,\mathrm{ramp}}_{m}$ represents the regulation capacity of the gas gate station; T_m represents the gas-fired power plant's gas generation ratio; χ represents the maximum load-shedding ratio; $\Pi^{\min}_{m}$ and $\Pi^{\max}_{m}$ represent the minimum and maximum pressure at node m, respectively; $F^{\max}_{mn}$ represents the maximum gas flow capacity of pipeline mn; $F^{a,\min}_{m}$ and $F^{a,\max}_{m}$ represents the minimum and maximum supply gas flow of the gas storage tank at node m; $F^{u,\min}_{m}$ and $F^{u,\max}_{m}$ represent the minimum and maximum supply gas flow of the gas-fired power plant at node m; $F^{w,\min}_{m}$ and $F^{w,\max}_{m}$ represent the minimum and maximum gas flow of the gas gate station at node m. $f^{a}_{n,t}$, $f^{w}_{n,t}$, and $f^{u}_{n,t}$ represent the gas flow rates at the gas storage tank, gas gate station, and gas-fired power plant nodes, respectively; $f_{mn,t}$ represents the gas flow through pipeline mn; $p^{u}_{i,t}$ represents the output of the gas-fired power plant; $U^{f}_{m,t}$ represents the gas user load shedding at node m; $\pi_{m,t}$ represents the gas pressure at node m. Additionally, the operator $\mathrm{sgn}(\cdot)$ denotes the unit step function.

Equation (5.17) represents the flow balance equation for gas network nodes, Eq. (5.18) corresponds to the steady-state gas flow equation for pipelines, Eq. (5.19) models the direction of gas flow, Eq. (5.20) limits the gas gate station's gas supply ramp-up capacity, Eq. (5.21) represents the energy conversion process of the gas-fired power plant, Eq. (5.22) sets the upper and lower limit for gas user load shedding, Eq. (5.23) provides the upper and lower limits for node pressure, Eq. (5.24) defines the pipeline gas transmission capacity, and Eqs. (5.25)–(5.27) corresponds to the upper and lower limits for the gas supply quantities at the gas storage tank, gas-fired power plant, and gas gate station, respectively.

5.3 Model Reformulation and Matrix Representation

The optimal electricity-and-gas flow model for the distribution-level IEGS constructed by Eqs. (5.1)–(5.27) is a mixed-integer nonlinear, non-convex optimization model, with its non-convexity primarily reflected in Eqs. (5.6), (5.18), and (5.19). The traditional approach is to convert this problem into a MISOCP problem through SOC relaxation, but this method does not guarantee the accuracy of the solution. Therefore, the solution proposed in this section is to first unify the non-convex constraints into SOC equality constraints of the form $z = \sqrt{x2 + y2}$, but without relaxing them. Section 5.3.1 reformulates the non-convex optimization into a SOC form, and Sect. 5.3.2 presents the matrix representation.

5.3.1 Second-Order Cone Formulation for Non-Convex Optimization

On the power system side, the SOC reformulation of the model mainly targets the branch flow Eq. (5.6). The SOC equality reformulation process is as follows:

$$\left\| p_{ij,t},\ q_{ij,t} \right\|_2 = s_{ij,t}, \quad \forall ij \in \Omega_l, \forall t \in \Omega_t \tag{5.28}$$

$$\left\| 2s_{ij,t},\ v_{i,t} - l_{ij,t} \right\|_2 = v_{i,t} + l_{ij,t}, \quad \forall ij \in \Omega_l, \forall t \in \Omega_t \tag{5.29}$$

where $s_{ij,t}$ is a newly defined intermediate variable, which physically corresponds to the apparent power of the transmission line. By introducing this intermediate variable, Eq. (5.6) can be equivalently transformed into the SOC equality constraints shown in Eqs. (5.28)–(5.29).

On the gas system side, to address the non-convexity introduced by Eqs. (5.18)–(5.19), this section first introduces a binary variable λ_{mn} to represent the gas flow direction in pipeline mn. Specifically, $\lambda_{mn} = 1$ indicates that the gas flow from node m to node n, where $\lambda_{mn} = 0$ represents the opposite direction. These constraints can be expressed as:

$$-M(1 - \lambda_{mn}) \le f_{mn,t} \le M\lambda_{mn}, \quad \forall mn \in \Omega_p, t \in \Omega_t. \tag{5.30}$$

Furthermore, for Eqs. (5.18)–(5.19), this section introduces auxiliary variables $\sigma_{mn,t}^{\max}$ and $\sigma_{mn,t}^{\min}$, which correspond to the maximum and minimum gas pressure values between the two ends of pipeline mn, i.e., $\pi_{m,t}$ and $\pi_{n,t}$, respectively:

$$\sigma_{mn,t}^{\max} = \max\{\pi_{m,t}, \pi_{n,t}\}, \sigma_{mn,t}^{\min} = \min\{\pi_{m,t}, \pi_{n,t}\}, \quad \forall mn \in \Omega_p, t \in \Omega_t. \tag{5.31}$$

Since gas in the pipeline always flows from the high-pressure side to the low-pressure side, the above expressions can be rewritten based on the newly defined gas flow direction variable λ_{mn} as:

$$\pi_{m,t} \le \sigma_{mn,t}^{\max} \le \pi_{m,t} + M(1 - \lambda_{mn}), \quad \forall mn \in \Omega_p, t \in \Omega_t \tag{5.32}$$

$$\pi_{n,t} \le \sigma_{mn,t}^{\min} \le \pi_{n,t} + M(1 - \lambda_{mn}), \quad \forall mn \in \Omega_p, t \in \Omega_t \tag{5.33}$$

$$\pi_{n,t} \le \sigma_{mn,t}^{\max} \le \pi_{n,t} + M\lambda_{mn}, \quad \forall mn \in \Omega_p, t \in \Omega_t \tag{5.34}$$

$$\pi_{m,t} \le \sigma_{mn,t}^{\min} \le \pi_{m,t} + M\lambda_{mn}, \quad \forall mn \in \Omega_p, t \in \Omega_t. \tag{5.35}$$

Based on these auxiliary variables, Eqs. (5.18)–(5.19) can be rewritten as:

$$\left\| f_{mn,t},\ \sqrt{W_{mn}}\sigma_{mn,t}^{\min} \right\|_2 = \sqrt{W_{mn}}\sigma_{mn,t}^{\max},\quad \forall mn \in \Omega_p,\ t \in \Omega_t. \tag{5.36}$$

Finally, by treating $\sqrt{W_{mn}}\sigma_{mn,t}^{\min}$ and $\sqrt{W_{mn}}\sigma_{mn,t}^{\max}$ as two independent variables, the original non-convex constraints Eqs. (5.18)–(5.19) can be equivalently reformulated.

5.3.2 Matrix Representation

For further simplification and ease of expression, the non-convex optimization model, reconstructed using SOC equality constraints, is now represented in a compact matrix form as follows:

$$\langle \boldsymbol{P} \rangle : \min_{\{x \in \mathbb{Z}, y \in \mathbb{R}\}} d^{\mathrm{T}} y \quad \text{s.t. } Ax + By \le f,\ \boldsymbol{L} \le y \le \boldsymbol{U} \tag{5.37}$$

$$\boldsymbol{y}_{r_n} = \sqrt{\boldsymbol{y}_{p_n}^2 + \boldsymbol{y}_{q_n}^2},\quad \forall n \in \Delta \tag{5.38}$$

where $\boldsymbol{x}$ and $\boldsymbol{y}$ correspond to integer variables and continuous variables, respectively, while $\boldsymbol{d}$, $\boldsymbol{A}$, $\boldsymbol{B}$, and $\boldsymbol{f}$ represent the objective function coefficients, integer variable coefficients, continuous variable coefficients, and constraint right-hand side terms, respectively. The variables $\boldsymbol{L}$ and $\boldsymbol{U}$ define the bounds for $\boldsymbol{y}$, and $\boldsymbol{y}_{rn}$, $\boldsymbol{y}_{pn}$, and $\boldsymbol{y}_{qn}$ form a subset of $\boldsymbol{y}$ that participates in the formulation of the SOC equality constraints, where the index triplet (p_n, q_n, r_n) is determined by the index set $n \in \Delta$. The linear constraints in Eq. (5.37) correspond to Eqs. (5.1)–(5.5), (5.7)–(5.17), (5.20)–(5.27), (5.30), and (5.32)–(5.35), while the non-convex constraints in Eq. (5.38) correspond to Eqs. (5.28), (5.29), and (5.36).

5.4 Design of a Non-convex Optimization Algorithm for Second-Order Cone Equality Constraints

SOC equality constraints are a common type of non-convex constraint in energy optimization problems. However, the academic community has yet to develop an efficient method for obtaining exact solutions to such non-convex optimization problems. While general solvers [55, 58] and traditional SB&B algorithms [82, 84] can be used to solve these problems, they require reformulating SOC equality constraints into multiple bilinear equality constraints before applying their respective SB&B algorithms. This transformation significantly increases the number of non-convex constraints in the problem. Taking the optimal electricity-and-gas flow problem of IEGS studied in this chapter as an example, each quadratic term must be replaced, meaning that reformulating Eqs. (5.6) and (5.18) requires three sets and two sets of bilinear equality constraints, respectively. If a spatial branching method specifically designed for SOC equality constraints can be constructed, it would eliminate the need for the mentioned reformulation process, significantly reducing the number of non-convex constraints and thereby improving the computational efficiency of the algorithm. Based on this idea, this section proposes a novel spatial branching method within the SB&B framework. Specifically, Sect. 5.4.1 introduces a spatial branching strategy for SOC surfaces, Sect. 5.4.2 presents the general algorithmic procedure, and Sect. 5.4.3 implements the approach using CPLEX-based programming.

5.4.1 Second-Order Cone Surface Spatial Branching Strategy

Solving non-convex optimization problems requires continuously approximating the optimal solution through multiple convex subproblems. In this section, the convex subproblem at the root node is obtained by directly applying convex relaxation to the original problem $\langle \mathbf{P} \rangle$, replacing the equality sign "=" in the non-convex constraints (5.38) with the inequality "$\geq$". For convenience, this convex relaxation problem is denoted as $\langle \mathbf{P_c} \rangle$. During the construction of the search tree, each convex subproblem corresponding to a child node is formulated by adding additional constraints to $\langle \mathbf{P_c} \rangle$, denoted as $\langle \mathbf{P_c} \rangle^{(s)}$, where s represents the index of the child node.

Since each subproblem $\langle \mathbf{P_c} \rangle^{(s)}$ is a MISOCP problem, it can be solved directly. However, the obtained solution may not satisfy the non-convex constraints (5.38), resulting in an infeasible relaxed solution, denoted as $\boldsymbol{y}^*$. Suppose this relaxed solution violates the k-th equality constraint in Eq. (5.38), which is expressed as:

$$y^*_{r_k} > \sqrt{(y^*_{P_k})^2 + (y^*_{q_k})^2}. \tag{5.39}$$

Additionally, it is assumed that in the current convex subproblem $\langle \mathbf{P_c} \rangle^{(s)}$, the upper and lower bounds of the variables $\boldsymbol{y}\ P_k$ and $\boldsymbol{y}\ q_k$ have been narrowed down to $[\boldsymbol{L}^{(s)}_{Pk}, \boldsymbol{U}^{(s)}_{Pk}]$ and $[\boldsymbol{L}^{(s)}_{qk}, \boldsymbol{U}^{(s)}_{qk}]$ during the branching process. The subsequent analysis will focus on the triplet variables in this infeasible solution. For clarity, the symbols mentioned above have been replaced, as shown in Table 5.1.

After the symbol substitution, the following summarizes the SOC relaxation error, variable bounds, and other properties of the infeasible relaxed solution corresponding to the current subproblem:

Table 5.1 Symbol substitution

Type	Variable name	Current solution	Upper and lower bounds
Before Substitution	$\boldsymbol{y}_{pk}, \boldsymbol{y}_{qk}, \boldsymbol{y}_{rk}$	$\boldsymbol{y}^*_{pk}, \boldsymbol{y}^*_{qk}, \boldsymbol{y}^*_{rk}$	$\boldsymbol{L}^{(s)}_{Pk}, \boldsymbol{U}^{(s)}_{Pk}, \boldsymbol{L}^{(s)}_{qk}, \boldsymbol{U}^{(s)}_{qk}$
After Substitution	x, y, z	x^*, y^*, z^*	l_x, u_x, l_y, u_y

$$\zeta = z^* - \sqrt{(x^*)^2 + (y^*)^2} > \varepsilon, \quad l_x \le x^* \le u_x, \quad l_y \le y^* \le u_y \tag{5.40}$$

where ζ represents the SOC relaxation error of the current infeasible solution, and ε represents the threshold of the SOC relaxation error. Based on the parameters in the above expression, this section proposes a novel spatial branching strategy.

Spatial Information Extraction: The schematic diagram of the SOC equality constraint and its infeasible relaxation solution is shown in Fig. 5.1a. In this diagram, the blue SOC surface represents the original feasible region corresponding to the non-convex constraint $z = \sqrt{x2 + y2}$; the green surface represents the feasible region of the current convex subproblem (a subset of the original feasible region). This surface is obtained by cutting the original feasible region with four planes corresponding to the upper and lower bounds of the current variables x and y. Therefore, the projection of this surface onto the x–y plane is a rectangle. The red point P_0 corresponds to the current infeasible solution (x^*, y^*, z^*), which is located above the blue/green SOC surface and does not satisfy the original non-convex constraint.

In this section, two points closest to the origin O are selected from the four vertices of the feasible region (green surface), which serve as waypoints: (l_x, l_y, $\|l_x, l_y\|_2$) and (l_x, u_y, $\|l_x, l_y\|_2$), as shown in Fig. 5.1a. From the origin O, rays are drawn through these two waypoints, and they intersect with the plane corresponding to the upper limit of the z-value at two points, denoted as P_1 and P_3 in Fig. 5.1b. Similarly, a ray originating from the origin O and passing through any point within the feasible region will also produce an intersection, which corresponds to a variable point that is not fixed, as shown by P_2 in Fig. 5.1b. The branching rules in this section will focus on these three reference points.

Since the projection of the feasible region (green surface) onto the x–y plane is a rectangle, it may cover all four quadrants or only one or two of them, so the following conditions need to be verified before establishing the branching rules:

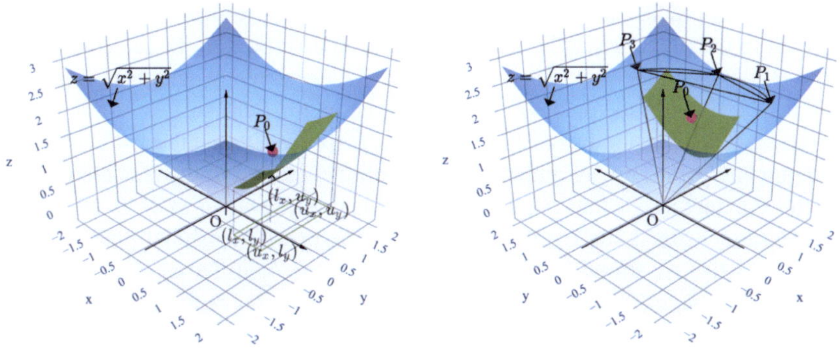

(a) feasible domain and infeasible solution diagram.

(b) spatial reference point diagram.

Fig. 5.1 Solution space visualization and information extraction

$$(l_x \cdot u_x < 0) \wedge \left(l_y \cdot u_y < 0\right). \tag{5.41}$$

In the above expression, the symbol "∧" represents the logical AND operation. If the condition is satisfied, it indicates that the projection of the feasible region onto the *x*–*y* plane covers all four quadrants. Otherwise, it implies that the projection of the feasible region onto the *x*–*y* plane only covers one to two quadrants. For these two different cases, the corresponding branching rules are defined as follows.

Half-Space Branching Rule: The half-space branching rule is mainly applicable in the case where the condition in Eq. (5.41) is not satisfied. In this case, the space branching depends on a triangular surface formed by points O, P_1, and P_3, as shown in Fig. 5.1b. It is important to note that this triangle is not a two-dimensional plane but a three-dimensional surface situated above the blue feasible region, representing a subset of the feasible region. Additionally, the triangular surface can also be viewed as a rotated surface formed by rotating the line segment OP_1 around the z-axis until it reaches the position of the line segment OP_3. In spherical coordinates,[1] the spherical coordinates of the two vertices P_1 and P_3 of the triangular surface can be represented as $(\rho_1, \theta_1, \varphi_1)$ and $(\rho_3, \theta_3, \varphi_3)$, respectively, with the following calculation methods:

$$\rho_1 = \rho_3 = \sqrt{2\left[\max(l_x^2, u_x^2) + \max(l_y^2, u_y^2)\right]} \tag{5.42}$$

$$\theta_1 = \theta_3 = \arctan(1) = \frac{\pi}{4} \tag{5.43}$$

$$\varphi_1 = \min_{w_y \in \{l_y, u_y\}, w_x \in \{l_x, u_x\}} \arctan \frac{w_y}{w_x} \tag{5.44}$$

$$\varphi_3 = \max_{w_y \in \{l_y, u_y\}, w_x \in \{l_x, u_x\}} \arctan \frac{w_y}{w_x} \tag{5.45}$$

where ρ_1 and ρ_3 represent the spherical radii of P_1 and P_3, respectively. Their values are equal to the maximum spherical radius of all the vertices in the feasible region. θ_1 and θ_3 represent the polar angles of P_1 and P_3, respectively, and their values are equal to $\pi/4$, because the polar angle of any point on the SOC surface is $\pi/4$. φ_1 and φ_3 represent the azimuthal angles of P_1 and P_3, respectively, and their values are the minimum and maximum azimuthal angles of all the vertices in the feasible region.

After connecting P_1 and P_3, a convex polyhedron, denoted as Γ, is formed by the three-dimensional surface triangle OP_1P_3, the two-dimensional plane triangle OP_1P_3, and the arc formed by the line segment P_1P_3 and the arc P_1P_3 (through point P_2). Although Γ is a convex hull of the current feasible region (green surface), it still contains many infeasible solutions. Therefore, based on the SB&B framework, this section will split Γ into two subspaces through space branching, ensuring that

[1] In this chapter, the spherical coordinate system is defined according to the conventions used in physics, where θ and ϕ represent the polar angle and the azimuthal angle, respectively, with $\theta \in [0, \pi]$, $\varphi \in [0, 2\pi]$.

feasible solutions are not eliminated while minimizing the elimination of infeasible solutions, including P_0. To achieve this, the coordinates of point P_2 are first fixed as follows:

$$\rho_2 = \rho_1 = \rho_3, \quad \theta_2 = \frac{\pi}{4}, \quad \varphi_2 = \arctan(\frac{y^*}{x^*}) \tag{5.46}$$

where ρ_2 and θ_2 are set to the same values as P_1 and P_3, while φ_2 is calculated based on the x and y coordinates of the current infeasible solution P_0.

Next, define the operator $C(\cdot)$ to convert coordinates from the spherical coordinate system to the Cartesian coordinate system. The specific calculation method is as follows:

$$\begin{aligned}(x, y, z) &= C(\rho, \theta, \varphi) \\ &= (\rho\cos\varphi\sin\theta,\ \rho\sin\varphi\sin\theta,\ \rho\cos\theta).\end{aligned} \tag{5.47}$$

Through the above operator, the spherical coordinates of P_1, P_2, and P_3 can be converted into Cartesian coordinates, resulting in the following coordinates for $P_1(x_1, y_1, z_1)$, $P_2(x_2, y_2, z_2)$, and $P_3(x_3, y_3, z_3)$:

$$\begin{cases}(x_1, y_1, z_1) = C(\rho_1, \theta_1, \varphi_1) \\ (x_2, y_2, z_2) = C(\rho_2, \theta_2, \varphi_2) \\ (x_3, y_3, z_3) = C(\rho_3, \theta_3, \varphi_3)\end{cases}. \tag{5.48}$$

Furthermore, based on the coordinates of the origin O, P_1, P_2, and P_3, the normal vectors of the planes containing the two-dimensional triangles ΔOP_1P_2 and ΔOP_2P_3 are calculated. These normal vectors are then used to construct two hyperplane equations for spatial branching of the convex polyhedron Γ, as shown below:

$$f_{\Delta OP_1P_2}(x, y, z) = \begin{vmatrix} x & y & z \\ x_1 - x_o & y_1 - y_o & z_1 - z_o \\ x_2 - x_o & y_2 - y_o & z_2 - z_o \end{vmatrix} = \begin{vmatrix} x & y & z \\ x_1 & y_1 & z_1 \\ x_2 & y_2 & z_2 \end{vmatrix} \leq 0 \tag{5.49}$$

$$f_{\Delta OP_2P_3}(x, y, z) = \begin{vmatrix} x & y & z \\ x_2 - x_o & y_2 - y_o & z_2 - z_o \\ x_3 - x_o & y_3 - y_o & z_3 - z_o \end{vmatrix} = \begin{vmatrix} x & y & z \\ x_2 & y_2 & z_2 \\ x_3 & y_3 & z_3 \end{vmatrix} \leq 0. \tag{5.50}$$

The coordinates of the origin O are given as $(x_o, y_o, z_o) = (0, 0, 0)$. The two hyperplane Eqs. (5.9)–(5.10) are both linear constraints expressed in determinant form, with the first row representing the variables and the second and third rows representing constants. A three-dimensional schematic of these hyperplanes is shown in Fig. 5.2.

Based on Eqs. (5.9)–(5.10), the following space branching rules for the current convex subproblem can be constructed:

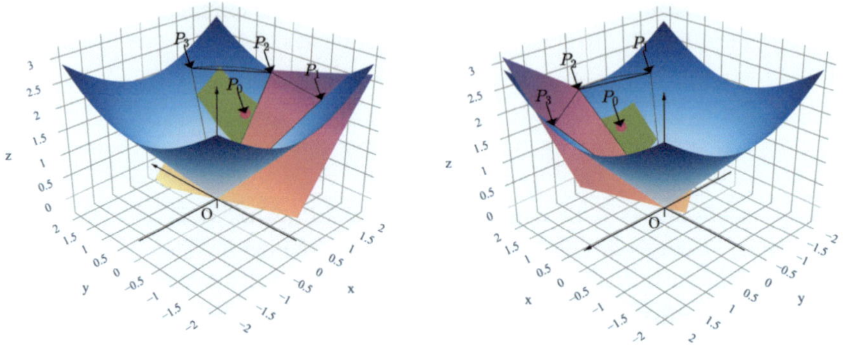

(a) diagram of the lower branch cutting plane. (b) diagram of the upper branch cutting plane.

Fig. 5.2 Spatial branching diagram

$$f_{\Delta OP_1P_2}(x, y, z) \leq 0 \vee f_{\Delta OP_2P_3}(x, y, z) \leq 0 \tag{5.51}$$

where the symbol "∨" represents the logical OR operation. Based on the above space branching rules, the convex polytope Γ can be divided into two subspaces, corresponding to the two subproblems of the current convex subproblem. Specifically, $f_{\Delta OP1P2} \leq 0$ is used to construct the lower branch, while $f_{\Delta OP2P3} \leq 0$ is used to construct the upper branch.

Proposition 5.1 *The infeasible solution P_0 does not exist in either of the two subspaces after branching in Eq. (5.51).*

Proof Since the proof process for the upper branch and the lower branch is similar, this section will only demonstrate the proof for the lower branch, where the cutting-plane equation is $f_{\Delta OP1P2}\ (x, y, z) \leq 0$. For the infeasible solution P_0 with coordinates (x^*, y^*, z^*), we will use a proof by contradiction, assuming that P_0 exists in the subspace of the lower branch after branching, meaning that P_0 satisfies:

$$f_{\Delta OP_1P_2}(x^*, y^*, z^*) \leq 0. \tag{5.52}$$

Substituting Eqs. (5.42)–(5.46) and (5.48) into Eq. (5.52) could obtain:

$$\begin{aligned} f_{\Delta OP_1P_2}(x^*, y^*, z^*) = & \frac{1}{2}\rho_1\rho_2(\sin\varphi_1 - \sin\varphi_2)x^* \\ & - \frac{1}{2}\rho_1\rho_2(\cos\varphi_1 - \cos\varphi_2)y^* \\ & + \frac{1}{2}\rho_1\rho_2(\cos\varphi_1\sin\varphi_2 - \sin\varphi_1\cos\varphi_2)z^* \leq 0. \end{aligned} \tag{5.53}$$

Since $\rho_1\rho_2 \neq 0$ and $\varphi_2 = \arctan(y^*/x^*)$, Eq. (5.53) can be further simplified as:

$$x^* \sin\varphi_1 - y^* \cos\varphi_1 \leq \frac{z^*}{\sqrt{(x^*)^2 + (y^*)^2}} \big(x^* \sin\varphi_1 - y^* \cos\varphi_1\big). \tag{5.54}$$

Based on the definition of Eq. (5.54), we have $\varphi_1 \leq \arctan(y^*/x^*)$, which implies that $x^* \sin\varphi_1 - y^*\cos\varphi_1 \leq 0$. Therefore, Eq. (5.54) can be further rewritten as:

$$\sqrt{(x^*)^2 + (y^*)^2} \geq z^*. \tag{5.55}$$

Noting that Eq. (5.55) violates the prerequisite condition for P_0 being an infeasible solution, we conclude that the assumption in Eq. (5.52) does not hold, meaning that the infeasible solution P_0 does not lie within the subspace after the lower branch. Similarly, it can be proven that P_0 does not exist in the upper branch's subspace either.

Proposition 5.2 *The two subspaces resulting from the branching in Eq. (5.51) will not eliminate any feasible solutions of the original problem.*

Proof Similar to Proposition 5.1, we only prove the case of the lower branch. Let (x^*, y^*, z^*) be any feasible solution of the original non-convex problem in the lower subspace, located on the green surface. Using proof by contradiction, assume that this solution is eliminated by the lower branch's cutting plane Eq. (5.49), i.e.,

$$f_{\Delta OP_1P_2}(x^*, y^*, z^*) > 0 \tag{5.56}$$

where (x^*, y^*, z^*) satisfies:

$$z^* = \sqrt{(x^*)^2 + (y^*)^2}, \quad \varphi 1 \leq \arctan\frac{y^*}{x^*} \leq \varphi 2. \tag{5.57}$$

Substituting Eqs. (5.46) and (5.48) into Eq. (5.56) could obtain:

$$\begin{aligned} f_{\Delta OP_1P_2}(x^*, y^*, z^*) = & \frac{1}{2}\rho_1\rho_2(\sin\varphi_1 - \sin\varphi_2)x^* \\ & - \frac{1}{2}\rho_1\rho_2(\cos\varphi_1 - \cos\varphi_2)y^* \\ & + \frac{1}{2}\rho_1\rho_2(\cos\varphi_1 \sin\varphi_2 - \sin\varphi_1 \cos\varphi_2)z^* > 0. \end{aligned} \tag{5.58}$$

Considering that $\rho_1\rho_2 \neq 0$ and $z^* \geq 0$, Eq. (5.58) can be further simplified as:

$$\frac{(\sin\varphi_2 - \sin\varphi_1)x^* - (\cos\varphi_2 - \cos\varphi_1)y^*}{\sin(\varphi_2 - \varphi_1)z^*} < 1. \tag{5.59}$$

Let $\varphi^* = \arctan(y^*/x^*)$ could have:

$$x^* = z^* \cos \varphi^*, \quad y^* = z^* \sin \varphi^*. \tag{5.60}$$

Substituting the above expression into Eq. (5.59) can get:

$$\sin(\varphi_2 - \varphi^*) + \sin(\varphi^* - \varphi_1) < \sin(\varphi_2 - \varphi_1) \tag{5.61}$$

For convenience, let A replace $\varphi_2 - \varphi^*$, and B replace $\varphi^* - \varphi_1$. The above expression can be rewritten as:

$$\begin{aligned} \sin A + \sin B &< \sin(A + B) \\ &= \sin A \cos B + \sin B \cos A. \end{aligned} \tag{5.62}$$

Note that the values of A, B, and $A + B$ are all within the range $[0, \pi]$, which means that the values of $\cos A$ and $\cos B$ are always less than or equal to 1, while the values of $\sin A$ and $\sin B$ are always greater than or equal to 0. As a result, the left-hand side of Eq. (5.62) is greater than the right-hand side, which contradicts the conclusion. Therefore, the assumption that Eq. (5.56) does not hold is incorrect, and the feasible solution (x^*, y^*, z^*) will not be cut off by the lower branch cutting plane Eq. (5.49). Similarly, we can prove that the feasible solution (x^*, y^*, z^*) will not be cut off by the upper branch cutting plane Eq. (5.50).

Based on Proposition 5.1 and Proposition 5.2, the validity of the space branching rule (5.51) can be proved. Specifically, this branching rule effectively removes infeasible solutions while preserving all feasible solutions. The above derivation establishes the space branching rule for SOC constraints under the half-space condition.

Full-Space Branching Rule: If Eq. (5.41) holds, it indicates that the projection of the feasible region onto the x–y plane covers all four quadrants. In this case, the half-space branching rule is no longer applicable. This is because the derivation of Proposition 5.2 only applies when the values of A, B, and $A + B$ lie within the range $[0, \pi]$. In full space, the range of azimuth angles is expanded to $[0, 2\pi]$, so Proposition 5.2 no longer applies. In other words, continuing to use Eq. (5.51) for space branching could result in the removal of feasible solutions.

In general, in full space, to ensure that infeasible solutions P_0 are removed from the generated subspaces while preserving feasible solutions, at least three sets of cutting-plane equations need to be introduced. However, current mainstream solvers typically support only binary branching operations. To address this issue, this section introduces the concept of pseudo-space branching, as illustrated in Fig. 5.3.

The concept of pseudo-space branching is to partition the space in such a way that it provides the necessary conditions for the half-space branching Eq. (5.51) without attempting to eliminate infeasible solutions. A conventional approach is to split the solution space directly using the planes $x = 0$ or $y = 0$, with the corresponding rule as follows:

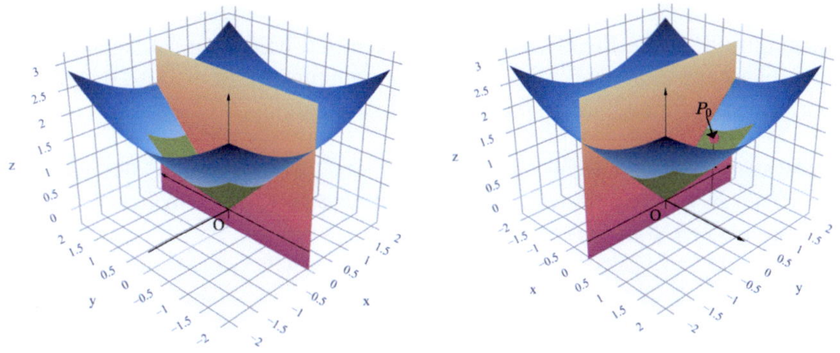

(a) lower branch diagram based on x ≤ 0 cut plane.

(b) upper branch diagram based on x ≥ 0 cut plane.

Fig. 5.3 Pseudo-space branching diagram

$$\begin{cases} x \geq 0 \vee x \leq 0, \text{ if } u_x - l_x \geq u_y - l_y \\ y \geq 0 \vee y \leq 0, \text{ if } u_y - l_y \geq u_x - l_x \end{cases}. \tag{5.63}$$

Under this rule, the algorithm will split the long edge of the projection of the feasible region onto the *x*–*y* plane, generating two subspaces. While this operation does not eliminate any solutions, it ensures that the feasible regions of the two generated convex subproblems are confined to half-space, thus allowing the continued application of the half-space branching rule (5.51) to handle the non-convex constraints.

5.4.2 General Algorithm Flow

Based on the proposed space branching rules, the solution framework of the traditional SB&B algorithm can be further improved to construct a non-convex optimization algorithm tailored for SOC equality constraints. The specific algorithm flow is out-lined as follows.

Space Branching Decision Process for SOC Equality Constraints: For any convex subproblem $\langle \mathbf{P_c} \rangle^{(s)}$ in the search tree, if the solution obtained does not satisfy the original non-convex constraints, space branching needs to be applied to this subproblem. The steps are described in Algorithm 5.1. First, select the variable triplet ($\boldsymbol{y}_{pk}$, $\boldsymbol{y}_{qk}$, $\boldsymbol{y}_{rk}$) corresponding to the set of constraints with the largest SOC relaxation error as the branching object. Then, gather model information from the three-dimensional space formed by these three variables, including the coordinates of the infeasible solution and the boundary of the feasible region. Next, based on Eq. (5.41), determine the type of branching rule to apply for the current subproblem. If Eq. (5.41) is not satisfied, use the half-space branching rule (5.51); otherwise, use the full-space branching rule (5.63). Finally, based on the selected branching rule, generate the cutting-plane equations and return them to the main function of the SB&B algorithm to perform space branching for the current subproblem.

Algorithm 5.1: Space Branching Decision Process for SOC Equality Constraints

```
Obtain the infeasible relaxed solution y* of the current convex subproblem (P_c)^(s), and
calculate its relaxation error vector ζ*.
Select the variable triplet (y_pk, y_qk, y_rk) corresponding to the largest error ζ*_k in ζ*.
Perform symbolic substitution to obtain the variables (x, y, z), infeasible solution (x*,
y*, z*), and their bounds [l_x, u_x], [l_y, u_y].
if Eq. (5.41) is not satisfied then
  |  Retrieve the coordinates of the spatial reference points P_1, P_2, and P_3.
  |  Based on the space branching rule (5.51), generate the cut-      ▷ Half-Space
  |  ting plane equation.                                                Branching
else
  |  Based on the space branching rule (5.63), generate the cut-      ▷ Full-Space
  |  ting plane equation.                                                Branching
end if
Return the cutting plane equation for space branching.
```

Complete Steps of SB&B Algorithm: By embedding Algorithm 5.1 into the general SB&B framework, we can construct a solution method for the original non-convex problem ⟨P⟩ with SOC equality constraints. The algorithm flow is as shown in Algorithm 5.2, where the space branching process for SOC equality constraints is located at lines 13 and 14.

Algorithm 5.2: SB&B Algorithm for SOC Equality Constraints

Set the relaxation error threshold ε and optimal error threshold ϵ based on engineering requirements.
Set index $k \leftarrow 0$, lower bound $lb \leftarrow -\infty$, and upper bound $ub \leftarrow +\infty$. ▷for the optimal solution bounds
Initialize the search tree sets $N \leftarrow \{\langle \mathbf{P}_c \rangle^{(0)}\}$ and $F \leftarrow \{lb\}$. ▷ for the search tree
if a known feasible solution $(\hat{x}, \hat{y})$ exists **then** ▷ warm start
  Solve the problem $\langle \mathbf{P}_c \rangle^{(0)}$, obtaining the solution $(\boldsymbol{x}^{(0)}, \boldsymbol{y}^{(0)})$ and the objective function value $f^{(0)}$ and update $lb \leftarrow f^{(0)}$.
  Based on the feasible solution $(\hat{x}, \hat{y})$, call Eq. (5.1) to update the upper bound ub.
Else ▷ cold start
  Initialize $(\hat{x}, \hat{y}) \leftarrow (0, 0)$.
end if
while $(ub - lb)/ub \geq \epsilon$ and $N \neq \varnothing$ **then**
  From the set N, select the convex subproblem $\langle \mathbf{P}_c \rangle^{(s)}$ with the smallest objective function value. ▷ accelerated lower bound increase
  Update $N \leftarrow N \setminus \{\langle \mathbf{P}_c \rangle^{(s)}\}$, $F \leftarrow F \setminus \{f^{(s)}\}$ ▷ remove this problem from the search tree
  Call Algorithm 5.1 to obtain the cutting plane equation for the selected subproblem
  Perform space branching based on the cutting plane information, creating two new subproblems: $\langle \mathbf{P}_c \rangle^{(k+1)}$ and $\langle \mathbf{P}_c \rangle^{(k+2)}$.
  for each $\langle \mathbf{P}_c \rangle^{(s)} \in \{\langle \mathbf{P}_c \rangle^{(k+1)}, \langle \mathbf{P}_c \rangle^{(k+2)}\}$ **then**
    if $\langle \mathbf{P}_c \rangle$ is solvable **then** ▷ Eliminated The infeasible subproblem node directly
      Solve $\langle \mathbf{P}_c \rangle^{(s)}$, obtaining the solution $(\boldsymbol{x}^{(s)}, \boldsymbol{y}^{(s)})$ and the objective function value $f^{(s)}$.
      if $(\boldsymbol{x}^{(s)}, \boldsymbol{y}^{(s)})$ is a feasible solution for the original non-convex problem **then**
        if $f^{(s)} \leq ub$ **then** ▷ find better solution
          update $ub \leftarrow f^{(s)}$, $(\hat{x}, \hat{y}) \leftarrow (\boldsymbol{x}^{(s)}, \boldsymbol{y}^{(s)})$ ▷ update best solution
        else
      else ▷ Re-add the subproblem to the search tree for further branching
        update $N \leftarrow N \cup \{\langle \mathbf{P}_c \rangle^{(s)}\}$, $F \leftarrow F\{f^{(s)}\}$.
      end if
    end if
  end while
  update $N \leftarrow N \setminus \{\langle \mathbf{P}_c \rangle^{(s)} | f^{(s)} \geq ub, \forall s\}$, $F \leftarrow F \setminus \{f^{(s)} \geq ub, \forall s\}$
  update lb to be the minimum value in the set F, $k \leftarrow k + 2$
end while
Return the optimal solution $(\hat{x}, \hat{y})$ and the optimal objective function value.

5.4.3 Programming Implementation Based on CPLEX

Implementing the above algorithm using conventional iterative programming methods is not the most efficient approach. This is because the main function from lines 10 to 29 treats each newly generated convex subproblem as an entirely new mathematical problem, without associating it with the previous subproblem-solving process. In reality, for a convex subproblem that needs to be branched (denoted as the parent node problem) and the two new convex subproblems generated after branching (denoted as child node problems), the child node problems only differ from the parent node problem by an additional cutting plane (either Eqs. (5.49) or (5.50)). Therefore, if the solution process of the parent node problem can be fully utilized in solving the child node problems, this can significantly reduce computational redundancy and improve the algorithm's solution efficiency.

To address this, this chapter continues to use CPLEX as the mathematical programming solver. It leverages the CPLEX Callback function interface to implement a unified solution process. In terms of model definition, the $\langle \mathbf{P_c} \rangle$ obtained from the original problem $\langle \mathbf{P} \rangle$ after SOC relaxation is passed into CPLEX as the problem at the root node. In terms of interface definition, this chapter implements the complete solution process of Algorithm 5.2 using three function interfaces "Heuristics Callback", "Incumbent Callback", and "Branch Callback". The entire process is achieved with only a single call to CPLEX.

5.5 Case Study Analysis

In this section, the effectiveness of the proposed SB&B algorithm is analyzed using three different scales of IEGS. Specifically, Sect. 5.5.1 analyzes optimal electricity-and-gas flow for a single time period, while Sect. 5.5.2 analyzes multi-period day-ahead scheduling. The case studies are as follows: Case A includes an 18-node power distribution network and a 6-node gas distribution network; Case B includes a 33-node power distribution network and a 37-node gas distribution network; and Case C includes a 136-node power distribution network and a 74-node gas distribution network. Detailed parameters can be found in Ref. [85].

To further illustrate the effectiveness of the proposed space branching strategy, this section compares and analyzes a total of 8 different non-convex optimization problem-solving methods, with their attributes shown in Table 5.2. The columns "feasibility" and "optimality" correspond to whether the methods have mechanisms to ensure the accuracy of the results. The "problem type" is set to $\langle \mathbf{P} \rangle$ for all methods, and the "approximation operations" include three types: SOC, linear, and no approximation. The last two columns provide the solvers used for each method and their corresponding references. From a method perspective, M1 uses the traditional SOC relaxation method, which is efficient but less accurate; M2 and M3 use piecewise approximation methods, where the accuracy depends on the number of segments

Table 5.2 Comparison of non-convex optimization algorithms

Number	Name	Optimality	Feasibility	Problem Type	Approximation	Solver	Source
M1	SOC convex relaxation	–	–	⟨**P**⟩	SOC	Gurobi	[72]
M2	Piecewise linear approximation	–	–	⟨**P**⟩	Linearity	Gurobi	[60]
M3	Piecewise McCormick linear envelope relaxation	–	–	⟨**P**⟩	Linearity	Gurobi	[61]
M4	Convex-concave method	–	✓	⟨**P**⟩	–	Gurobi	[73]
M5	Gurobi 10.0.2	✓	✓	⟨**P**⟩	–	Gurobi	[55]
M6	SCIP 8.0.3	✓	✓	⟨**P**⟩	–	SCIP	[58]
M7	SB&B algorithm proposed in [81]	✓	✓	⟨**P**⟩	–	CPLEX	
M8	SB&B algorithm proposed in this chapter	✓	✓	⟨**P**⟩	–	CPLEX	

chosen; M4 uses a convex-concave approach that guarantees feasibility but does not guarantee optimality; M5 and M6 use the Gurobi and SCIP solvers, respectively, representing the simplest and most direct strategies for solving non-convex optimization problems; M7 is the SB&B algorithm proposed in Ref. [81] for bilinear equality constraints, and M8 is the SB&B algorithm proposed in this chapter for SOC equality constraints. The comparison between these two methods (M7 and M8) highlights the effectiveness of the proposed method for solving specific non-convex constraints. For the case study analysis, the feasibility and optimality thresholds are both set to 0.5%. Four typical load scenarios are selected, with each scenario containing 24 time intervals of 1-h load data.

5.5.1 Single-Period Optimal Electricity-Gas Flow

To test the performance of the algorithm in solving single-period OEGF problems, this section sets $|\Omega_t|$ to 1. Each time period's OEGF in a load scenario is treated as an independent optimization problem. The results of 96 single-period optimal power flow problems ($24 \times 4 = 96$) are compared and analyzed across 8 different solution methods.

Accuracy: In this subsection, the accuracy of the algorithm is evaluated from two aspects: the "optimality indicator" and the "feasibility indicator". The optimality indicator corresponds to the relative error between the objective function value obtained by the algorithm and the result from M6 (SCIP), while the feasibility indicator corresponds to the maximum imbalance in the non-convex constraint (5.38). Table 5.3 shows a comparison of the accuracy indicators for the 8 solution methods across three case studies.

Overall, M5–M8 exhibit higher accuracy than M1–M4, as the former employ global optimization algorithms capable of handling non-convex constraints, whereas the latter fail to simultaneously guarantee both optimality and feasibility. Specifically, while M1 performs well in terms of the optimality indicator, its feasibility indicator exceeds 3% in all three cases, indicating that although the obtained objective function value is close to the actual value, the variable assignments generally fail to satisfy the non-convex constraints with significant errors. M2 and M3, which rely on piecewise linear approximation, perform well in terms of feasibility, particularly in Case A, where the feasibility indicator is nearly zero. However, as the system scale increases, their feasibility indicators exceed 1% in Cases B and C. M4, which applies CCP, ensures convergence to the feasible region through iterative operations but lacks optimality guarantees, with its deviation from the benchmark value exceeding 10% in Case C. M5 and M6, both based on general-purpose solvers, can guarantee both optimality and feasibility, making them relatively reliable solution methods. M7, which implements the SB&B algorithm proposed in Ref. [81], reformulates SOC equality constraints as bilinear equality constraints during the solution process, but this does not affect result accuracy. M8, which applies the SB&B algorithm proposed in this chapter, directly operates in the space of SOC equality constraints, achieving comparable optimality to M7 and obtaining the global optimum in Cases B and C. Moreover, M8 slightly improves feasibility over M7, with a maximum relaxation error of only 0.2%.

Table 5.3 Comparison of accuracy index of M1–M8 algorithm to solve single-period optimal electricity-gas flow problem

Algorithm	Case A		Case B		Case C	
	Optimality (%)	Feasibility (%)	Optimality (%)	Feasibility (%)	Optimality (%)	Feasibility (%)
M1	↓ 0.34	3.00	↓ 0.00	4.00	↓ 0.00	4.00
M2	↓ 0.49	0.00	↓ 0.39	1.62	↓ 1.22	1.65
M3	↓ 0.29	0.01	↑ 0.00	1.24	↑ 0.06	1.55
M4	↑ 0.34	0.00	↑ 8.83	0.24	↑ 10.7	0.24
M5	↑ 0.28	0.02	↑ 0.00	0.48	↑ 0.11	0.50
M6	Benchmark	0.50%	Benchmark	0.50	Benchmark	0.36
M7	↑ 0.38%	0.00%	↑ 0.00%	0.25	↑ 0.00	0.24
M8	↑ 0.34%	0.00%	↑ 0.00%	0.20	↑ 0.00	0.20

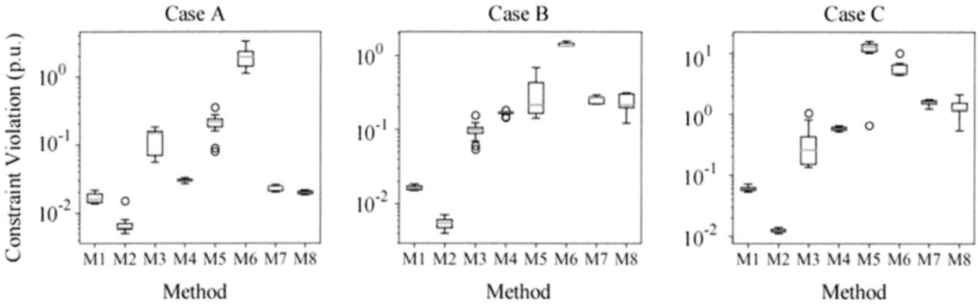

Fig. 5.4 Calculation time comparison of M1–M8 algorithm for single-period optimal energy flow problem

Efficiency: To further demonstrate the efficiency of the proposed algorithm, this subsection compares the computation time of M1–M8 for solving the single-period optimal electricity-gas flow problem. The results for the 24 time intervals in each case are aggregated and presented in a boxplot, as shown in Fig. 5.4 (with the vertical axis in logarithmic scale).

A horizontal comparison among the algorithms reveals a certain negative correlation between computational efficiency and accuracy. M1 and M2 exhibit the highest computational efficiency but at the cost of lower accuracy. M3 is slightly less efficient than M1 and M2 because its piecewise McCormick convex envelope approach requires additional 0–1 variables and four linear constraints per segment, increasing computational complexity. The efficiency of M4 largely depends on the number of iterations. For single-period optimal electricity-gas flow problems, which are relatively small in scale, the initial relaxed solution is often close to the feasible region, resulting in fewer iterations and relatively high efficiency. Among the general-purpose solvers, M5 and M6 perform efficiently for small-scale optimization problems, but as the problem size increases, M6 gains an advantage and outperforms M5 in Case C. In comparison, M7 and M8 demonstrate lower and more stable computation times. Specifically, M8 achieves an average runtime of 0.007, 0.081, and 0.451 s across the three cases, highlighting the effectiveness of the spatial branching strategy proposed in this chapter.

5.5.2 Day-Ahead Multi-period Scheduling

To further assess the effectiveness of the proposed algorithm in solving the multi-period day-ahead scheduling problem, this section analyzes cases where $|\Omega_t| = 24$. The accuracy and computational efficiency of algorithms M1–M8 are evaluated under four sets of typical load curves, and the results are presented as follows.

Accuracy: Similar to the single period optimal electricity-gas flow problem, the accuracy of the multi-period day-ahead scheduling problem is assessed based on two criteria: "optimality metric" and "feasibility metric". The results are summarized in Table 5.4.

Compared to Table 5.3, as the problem size increases, the optimality and feasibility metrics of all methods deteriorate relative to their performance in the single-period optimal electricity-gas flow problem. Specifically, M1–M3 still fail to guarantee solution feasibility, with the imbalance error in non-convex constraints increasing compared to the single-period case. M4 is one of the few methods that performs better in solving the multi-period day-ahead scheduling problem than in the single-period case. Although its relative error to the global optimal solution decreases, it still does not meet the 5% accuracy requirement. This is because the concave-convex procedure used in M4 is sensitive to the initial solution, leading to stochastic performance in optimization results. M5 successfully solves Cases A and B but fails to find a feasible solution for Case C within the given time limit. Compared to M6, which serves as the benchmark algorithm, M7 and M8 achieve the best optimality metrics among all methods when solving the multi-period day-ahead scheduling problem. Additionally, their feasibility metrics meet the required accuracy threshold.

Efficiency: Similar to Fig. 5.4, Fig. 5.5 also presents a box plot illustrating the computation time of M1–M8 for solving the multi-period day-ahead scheduling problem, with the vertical axis in logarithmic scale. It is important to note that since M5 failed to obtain a feasible solution for Case C, its data is not included in Fig. 5.5c.

It can be observed that the expansion of the problem size significantly increases the computation time for all methods, while the presence of non-convex constraints further amplifies the problem complexity and the required computational effort. This results in global optimization algorithms, represented by M5–M8, being more than an order of magnitude slower than M1–M4 in terms of solution time. However, in practical engineering applications, solution accuracy is critical for the safe operation of urban energy systems. Therefore, an increase in computation time is acceptable as long as solution accuracy is ensured. Among M5–M8, thanks to the spatial branching

Table 5.4 Comparison of accuracy index of M1–M8 algorithm for multi-period day-ahead scheduling problem

Algorithm	Case A		Case B		Case C	
	Optimality (%)	Feasibility (%)	Optimality (%)	Feasibility (%)	Optimality (%)	Feasibility (%)
M1	↑ 0.00	1.87	↑ 0.00	4.00	↑ 0.00	4.00
M2	↓ 0.16	0.00	↓ 0.40	2.00	↓ 1.09	2.00
M3	↓ 0.01	0.01	↑ 0.00	1.39	↑ 0.00	1.69
M4	↑ 0.00	0.00	↑ 3.48	0.41	↑ 4.22	0.41
M5	↑ 0.09	0.39	↑ 0.00	0.48	–	–
M6	Benchmark	0.48	Benchmark	0.46	Benchmark	0.47
M7	↓ 0.04	0.43	↑ 0.00	0.45	↑ 0.00	0.46
M8	↑ 0.00	0.46	↑ 0.00	0.44	↑ 0.00	0.42

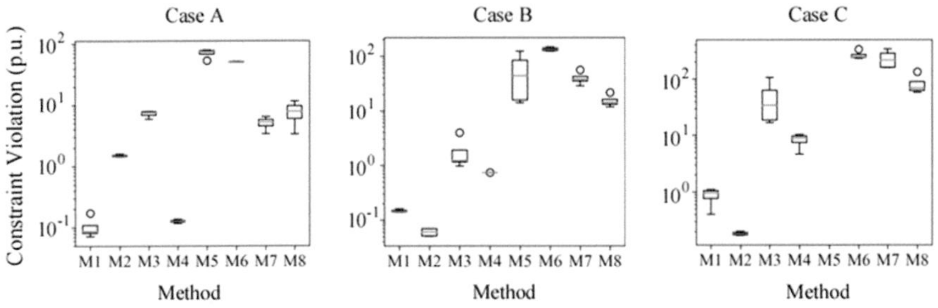

Fig. 5.5 Comparison of accuracy index of M1–M8 algorithm for multi-period day-ahead scheduling problem

strategy specifically designed for SOC equality constraints proposed in this chapter, M8 demonstrates a significant efficiency advantage in computation time compared to the other methods, making it highly promising and valuable for practical engineering applications.

5.6 Conclusions

This chapter investigates the optimal electricity-gas flow problem of urban distribution-level integrated IEGS from both modeling and algorithmic perspectives. First, based on the branch flow model of the distribution network and the steady-state transmission model of the gas network, an optimal electricity-gas flow model for distribution-level IEGS was established. Next, a reformulation method is proposed to transform the model's non-convex constraints into a standard form of SOC equality constraints. Subsequently, a spatial branching strategy specifically tailored for this type of constraint is designed and integrated into the SB&B algorithm framework. Finally, through comparative analyses across multiple systems and various solution methods, the effectiveness of the proposed algorithm is demonstrated. The key contributions and innovations of this chapter are summarized as follows:

(1) For the non-convex and nonlinear optimal electricity-gas flow model of urban distribution-level IEGS, a standardization process for non-convex constraints based on SOC equality reformulation is proposed. Compared to the bilinear constraint-based reformulation method, this approach effectively reduces the number of non-convex constraints and improves the efficiency of individual branching operations; (2) for SOC equality constraints, which have specific physical significance and mathematical forms, a specialized spatial branching strategy is designed by introducing a series of reference points within the three-dimensional space of SOC, and the effectiveness of this strategy is proven; (3) from the perspectives of accuracy and efficiency, a comparative analysis is conducted between the proposed method and seven other methods across multiple systems, verifying the effectiveness and superiority of the proposed method.

References

1. Kirchmayer, L. K. (1958). *Economic operation of power systems.* Wiley.
2. Carpentier, J. (1979). Optimal power flows. *International Journal of Electrical Power & Energy Systems*, 3–15.
3. Bukhsh, W. A., Zhang, C., & Pinson, P. (2016). An integrated multiperiod OPF model with demand response and renewable generation uncertainty. *IEEE Transactions on Smart Grid, 7*(3), 1495–1503.
4. Khanabadi, M., Ghasemi, H., & Doostizadeh, M. (2013). Optimal transmission switching considering voltage security and N-1 contingency analysis. *IEEE Transactions on Power Systems, 28*(1), 542–550.
5. Wang, H., Murillo-Sanchez, C. E., Zimmerman, R. D., & Thomas, R. J. (2007). On computational issues of market-based optimal power flow. *IEEE Transactions on Power Systems, 22*(3), 1185–1193.
6. Chondrogiannis, S., & Blanco, M. P. (2016). Market integration scheme of a multi-terminal HVDC grid in the North Seas. *IEEE Transactions on Power Systems, 31*(3), 2415–2422.
7. Fatemi, S. M., Abedi, S., Gharehpetian, G. B., & Liu, W. (2015). Introducing a novel DC power flow method with reactive power considerations. *IEEE Transactions on Power Systems, 30*(6), 3012–3023.
8. Yang, J., Zhang, N., Kang, C., & Xia, Q. (2017). A state-independent linear power flow model with accurate estimation of voltage magnitude. *IEEE Transactions on Power Systems, 32*(5), 3607–3617.
9. Yang, Z., Zhong, H., Bose, A., & Xia, Q. (2018). A linearized OPF model with reactive power and voltage magnitude: A pathway to improve the MW-only DC OPF. *IEEE Transactions on Power Systems, 33*(2), 1734–1745.
10. Farivar, M., & Low, S. H. (2013). Branch flow model: Relaxations and convexification—Part I. *IEEE Transactions on Power Systems, 28*(3), 2554–2564.
11. Liu, Y., Wu, W., Zhang, B., & Li, Y. (2014). Active power-reactive power coordinated multi-period optimal operation of active distribution networks based on mixed-integer second-order cone programming. *Proceedings of the CSEE, 34*(16), 2575–2583.
12. Ma, X., Guo, R., Wang, L., & Wu, W. (2018). Day-ahead scheduling model for AC/DC active distribution networks based on second-order cone programming. *Automation of Electric Power Systems, 42*(22), 144–150.
13. Yuan, Z., & Hesamzadeh, M. R. (2019). A modified Benders decomposition algorithm to solve second-order cone AC optimal power flow. *IEEE Transactions on Smart Grid, 10*(2), 1713–1724.
14. Yang, T., Guo, Y., Deng, L., & Xu, B. (2021). A linear branch flow model for radial distribution networks and its application to reactive power optimization and network reconfiguration. *IEEE Transactions on Smart Grid, 12*(3), 2027–2036.
15. Tian, J., Wei, H., & Bai, X. (2014). Parallel computing method for semidefinite programming optimal power flow. *Power System Technology, 38*(362), 175–180.
16. Li, C., Wang, Y., Chen, Z., & Zhang, X. (2008). *Natural gas pipeline transportation.* China Petrochemical Press.
17. Martinez-Mares, A., & Fuerte-Esquivel, C. R. (2012). A unified gas and power flow analysis in natural gas and electricity coupled networks. *IEEE Transactions on Power Systems, 27*(4), 2156–2166.
18. Osiadacz, A. J., & Chaczykowski, M. (2020). Modeling and simulation of gas distribution networks in a multienergy system environment. *Proceedings of the IEEE, 108*(9), 1580–1595.
19. Gu, C., Zhang, Y., Wang, J., & Zhang, W. (2021). Joint planning of electrical storage and gas storage in power-gas distribution network considering high-penetration electric vehicle and gas vehicle. *Applied Energy, 301*, 117447.
20. Xie, H., Sun, X., Chen, C., & Zhang, J. (2022). Resilience metrics for integrated power and natural gas systems. *IEEE Transactions on Smart Grid, 13*(3), 2483–2486.

21. Zhuang, W., Zhou, S., Gu, W., & Zhang, X. (2021). Optimized dispatching of city-scale integrated energy system considering the flexibilities of city gas gate station and line packing. *Applied Energy, 290*, 116689.
22. Chen, S., Conejo, A. J., Sioshansi, R., & Wei, Z. (2020). Operational equilibria of electric and natural gas systems with limited information interchange. *IEEE Transactions on Power Systems, 35*(1), 662–671.
23. Wei, Z., Zhang, S., Sun, G., & Li, J. (2017). Peak shaving and valley filling in integrated electricity-gas energy systems considering power-to-gas. *Proceedings of the CSEE, 37*(16), 4601–4609.
24. Liu, W., Li, P., Yang, W., & Zhang, H. (2019). Optimal energy flow for integrated energy systems considering gas transients. *IEEE Transactions on Power Systems, 34*(6), 5076–5079.
25. Fang, J., Zeng, Q., Ai, X., & Zhang, J. (2018). Dynamic optimal energy flow in the integrated natural gas and electrical power systems. *IEEE Transactions on Sustainable Energy, 9*(1), 188–198.
26. Bayani, R., & Manshadi, S. D. (2022). Natural gas short-term operation problem with dynamics: A rank minimization approach. *IEEE Transactions on Smart Grid, 13*(4), 2761–2773.
27. Ameli, H., Qadrdan, M., & Strbac, G. (2017). Value of gas network infrastructure flexibility in supporting cost-effective operation of power systems. *Applied Energy, 202*, 571–580.
28. Huang, W., Zhang, N., Cheng, Y., & Liu, K. (2020). Multienergy networks analytics: Standardized modeling, optimization, and low carbon analysis. *Proceedings of the IEEE, 108*(9), 1411–1436.
29. Rastegar, M., Fotuhi-Firuzabad, M., Zareipour, H., & Khatami, R. (2017). A probabilistic energy management scheme for renewable-based residential energy hubs. *IEEE Transactions on Smart Grid, 8*(5), 2217–2227.
30. Correa-Posada, C. M., & Sanchez-Martin, P. (2014). Security-constrained optimal power and natural-gas flow. *IEEE Transactions on Power Systems, 29*(4), 1780–1787.
31. Sayed, A. R., Wang, C., Zhao, J., & Wu, J. (2020). Distribution-level robust energy management of power systems considering bidirectional interactions with gas systems. *IEEE Transactions on Smart Grid, 11*(3), 2092–2105.
32. Qi, F., Shahidehpour, M., Wen, F., & Wang, X. (2020). Decentralized privacy-preserving operation of multi-area integrated electricity and natural gas systems with renewable energy resources. *IEEE Transactions on Sustainable Energy, 11*(3), 1785–1796.
33. Chen, S., Conejo, A. J., & Wei, Z. (2022). Gas-power coordination: From day-ahead scheduling to actual operation. *IEEE Transactions on Power Systems, 37*(2), 1532–1542.
34. Li, G., Yan, K., Zhang, R., & Zhao, Y. (2022). Resilience-oriented distributed load restoration method for integrated power distribution and natural gas systems. *IEEE Transactions on Sustainable Energy, 13*(1), 341–352.
35. Jin, X. (2018). *Modeling and operation optimization of regional integrated energy systems incorporating intelligent buildings*. Tianjin University.
36. Liu, P., Wu, Z., Gu, W., & Zhang, C. (2022). Security-constrained AC-DC hybrid distribution system expansion planning with high penetration of renewable energy. *International Journal of Electrical Power & Energy Systems*, 142.
37. Wen, K., Lu, Y., Lu, M., et al. (2022). Multi-period optimal infrastructure planning of natural gas pipeline network system integrating flowrate allocation. *Energy, 257*, 124745.
38. Unsihuay-Vila, C., Marangon-Lima, J. W., de Souza, A. C. Z., et al. (2010). A model to long-term, multi-area, multistage, and integrated expansion planning of electricity and natural gas systems. *IEEE Transactions on Power Systems, 25*(2), 1154–1168.
39. Qiu, J., Dong, Z. Y., Zhao, J. H., et al. (2015). Low carbon oriented expansion planning of integrated gas and power systems. *IEEE Transactions on Power Systems, 30*(2), 1035–1046.
40. Yang, W., Liu, W., Chung, C. Y., et al. (2020). Coordinated planning strategy for integrated energy systems in a district energy sector. *IEEE Transactions on Sustainable Energy, 11*(3), 1807–1819.
41. Qiu, J., Dong, Z. Y., Zhao, J. H., et al. (2015). Multi-stage flexible expansion co-planning under uncertainties in a combined electricity and gas market. *IEEE Transactions on Power Systems, 30*(4), 2119–2129.

42. Shao, C., Shahidehpour, M., Wang, X., et al. (2017). Integrated planning of electricity and natural gas transportation systems for enhancing the power grid resilience. *IEEE Transactions on Power Systems, 32*(6), 4418–4429.
43. Shao, C., Li, K., Hu, Z., et al. (2022). Coordinated planning of electric power and natural gas distribution systems with refueling stations for alternative fuel vehicles in transportation systems. *IEEE Transactions on Smart Grid, 13*(5), 3558–3569.
44. Zhan, Y., Zheng, Q. P., Wang, J., et al. (2017). Generation expansion planning with large amounts of wind power via decision-dependent stochastic programming. *IEEE Transactions on Power Systems, 32*(4), 3015–3026.
45. Li, C., Conejo, A. J., Siirola, J. D., et al. (2022). On representative day selection for capacity expansion planning of power systems under extreme operating conditions. *International Journal of Electrical Power & Energy Systems, 137*, 107697.
46. Wang, C., Wei, W., Wang, J., et al. (2017). Robust defense strategy for gas-electric systems against malicious attacks. *IEEE Transactions on Power Systems, 32*(4), 2953–2965.
47. He, C., Dai, C., Wu, L., et al. (2018). Robust network hardening strategy for enhancing resilience of integrated electricity and natural gas distribution systems against natural disasters. *IEEE Transactions on Power Systems, 33*(5), 5787–5798.
48. Murty, K. (1984). *Linear programming.* Wiley. https://books.google.com/books?id=ibQJvAEACAAJ
49. Dikin, I. I. (1967). Iterative solution of problems of linear and quadratic programming. *Soviet Mathematics Doklady, 8*, 674–675.
50. Boyd, S., & Vandenberghe, L. (2004). *Convex optimization.* Cambridge University Press.
51. Land, A. H., & Doig, A. G. (1960). An automatic method of solving discrete programming problems. *Econometrica, 28*(3), 497–520.
52. Kelley, J. E. (1960). The cutting-plane method for solving convex programs. *Journal of the Society for Industrial and Applied Mathematics, 8*(4), 703–712.
53. Geoffrion, A. M. (1972). Generalized Benders decomposition. *Journal of Optimization Theory and Applications, 10*(4), 237–260.
54. Duran, M. A., & Grossmann, I. E. (1986). An outer-approximation algorithm for a class of mixed-integer nonlinear programs. *Mathematical Programming, 36*(3), 307–339.
55. Gurobi Optimization, LLC. (2023). *Gurobi Optimizer Reference Manual.* https://www.gurobi.com
56. IBM. (2022). *IBM ILOG CPLEX Optimization Studio 22.1.0 User's Manual.* https://www.ibm.com/docs/en/icos/22.1.0?topic=callbacks-what-are
57. Wächter, A., & Biegler, L. T. (2006). On the implementation of an interior-point filter line-search algorithm for large-scale nonlinear programming. *Mathematical Programming, 106*(1), 25–57.
58. Vigerske, S., & Gleixner, A. (2018). SCIP: Global optimization of mixed-integer nonlinear programs in a branch-and-cut framework. *Optimization Methods and Software, 33*(3), 563–593.
59. Chen, S., Wei, Z., Sun, G., et al. (2018). Optimal power and gas flow with a limited number of control actions. *IEEE Transactions on Smart Grid, 9*(5), 5371–5380.
60. Ding, T., Xu, Y., Wei, W., et al. (2020). Energy flow optimization for integrated power-gas generation and transmission systems. *IEEE Transactions on Industrial Informatics, 16*(3), 1677–1687.
61. Wang, C., Ju, P., Wu, F., et al. (2022). Long-term voltage stability-constrained coordinated scheduling for gas and power grids with uncertain wind power. *IEEE Transactions on Sustainable Energy, 13*(1), 363–377.
62. Sun, G., Chen, S., Wei, Z., et al. (2020). Corrective security-constrained optimal power and gas flow with binding contingency identification. *IEEE Transactions on Sustainable Energy, 11*(2), 1033–1042.
63. Wang, S., Zhai, J., & Hui, H. (2023). Optimal energy flow in integrated electricity and gas systems with injection of alternative gas. *IEEE Transactions on Sustainable Energy, 14*(3), 1540–1557.

64. Manshadi, S. D., & Khodayar, M. E. (2018). A tight convex relaxation for the natural gas operation problem. *IEEE Transactions on Smart Grid, 9*(5), 5467–5469.
65. Chen, S., Conejo, A. J., Sioshansi, R., et al. (2019). Unit commitment with an enhanced natural gas-flow model. *IEEE Transactions on Power Systems, 34*(5), 3729–3738.
66. Yang, L., Xu, Y., & Sun, H. (2020). A dynamic linearization and convex relaxation-based approach for a natural gas optimal operation problem. *IEEE Transactions on Smart Grid, 11*(2), 1802–1804.
67. Gan, L., Li, N., Topcu, U., et al. (2015). Exact convex relaxation of optimal power flow in radial networks. *IEEE Transactions on Automatic Control, 60*(1), 72–87.
68. Huang, S., Wu, Q., Wang, J., et al. (2017). A sufficient condition on convex relaxation of AC optimal power flow in distribution networks. *IEEE Transactions on Power Systems, 32*(2), 1359–1368.
69. Eiselt, H., Spielberg, K., Richards, E., et al. (2013). *Integer programming and network models.* Springer.
70. Danna, E., Rothberg, E., & Pape, C. L. (2005). Exploring relaxation induced neighborhoods to improve MIP solutions. *Mathematical Programming, 102*(1), 71–90.
71. Lipp, T., & Boyd, S. (2016). Variations and extension of the convex-concave procedure. *Optimization and Engineering, 17*(2), 263–287.
72. Wang, C., Wei, W., Wang, J., et al. (2018). Convex optimization based distributed optimal gas-power flow calculation. *IEEE Transactions on Sustainable Energy, 9*(3), 1145–1156.
73. Yang, L., Xu, Y., Sun, H., et al. (2020). Two-stage convexification-based optimal electricity-gas flow. *IEEE Transactions on Smart Grid, 11*(2), 1465–1475.
74. Wang, C., Gao, R., Wei, W., et al. (2019). Risk-based distributionally robust optimal gas-power flow with Wasserstein distance. *IEEE Transactions on Power Systems, 34*(3), 2190–2204.
75. Mhanna, S., Saedi, I., & Mancarella, P. (2022). Iterative LP-based methods for the multiperiod optimal electricity and gas flow problem. *IEEE Transactions on Power Systems, 37*(1), 153–166.
76. Chen, C., Atamtürk, A., & Oren, S. S. (2016). A spatial branch-and-cut method for non-convex QCQP with bounded complex variables. *Mathematical Programming, 165*(2), 549–577.
77. Son, T. Q., Strodiot, J. J., & Nguyen, V. H. (2009). ε-optimality and ε-Lagrangian duality for a non-convex programming problem with an infinite number of constraints. *Journal of Optimization Theory and Applications, 141*(2), 389–409.
78. Fischetti, M., & Monaci, M. (2020). A branch-and-cut algorithm for mixed-integer bilinear programming. *European Journal of Operational Research, 282*(2), 506–514.
79. Nagarajan, H., Lu, M., Wang, S., et al. (2019). An adaptive, multivariate partitioning algorithm for global optimization of non-convex programs. *Journal of Global Optimization, 74*(4), 639–675.
80. Kocuk, B., Dey, S. S., & Sun, X. A. (2018). Matrix minor reformulation and SOCP-based spatial branch-and-cut method for the AC optimal power flow problem. *Mathematical Programming Computation, 10*(4), 557–596.
81. Liu, P., Wu, Z., Gu, W., et al. (2022). An improved spatial branch-and-bound algorithm for non-convex optimal electricity-gas flow. *IEEE Transactions on Power Systems, 37*(2), 1326–1339.
82. Liu, P., Wu, Z., Gu, W., et al. (2023). Spatial branching for conic non-convexities in optimal electricity-gas flow. *IEEE Transactions on Power Systems, 38*(1), 972–975.
83. Martinez Cesena, E. A., Loukarakis, E., Good, N., et al. (2020). Integrated electricity-heat-gas systems: Techno-economic modeling, optimization, and application to multienergy districts. *Proceedings of the IEEE, 108*(9), 1392–1410.
84. Gerard, D., Köppe, M., & Louveaux, Q. (2017). Guided dive for the spatial branch-and-bound. *Journal of Global Optimization, 65*(1), 685–711.
85. Liu, P. (2023). *Data.* Available at: https://pan.seu.edu.cn:443/link/65F657F40287AE18CE67416-BFD877E27

Chapter 6
Accelerated Resilience Assessment for IES Considering Non-convex Gas Flow Dynamics (Part II): Accelerated Evaluation Algorithm

Abstract Building on the previous chapter, this chapter proposes an improved spatial branch-and-cut (B&C) method for the rapid resilience assessment of integrated energy systems (IES). To accelerate fault recovery solutions, the proposed method introduces two key improvements to the traditional B&C algorithm. First, to address the non-convex characteristics of the integrated electricity-gas flow (OEGF), a novel cone-specific spatial branching strategy is proposed. Compared to existing strategies that can only handle bilinear equality constraints, the proposed method reduces the scale of branching variables by two-thirds without compromising solution accuracy. Second, a combined cut generation strategy is introduced, transforming fault scenario enumeration into a guided generation process within a single search tree. The core idea is to formulate the solution process for each fault scenario as a subset of the search tree, thereby reducing the load restoration problem to a single optimization process. The proposed algorithm is tested on systems of three different scales. Comparative analysis with seven existing algorithms further validates its superiority in both solution accuracy and computational efficiency.

Keywords Integrated electricity-gas system resilience · Non-convex optimization · Spatial branching · Combinatorial cut

Nomenclature

v	Index for the fault scenarios
t	Index for the scheduling time period
i	Index for electricity grid nodes
m	Index for gas grid nodes
ij	Index for power line index
mn	Index for natural gas pipeline index
s	Index for the spatial point
$\mathcal{V}$	Set of failure scenarios
$\mathcal{T}$	Set of failure periods

Z. Wu et al., *Integrated Energy System Resilience*,
https://doi.org/10.1007/978-981-95-2363-4_6

$\mathcal{B}$	Set of buses in EPS
$\mathcal{N}$	Set of nodes in NGS
$\mathcal{L}$	Set of lines in EPS
$\mathcal{C}$	Set of CPGs
$\mathcal{U}$	Set of GFUs
$\mathcal{R}$	Set of RESs
$\mathcal{C}_i$	Set of CPGs at bus i
$\mathcal{U}_i$	Set of GFUs at bus i
$\mathcal{R}_i$	Set of RESs at bus i
$\mathcal{P}$	Set of pipelines in NGS
$\mathcal{W}$	Set of gas suppliers
$\mathcal{K}$	Set of gas compressors
$\mathcal{W}_m$	Set of gas suppliers at node m
$\mathcal{U}_m$	Set of GFUs at node m
K_v	Probability of failure scenarios
LS_v	Load-shedding amount at the load node for fault scenario
Δt	Time slot duration
Q	Calorific value of natural gas
$p_{i,t}^d$	Active power load shedding at bus i
$f_{m,t}^d$	Gas load shedding at node m
$A_{ij,t,v}$	Operating status of lines
M	A sufficient large number
$x_{ij,t}$	Operating status of line ij
$x_{mn,t}$	Direction of gas flow in pipeline mn
x_v	Selection sign of failure scenario v
$\mathcal{L}_i^+$	Set of lines with bus i as the from bus
$\mathcal{L}_i^-$	Set of lines with bus i as the to bus
$LD_{i,t}^p$	Power load demand
G_{ij}	Conductance of lines
B_{ij}	Susceptance of lines
$P_{ij}^{\max}$	Maximum active power flow of lines
$p_{ij,t}^+$	Power flow at the from bus of line ij
$p_{ij,t}^-$	Power flow at the to bus of line ij
$p_{i,t}^c$	Power generation of CPG i
$p_{i,t}^u$	Power generation of GFU i
$p_{i,t}^r$	Power generation of RES i
$\theta_{ij,t}$	Voltage angle difference along line ij
$\theta_{i,t}$	Voltage angle of bus i
$\theta_{j,t}$	Voltage angle of bus j
$P_{i,t}^{r,\mathrm{for}}$	Forecasted power generation of RESs
$P_i^{c,\max}$	Maximum power generation of CPGs
$P_i^{u,\max}$	Maximum power generation of GFUs
$P_i^{c,\min}$	Minimum power generation of CPGs
$P_i^{u,\min}$	Minimum power generation of GFUs

$P_i^{c,rd}$	Ramp down rates of CPGs
$P_i^{u,rd}$	Ramp down rates of GFUs
$P_i^{c,ru}$	Ramp up rates of CPGs
$P_i^{u,ru}$	Ramp up rates of GFUs
$\Theta_i^{\min}$	Minimum voltage angle of buses
$\Theta_i^{\max}$	Maximum voltage angle of buses
C	Speed of sound
D/D_{mn}	Diameter of pipelines
Λ/Λ_{mn}	Friction factor of pipelines
$\mathcal{K}_m^+$	Set of gas compressors with node m as the inlet node
$\mathcal{K}_m^-$	Set of gas compressors with node m as the outlet node
$\mathcal{P}_m^+$	Set of pipelines with node m as predefined inlet node
$\mathcal{P}_m^-$	Set of pipelines with node m as predefined outlet node
E_{mn}	Conversion ratio of gas compressors
U_m	Energy conversion coefficient of GFUs
Δx	Length of pipeline segments
$F_{mn}^{\max}$	Maximum gas flow of pipelines
$F_{mn}^{w,\min}$	Minimum output of gas suppliers
$F_{mn}^{w,\max}$	Maximum output of gas suppliers
$F_{mn}^{u,\min}$	Minimum gas consumption of GFUs
$F_{mn}^{u,\max}$	Maximum gas consumption of GFUs
$F_{mn}^{k,\min}$	Minimum gas flow of compressors
$F_{mn}^{k,\max}$	Maximum gas flow of compressors
$H_{mn}^{\min}$	Minimum compression ratio of compressors
$H_{mn}^{\max}$	Maximum compression ratio of compressors
$LD_{m,t}^f$	Gas load demand
$\Psi_{mn}^{\min}$	Lower limit of gas pressure
$\Psi_{mn}^{\max}$	Upper limit of gas pressure
$f_{m,t}^k$	Gas flow of compressor m
$f_{m,t}^w$	Gas injection from gas supplier m
$f_{m,t}^u$	Gas consumption of GFU m
$f_{m,t}^+$	Gas flow at inlet of pipeline mn
$f_{m,t}^-$	Gas flow at outlet of pipeline mn
$f_{m,t}^{s,t}$	Gas flow at point s of pipeline mn
$\Psi_{mn}^{s,t}$	Gas pressure at point s of pipeline mn
$\tau_{m,t}$	Gas consumption of compressor mn
$\Psi_{m,t}$	Gas pressure of node m
B&C	Branch-and-cut
IES	Integrated energy systems
EPSs	Electrical power systems
NGSs	Natural gas systems
IEGSs	Integrated electricity-gas systems
GFUs	Gas-fired units
HILP	High-impact, low-probability

OEGF	Optimal electricity-and-gas flow
PDEs	Partial differential equations
FDM	Finite difference method
MINLP	Mixed integer nonlinear programming
NP-hard	Non-deterministic polynomial-time hard
SOC	Second-order cone
CCP	Convex-concave procedure
SLP	Sequential linear programming
B&B	Branch-and-bound
EENS	Expected energy-not-served
DC	Direct current
OPF	Optimal power flow
CPGs	Conventional generators
RESs	Renewable energy sources
MISOCP	Mixed-integer second-order cone programming
SB&B	Spatial branch-and-bound
AC OPF	Alternating current optimal power flow

6.1 Introduction

As the coupling between electrical power systems (EPSs) and natural-gas systems (NGSs) becomes increasingly tight, electricity-gas coordination plays a crucial role in enhancing energy supply flexibility [1]. However, this interdependence also introduces new challenges for the resilience assessment of integrated electricity-gas systems (IEGSs). Specifically, faults in one system may propagate through interconnected components, such as gas-fired units (GFUs) [2], and impact the other system. Therefore, resilience assessment serves as a fundamental basis for disaster preparedness planning [3], post-disaster scheduling [4], and reinforcement strategies [5] in IEGSs. The accuracy and computational efficiency of resilience assessment are critical to optimizing the coordinated operation of the integrated systems.

The term "resilience" describes an energy system's ability to withstand high-impact, low-probability (HILP) events [6, 7]. Existing resilience assessment methods for energy systems can be broadly categorized into three approaches: machine learning, topological analysis, and operational optimization. Machine learning, as a model-free approach, utilizes black-box neural networks to derive resilience metrics [8]. While data-driven methods offer significant computational advantages, fully trained neural networks still suffer from limitations in generalizability and interpretability [9]. Topological analysis, on the other hand, assesses IEGS resilience based on simplified power flow models [10, 11]. Although this method has a lower computational complexity for large-scale networks, its accuracy is constrained by the omission of key energy flow characteristics, potentially leading to overly optimistic assessments [12]. In contrast, operational optimization-based approaches are widely applied in IEGS resilience assessment. For a given fault scenario, these methods derive resilience metrics by solving load restoration models [13]. When faults are caused by stochastic events such as extreme weather, Monte Carlo simulation is commonly used to generate numerous fault scenarios for resilience evaluation [14, 15]. However, performing integrated energy flow optimization across all fault scenarios significantly increases the computational burden of resilience assessment [16].

The non-convexity of optimal electricity-gas flow problem is the primary cause of its computational burden. Due to the dynamic characteristics of gas flow in NGS, this problem is typically formulated by using partial differential equations (PDEs) and discretized via the finite difference method (FDM) [17, 18]. Consequently, the OEGF problem is classified as a non-convex mixed-integer nonlinear programming (MINLP). Since solving such non-convex MINLP to obtain a globally optimal solution is non-deterministic polynomial-time hard (NP-hard), developing efficient solution methods has become a key research focus. One commonly used approach is second-order cone (SOC) relaxation, which transforms non-convex quadratic constraints into SOC constraints [19], thereby convexifying the problem. However, if certain conditions are not met, this method may render the problem infeasible. Another alternative is piecewise approximation, which approximates non-convex constraints using a set of linear functions [20] or McCormick envelopes [21]. Nevertheless, even with conservative accuracy settings, introducing piecewise variables significantly increases the model size and reduces computational efficiency. Therefore, an efficient solution algorithm for the OEGF problem is urgently needed.

When solving non-convex optimization problems, feasibility and optimality are the two core metrics for evaluating the performance of an optimization algorithm. Under normal operating conditions, obtaining a feasible solution that is near-optimal is often sufficient for practical applications. Since local optimum is more accessible than the global optimum, most studies adopt local optimization methods to handle non-convex constraints. For example, convex-concave procedure (CCP) is used in Ref. [22] to recover feasible solutions from non-tight SOC relaxation, while sequential linear programming (SLP) is explored in Ref. [23], where two SLP-based algorithms are proposed to obtain high-quality feasible solutions for the OEGF problem. The effectiveness of these methods largely stems from extensive empirical observations: Under normal operating conditions, the relaxed solution of the OEGF problem is often very close to the local (could be even global) optimal, enabling efficient computation by solving a sequence of convex optimization problems [24]. However, the applicability of local optimization methods diminishes significantly in resilience assessment. Extreme events may push the IEGS operating point to the boundary of the feasible region, causing local optimization methods to converge to solutions far from the global optimum (or even fail to converge). Therefore, to ensure optimality in OEGF modeling for resilience assessment, global optimization techniques are required. Currently, commercial and general-source solvers integrating various mathematical programming algorithms, such as Gurobi [25] and SCIP [26], offer solutions for non-convex OEGF problems. However, their built-in branch-and-bound (B&B) algorithms still exhibit performance limitations. As a result, several studies have focused on optimizing embedded spatial branching strategies to improve computational efficiency [27, 28].

Another major source of computational burden in resilience assessment is the need to evaluate all fault scenarios, particularly when employing Monte Carlo simulation method to capture the probabilistic characteristics of system failures. Due to the lack of effective optimization strategies, most existing studies [14, 29] solve the OEGF problem independently for each scenario, leading to a computational complexity that scales linearly with the number of scenarios. Currently, redundancy reduction is the primary approach to addressing this issue. The most common method is scenario simplification via clustering [30], where only the most representative scenarios are selected for resilience assessment. Another approach is constraint filtering within each scenario, as certain key constraints may be sufficient for evaluating system resilience [31]. However, while these methods improve computational efficiency, they come at the cost of reduced assessment accuracy.

To address this challenge, this chapter proposes a fast resilience assessment method within the classical B&C algorithm framework. The objective is to reduce the computational complexity of resilience assessment while ensuring that the obtained solutions meet predefined accuracy thresholds. It is important to emphasize that the B&C algorithm is a framework designed for solving NP-hard problems, with its performance largely dependent on the design of the branching and cutting strategies [27, 32]. Table 6.1 summarizes relevant studies in this field, and the key contributions of this chapter are outlined as follows:

(1) An improved spatial branching strategy is proposed for the conic non-convex SOC constraints in post-contingency OEGF problems. This chapter proposes an enhanced spatial branching strategy to address the non-convex SOC constraints in the OEGF problem. Existing global solvers implement built-in spatial branching strategies primarily designed for bilinear terms ($z = xy$ with $(x, y, z) \in \mathbb{R}^3$) [25, 26], leading to suboptimal performance when applied to such problems. However, in the OEGF model, non-convex forms ($z = \sqrt{x^2 + y^2}$ with $(x, y, z) \in \mathbb{R}^3$) are more prevalent. Current global solvers require decomposing these constraints into three bilinear terms [27], which results in a threefold increase in the number of branching variables. Additionally, existing cone-specific spatial branching strategies for SOC [28] perform poorly in handling the bi-directional nature of gas flow, thereby restricting the feasible region of branching variables. To address this issue, this study introduces a generalized spatial branching strategy and integrates it into the *Branch* process of the B&C algorithm. This strategy ensures that, for any given SOC equality constraint, only a single branching operation is performed within the three-dimensional solution space, as opposed to three times in conventional methods. Consequently, the computational complexity of handling non-convex constraints is significantly reduced.

(2) An improved cut generation strategy is proposed to restructure the iterative solution process across all failure scenarios into a single traversal of a specialized B&C search tree. The key observation is that OEGF models under different fault

Table 6.1 Overview of previous studies

References	Gas flow modeling			Local optimization		Global optimization		Assessment	
	Steady-state	Quasi-dynamic	Dynamic	Convexify	Heuristic	Bilinear	Quadratic-cone	Enumerate all	Once-for-all
[1, 5, 13, 21]	✓	×	×	✓	×	×	×	–	–
[2, 4]	✓	×	×	×	✓	×	×	✓	×
[3, 8, 14, 29]	✓	×	×	✓	×	×	×	✓	×
[15, 30]	–	–	–	✓	×	×	×	✓	×
[17, 18]	×	×	✓	✓	×	×	×	–	–
[19]	×	✓	×	✓	×	×	×	✓	×
[20]	×	✓	×	✓	×	×	×	–	–
[22]	✓	×	×	×	✓	×	×	–	–
[23]	×	✓	×	×	✓	×	×	–	–
[24–-[26]	–	–	–	×	×	✓	×	–	–
[27]	×	✓	×	×	×	✓	×	–	–
[28]	✓	×	×	×	×	×	✓	–	–
Proposed method	×	×	✓	×	×	×	✓	×	✓

1. Symbol "–" denotes missing or unavailable information
2. Review papers are not included

scenarios exhibit a high degree of structural similarity, with variations occurring only in the timing and location of faults. As a result, the corresponding search tree structures for solving each OEGF model are also similar. Conventional enumeration methods [14, 29], which solve each model from the root node of the search tree, lead to excessive computational redundancy, as previously explored nodes are revisited in successive iterations. To eliminate such redundant computations, a composite cutting-plane generation strategy is introduced to construct a search tree equivalent to the union of all fault scenario trees. This strategy embeds the composite cutting planes into the *Cut* process of the B&C algorithm using a lazy constraint formulation [33]. By consolidating all composite cutting planes into a single search tree, this approach eliminates a significant amount of redundant computation, thereby enhancing computational efficiency while maintaining solution accuracy.

To address the existing challenges, this chapter proposes a fast resilience assessment method for integrated energy systems, considering the non-convexity of dynamic gas flow. Section 6.2 provides a detailed mathematical formulation of non-convex constraints, followed by Sect. 6.3, which introduces acceleration strategies for resilience assessment. Section 6.4 presents case studies for performance evaluation, while Sect. 6.5 conducts an in-depth analysis and discussion of the results. Finally, Sect. 6.6 concludes the chapter with a summary of key findings.

6.2 Mathematical Modeling Approach

In this section, Sect. 6.2.1 introduces the model assumptions, Sect. 6.2.2 describes the resilience indicators of the IEGS, Sect. 6.2.3 presents the proposed generalized fault recovery phase OEGF model, Sect. 6.2.4 discusses the reconstruction of the proposed model, and Sect. 6.2.5 provides the matrix representation of the model.

6.2.1 Assumptions

Before developing the mathematical model, the following assumptions are made regarding the construction of the optimal power flow model for the IEGS:

- This resilience assessment method focuses on extreme weather events (e.g., hurricanes) and considers only the faults on overhead transmission lines [14, 15].
- The fuel consumption of gas compressor is represented by a simplified linear model [20, 27].
- The voltage magnitude at each bus is close to 1 p.u., and the phase angle difference between the two ends of each transmission line is minimal [22].

6.2.2 Resilience Evaluation Metrics

In this section, the expected energy-not-served (EENS) [3, 14] is used as the quantifiable indicator for IEGS resilience assessment, defined as follows:

$$R_{\text{EENS}} = E(\text{LS}) = \sum_{v \in \mathcal{V}} K_v \cdot \text{LS}_v, \tag{6.1}$$

where $\mathcal{V}$ represents the set of fault scenarios. In terms of coefficients: K_v represents the probability of the fault scenario, and LS_v represents the load-shedding amount at the load node. For each HILP fault scenario, flexibility resources are mobilized, and energy recovery strategies are implemented to minimize the load-shedding amount. The indicator R_{EENS} is calculated by computing the weighted sum of LS_v across all fault scenarios, as shown in Fig. 6.1. Figure 6.1 illustrates the resilience recovery process under HILP fault scenarios.

Traditional methods calculate and verify the load-shedding amount for each HILP fault scenario separately [14, 29], leading to a computational complexity that is proportional to the scale of the scenarios. To address this issue, this section proposes a generalized fault recovery phase OEGF model, whose feasible region encompasses all HILP fault scenarios. During the solution process, this model significantly reduces the computational burden by utilizing the cutting-plane generation strategy proposed in this chapter (details of which will be introduced in the next section).

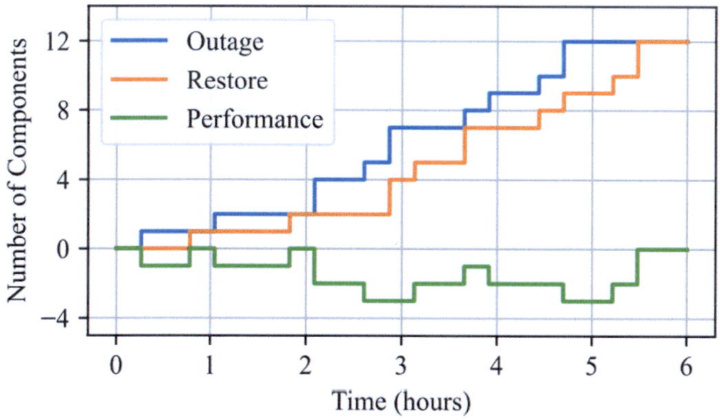

Fig. 6.1 Schematic of resilience-oriented actions for HILP events

6.2.3 Generalized OEGF Model for Post-Contingency Analysis

Objective: The objective is to minimize the load-shedding amount, as shown below:

$$\min\ \text{LS} = \sum_{t\in\mathcal{T}} \Delta t \left[\sum_{i\in\mathcal{B}} p_{i,t}^{d} + Q \sum_{m\in\mathcal{N}} f_{m,t}^{d} \right], \tag{6.2}$$

where t denotes the scheduling time period index, and i, m are node indices. $\mathcal{T}$ represents the set of fault periods, $\mathcal{B}$ represents the set of electricity grid nodes, and $\mathcal{N}$ represents the set of gas grid nodes. Δt represents the fault duration, and Q represents the heating value of natural gas. $p_{i,t}^{d}$ represents the active power load shedding at electricity grid node i during time period t, while $f_{m,t}^{d}$ represents the load shedding at gas load node m during time period t. The objective function (6.2) primarily includes the load-shedding cost in both the electricity grid and the gas network.

Failure Representation: In this chapter, the operational state of the lines in each fault scenario v is pre-determined. Since the specific fault scenario selection is uncertain, the following constraints apply:

$$\left| x_{ij,t} - A_{ij,t,v} \right| \leq M(1 - x_v), \quad \forall ij \in \mathcal{L}, v \in \mathcal{V} \tag{6.3}$$

$$\sum_{v\in\mathcal{V}} x_v = 1, \tag{6.4}$$

where ij refers to the power line index. In terms of sets, $\mathcal{L}$ represents the set of power lines in the grid. In terms of coefficients, $A_{ij,t,v}$ represents the operational status of the line at time t under fault scenario v, and M is a sufficiently large number. In terms of variables, $x_{ij,t}$ denotes the operational status of line ij at time t, and x_v is the selection flag for fault scenario v.

The constraint (6.3) indicates that when fault scenario v is selected ($x_v = 1$), the variable $x_{ij,t}$ is replaced by $A_{ij,t,v}$. When fault scenario v is not selected ($x_v = 0$), $x_{ij,t}$ remains as a decision variable. The constraint (6.4) ensures that only one fault scenario can be selected for analysis. Unlike traditional methods that set $x_v = 1$ for each $\forall v \in \mathcal{V}$ to calculate the load-shedding index LS_v, this section treats x_v as a decision variable, ensuring that the optimal solutions for all fault scenarios lie within the feasible domain of the proposed model. Consequently, the optimal solution can be obtained through observation and optimization of the solving process.

EPS Failure Modeling: An improved direct current (DC) optimal power flow (OPF) model is used to approximate the line power loss through a second-order approximation [23], as follows:

$$\sum_{ij\in\mathcal{L}_i^-} p_{ij,t}^- - \sum_{ij\in\mathcal{L}_i^+} p_{ij,t}^+ + \sum_{j\in\mathcal{R}_i} p_{i,t}^r + \sum_{j\in\mathcal{C}_i} p_{i,t}^c + \sum_{j\in\mathcal{U}_i} p_{i,t}^u = LD_{i,t}^p - p_{i,t}^d, \quad \forall i \in \mathcal{B} \tag{6.5}$$

$$p_{ij,t}^+ = G_{ij}\big(1 - \cos\theta_{ij,t}\big) - B_{ij}\sin\theta_{ij,t}, \quad \forall ij \in \mathcal{L} \tag{6.6}$$

$$p_{ij,t}^- = G_{ij}\big(1 - \cos\theta_{ij,t}\big) + B_{ij}\sin\theta_{ij,t}, \quad \forall ij \in \mathcal{L} \tag{6.7}$$

$$-P_{ij}^{\max} x_{ij,t} \le p_{ij,t}^+ \le P_{ij}^{\max} x_{ij,t}, \quad \forall ij \in \mathcal{L} \tag{6.8}$$

$$-P_{ij}^{max} x_{ij,t} \le p_{ij,t}^- \le P_{ij}^{\max} x_{ij,t}, \quad \forall ij \in \mathcal{L} \tag{6.9}$$

$$\left|\theta_{ij,t} - (\theta_{i,t} - \theta_{j,t})\right| \le M(1 - x_{ij,t}), \quad \forall ij \in \mathcal{L} \tag{6.10}$$

$$-Mx_{ij,t} \le \theta_{ij,t} \le Mx_{ij,t}, \quad \forall ij \in \mathcal{L} \tag{6.11}$$

$$0 \le p_{i,t}^r \le P_{i,t}^{r,\text{for}}, \quad \forall i \in \mathcal{R} \tag{6.12}$$

$$P_i^{c,\min} \le p_{i,t}^c \le P_i^{c,\max}, \quad \forall i \in \mathcal{C} \tag{6.11}$$

$$P_i^{u,\min} \le p_{i,t}^u \le P_i^{u,\max}, \quad \forall i \in \mathcal{U} \tag{6.14}$$

$$-P_i^{c,rd}\Delta t \le p_{i,t}^c - p_{i,t-1}^c \le P_i^{c,ru}\Delta t, \quad \forall i \in \mathcal{C} \tag{6.15}$$

$$-P_i^{u,rd}\Delta t \le p_{i,t}^u - p_{i,t-1}^u \le P_i^{u,ru}\Delta t, \quad \forall i \in \mathcal{U} \tag{6.16}$$

$$\Theta_i^{\min} \le \theta_{i,t} \le \Theta_i^{\max}, \quad \forall i \in \mathcal{B}, \tag{6.17}$$

where $\mathcal{L}_i^+/\mathcal{L}_i^-$ represent the sets of lines with node i as the start or end node, respectively. $\mathcal{C}/\mathcal{U}/\mathcal{R}$ represent the sets of conventional generators (CPGs), GFUs, and renewable energy sources (RESs), respectively. $\mathcal{C}_i/\mathcal{U}_i/\mathcal{R}_i$ represent the sets of CPGs, GFUs, and RESs at bus i, respectively. In terms of coefficients, $\text{LD}_{i,t}^p$ denotes the electrical load demand, G_{ij}/B_{ij} represent the conductance and susceptance of line ij, respectively, and $P_{ij}^{\max}$ represents the maximum active power flow of line ij. The variables include $p_{ij,t}^+/p_{ij,t}^-$, representing the active power flow at the start and end of line ij, respectively, $p_{i,t}^c,p_{i,t}^u/p_{i,t}^r$ denoting the power generation from CPGs, GFUs, and

RESs, respectively. $\theta_{ij,t}$ represents the voltage angle along line ij, and $\theta_{i,t}$ /$\theta_{j,t}$ represent the voltage angles at nodes i and j, respectively. Additionally, $P_{i,t}^{r,\text{for}}$ represents the forecasted power generation from renewable sources, and $P_i^{c,\max}$/$P_i^{u,\max}$ denote the upper limits of active power output for CPGs and GFUs. $P_i^{c,\min}$/$P_i^{u,\min}$ represent the lower limits, while $P_i^{c,rd}$/$P_i^{u,rd}$ represent the downward ramp rates, and $P_i^{c,ru}$/$P_i^{u,ru}$ denote the upward ramp rates for CPGs and GFUs, respectively. Finally, $\Theta_i^{\min}$/$\Theta_i^{\max}$ represent the lower and upper voltage angle limits at node i, respectively.

Equation (6.5) represents the node power balance constraint. Equations (6.6)–(6.7) express the power flow at the start and end of line ij, where the node voltage magnitude is close to 1 in the per-unit system. Equations (6.8)–(6.9) specify the upper and lower limits of the power flow on the line. In constraints (6.10)–(6.11), the big-M method is used to assign the value of $\theta_{ij,t}$, and binary variable $x_{ij,t}$ is introduced to represent the operational state of line ij at time t: $x_{ij,t} = 1$ indicates the line ij is in operation, while $x_{ij,t} = 0$ means it is not in service. Equations (6.12)–(6.16) represent the upper and lower power limits for RESs, CPGs, and GFUs, respectively. Specifically, Eqs. (6.15)–(6.16) impose ramping power constraints on CPGs and GFUs. Equation (6.17) defines the voltage angle limits for the system.

NGS Failure Modeling: The dynamic behavior of gas flow can be modeled using the following two partial differential equations (PDEs) [17].

$$\frac{\partial \psi}{\partial t} + \frac{4C^2}{\pi D^2}\frac{\partial f}{\partial x} = 0 \tag{6.18}$$

$$\frac{\partial \psi}{\partial x} + \frac{4}{\pi D^2}\frac{\partial f}{\partial t} + \frac{8\Lambda C^2}{\pi^2 D^5}\frac{f|f|}{\psi} = 0, \tag{6.19}$$

where C represents the speed of sound, D is the pipeline diameter, and Λ is the friction coefficient of the pipeline. ψ and f represent the gas pressure and mass flow rate through the pipeline, respectively.

Equations (6.18)–(6.19) represent the continuity equation and the momentum equation. By employing the finite difference method [17, 18], the above two partial differential Eqs. (6.18) and (6.19) are approximated into a set of algebraic equations, thereby constructing a transient gas flow model. The specific constraints are as follows:

$$\begin{aligned}&\sum_{mn\in\mathcal{K}_m^-}(f_{mn,t}^k - \tau_{mn,t}) - \sum_{mn\in\mathcal{K}_m^+} f_{mn,t}^k + \sum_{n\in\mathcal{W}_m} f_{n,t}^w \\ &\quad - \sum_{mn\in\mathcal{P}_m^+} f_{mn,t}^+ + \sum_{mn\in\mathcal{P}_m^-} f_{mn,t}^- - \sum_{n\in\mathcal{U}_m} f_{n,t}^u \\ &= LD_{m,t}^f - f_{m,t}^d, \quad \forall m\in\mathcal{N}\end{aligned} \tag{6.20}$$

$$\frac{\psi_{mn}^{s+1,t+1} - \psi_{mn}^{s+1,t}}{2\Delta t} + \frac{\psi_{mn}^{s,t+1} - \psi_{mn}^{s,t}}{2\Delta t}$$

$$+\frac{4C^2}{\pi D_{mn}^2}\left(\frac{f_{mn}^{s+1,t+1}-f_{mn}^{s,t+1}}{2\Delta x}+\frac{f_{mn}^{s+1,t}-f_{mn}^{s,t}}{2\Delta x}\right)$$
$$=0, \quad \forall mn \in \mathcal{P}, s \in \mathcal{S}_{mn} \tag{6.21}$$

$$\frac{\psi_{mn}^{s+1,t+1}-\psi_{mn}^{s,t+1}}{2\Delta x}+\frac{\psi_{mn}^{s+1,t}-\psi_{mn}^{s,t}}{2\Delta x}$$
$$+\frac{4}{\pi D_{mn}^2}\left(\frac{f_{mn}^{s+1,t+1}-f_{mn}^{s+1,t}}{2\Delta t}+\frac{f_{mn}^{s,t+1}-f_{mn}^{s,t}}{2\Delta t}\right)$$
$$+\frac{8\Lambda_{mn}C^2}{\pi^2 D_{mn}^5}\frac{f_{mn}^{s,t}\left|f_{mn}^{s,t}\right|}{\psi_{mn}^{s,t}}=0, \quad \forall mn \in \mathcal{P}, s \in \mathcal{S}_{mn} \tag{6.22}$$

$$\psi_{mn}^{1,t}=\psi_{m,t},\ \psi_{mn}^{|S_{mn}|,t}=\psi_{n,t}, \quad \forall mn \in \mathcal{P} \tag{6.23}$$

$$f_{mn}^{1,t}=f_{mn,t}^{+},\ f_{mn}^{|S_{mn}|,t}=f_{mn,t}^{-}, \quad \forall mn \in \mathcal{P} \tag{6.24}$$

$$\tau_{mn,t}=E_{mn}f_{mn,t}^{k}, \quad \forall mn \in \mathcal{K} \tag{6.25}$$

$$p_{i,t}^{u}=U_m f_{m,t}^{u}, \quad \forall i, m \in \mathcal{U} \tag{6.26}$$

$$H_{mn}^{\min}\psi_{m,t} \le \psi_{n,t} \le H_{mn}^{\max}\psi_{m,t}, \quad \forall mn \in \mathcal{K} \tag{6.27}$$

$$\Psi_m^{\min} \le \psi_{m,t} \le \Psi_m^{\max}, \quad \forall m \in \mathcal{N} \tag{6.28}$$

$$-F_{mn}^{\max} \le f_{mn,t}^{-} \le F_{mn}^{\max}, \quad \forall mn \in \mathcal{P} \tag{6.29}$$

$$-F_{mn}^{\max} \le f_{mn,t}^{+} \le F_{mn}^{\max}, \quad \forall mn \in \mathcal{P} \tag{6.30}$$

$$F_{mn}^{k,\min} \le f_{mn,t}^{k} \le F_{mn}^{k,\max}, \quad \forall mn \in \mathcal{K} \tag{6.31}$$

$$F_m^{u,\min} \le f_{m,t}^{u} \le F_m^{u,\max}, \quad \forall m \in \mathcal{U} \tag{6.32}$$

$$F_m^{w,\min} \le f_{m,t}^{w} \le F_m^{w,\max}, \quad \forall m \in \mathcal{W}, \tag{6.33}$$

where *mn* refers to the natural gas pipeline index. The sets are defined as follows: $\mathcal{K}_m^+/\mathcal{K}_m^-$ represent the sets of gas compressors with node m as the inlet or outlet node, $\mathcal{P}$ represents the gas system pipeline set, and $\mathcal{P}_m^+/\mathcal{P}_m^-$ represent the pipelines with node m as the predefined inlet or outlet node. The set $\mathcal{S}_{mn}$ refers to the spatial grid points of pipeline *mn*, and $\mathcal{W}$ denotes the natural gas supply equipment set, while $\mathcal{K}$

denotes the gas compressor set. In terms of coefficients, D_{mn} denotes the gas pipeline diameter, E_{mn} represents the gas compressor conversion ratio, U_m refers to the energy conversion factor for gas turbines, Δx is the length of the gas pipeline segment, and $F_{mn}^{\max}$ represents the maximum gas flow through the pipeline. The variables $F_{mn}^{w,\min}/F_{mn}^{w,\max}$ represent the minimum and maximum gas supply volumes at the gas gate stations, respectively, while $F_{mn}^{u,\min}/F_{mn}^{u,\max}$ represent the minimum and maximum gas supply volumes at the gas power plants. The variables $F_{mn}^{k,\min}/F_{mn}^{k,\max}$ correspond to the minimum and maximum gas supply volumes at the gas compressors, and $H_{mn}^{\min}/H_{mn}^{\max}$ denote the compression ratio limits of the gas compressors. Further, $\mathrm{LD}_{m,t}^{f}$ represents the gas load demand at time t, while $\Psi_{mn}^{\min}/\Psi_{mn}^{\max}$ denote the lower and upper pressure limits at node m, respectively. Λ_{mn} is the pipeline friction coefficient. Regarding the variables, $f_{m,t}^{k}$, $f_{m,t}^{w}$, $f_{m,t}^{u}$ represent the gas flow at the gas compressor, gas supply device, and gas turbine nodes, respectively. $f_{m,t}^{+}/f_{m,t}^{-}$ represent the incoming and outgoing gas flow at pipeline mn, respectively, while $f_{m,t}^{s,t}$ refers to the gas flow at the spatial point s in pipeline mn. The gas pressure at spatial point s along pipeline mn is denoted by $\psi_{mn}^{s,t}$, and $\tau_{m,t}$ refers to the gas consumption at compressor mn. Finally, $\psi_{m,t}$ denotes the gas pressure at node m at time t.

Equation (6.20) represents the gas balance equation at the node. Equations (6.21)–(6.22) discretize the continuity and momentum partial differential equations (PDEs) of the natural gas pipeline dynamics using the Wendroff difference scheme [34]. Equations (6.23)–(6.24) define the gas pressure and mass flow rate at the start and end of each pipeline, which are equal to the pressure and flow rate at the neighboring nodes. Equation (6.25) represents the linear relationship between the gas consumption and input flow rate at the compressor. Equation (6.26) linearizes the power generation model of the gas-fired unit. Equation (6.27) represents the compressor's compression ratio constraint, and Eqs. (6.28)–(6.33) represent the upper and lower limits on node pressure and pipeline flow rates.

6.2.4 Model Refinement

The non-convexity of the proposed OEGF model mainly arises from the constraints (6.6)–(6.7) and (6.22). To facilitate solving, the model needs to be reconstructed to improve its computational feasibility. The specific reconstruction process is as follows.

Power Flow Reformulation: Based on the assumption that θ_{ij} is very small, the following approximation can be applied:

$$\cos\theta_{ij,t} \approx 1 - \frac{1}{2}\theta_{ij,t}^{2}, \ \sin\theta_{ij,t} \approx \theta_{ij,t}, \quad \forall ij \in \mathcal{L}. \tag{6.34}$$

Based on Eq. (6.34), the constraints (6.6)–(6.7) can be reconstructed as follows:

$$p_{ij,t}^{+} = \frac{1}{2} G_{ij} w_{ij,t} - B_{ij} \theta_{ij,t}, \quad \forall ij \in \mathcal{L} \tag{6.35}$$

$$p_{ij,t}^{-} = \frac{1}{2} G_{ij} w_{ij,t} + B_{ij} \theta_{ij,t}, \quad \forall ij \in \mathcal{L} \tag{6.36}$$

$$w_{ij,t} = \theta_{ij,t}^{2}, \quad \forall ij \in \mathcal{L}, \tag{6.37}$$

where the auxiliary variable $w_{ij,t}$ is introduced to replace $\theta 2\ ij,t$. Therefore, the bilinear Eq. (6.37) can be reconstructed as a quadratic equality constraint, expressed as follows:

$$w_{ij,t} + \frac{1}{4} = \sqrt{(w_{ij,t} - \frac{1}{4})^2 + \theta_{ij,t}^{2}}, \quad \forall ij \in \mathcal{L}, \tag{6.38}$$

where $w_{ij,t} + 1/4$ and $w_{ij,t} - 1/4$ are both decision variables.

Gas Flow Reformulation: The last term in Eq. (6.22) is the primary source of non-convexity. To address this issue, a substitution variable $z_{mn}^{s,t}$ is introduced, which is specifically defined as follows:

$$z_{mn}^{s,t} = \frac{f_{mn}^{s,t} \left| f_{mn}^{s,t} \right|}{\psi_{mn}^{s,t}}, \quad \forall mn \in \mathcal{P}, s \in \mathcal{S}_{mn}. \tag{6.39}$$

Thus, Eq. (6.22) can be reformulated as:

$$\begin{aligned} &\frac{\psi_{mn}^{s+1,t+1} - \psi_{mn}^{s,t+1}}{2\Delta x} + \frac{\psi_{mn}^{s+1,t} - \psi_{mn}^{s,t}}{2\Delta x} \\ &+ \frac{4}{\pi D_{mn}^2} \left(\frac{f_{mn}^{s+1,t+1} - f_{mn}^{s+1,t}}{2\Delta t} + \frac{f_{mn}^{s,t+1} - f_{mn}^{s,t}}{2\Delta t} \right) \\ &+ \frac{8 \Lambda_{mn} C^2}{\pi^2 D_{mn}^5} z_{mn}^{s,t} = 0, \quad \forall mn \in \mathcal{P}, s \in \mathcal{S}_{mn}. \end{aligned} \tag{6.40}$$

It should be noted that Eq. (6.39) remains a complex non-convex equation due to the presence of absolute value terms and requires further reformulation. To address this issue, a new set of binary variables $x_{mn,t}$ is introduced to characterize the gas flow direction. Specifically, $x_{mn,t} = 1$ indicates that the gas in pipeline mn flows from node m to node n, while $x_{mn,t} = 0$ signifies that the flow direction is from node n to node m. Based on the big-M method, Eq. (6.39) can be reformulated as follows:

$$-M(1 - x_{mn,t}) \leq z_{mn}^{s,t} \leq M x_{mn,t} \tag{6.41}$$

$$\left| u_{mn}^{s,t} - 0.5\left(\psi_{mn}^{s,t} - z_{mn}^{s,t}\right) \right| \leq M(1 - x_{mn,t}) \tag{6.42}$$

$$\left|b_{mn}^{s,t} - 0.5\left(\psi_{mn}^{s,t} + z_{mn}^{s,t}\right)\right| \le M\left(1 - x_{mn,t}\right) \tag{6.43}$$

$$\left|a_{mn}^{s,t} - 0.5\left(\psi_{mn}^{s,t} + z_{mn}^{s,t}\right)\right| \le Mx_{mn,t} \tag{6.44}$$

$$\left|b_{mn}^{s,t} - 0.5\left(\psi_{mn}^{s,t} - z_{mn}^{s,t}\right)\right| \le Mx_{mn,t} \tag{6.45}$$

$$b_{mn}^{s,t} = \sqrt{\left(a_{mn}^{s,t}\right)^2 + \left(f_{mn}^{s,t}\right)^2}, \tag{6.46}$$

where $a_{mn}^{s,t}$ and $b_{mn}^{s,t}$ are auxiliary variables whose values are determined based on the gas flow direction. Eq. (6.41) specifies that when the gas flows in the positive direction, $z_{mn}^{s,t}$ takes a positive value, whereas when the gas flows in the reverse direction, $z_{mn}^{s,t}$ takes a negative value. When $x_{mn,t} = 1$, Eqs. (6.42)–(6.43) are activated, setting $a_{mn}^{s,t}$ and $b_{mn}^{s,t}$ to $0.5(\psi_{mn}^{s,t} \pm z_{mn}^{s,t})$, respectively. Conversely, when $x_{mn,t} = 0$, Eqs. (6.44)–(6.45) are activated, and the values of $a_{mn}^{s,t}$ and $b_{mn}^{s,t}$ are reversed. Additionally, by introducing constraint (6.46), the non-convex equation in Eq. (6.39) is further standardized.

6.2.5 Matrix Formulation

For clarity in subsequent discussions, the proposed model is reformulated in a matrix representation as follows:

$$\langle \mathbf{P} \rangle : \min \boldsymbol{d}^T \boldsymbol{y} \quad \text{s.t. } \boldsymbol{Ax} + \boldsymbol{By} \le \boldsymbol{f}, \quad \boldsymbol{x} \in \mathbb{Z}, \quad \boldsymbol{y} \in \mathbb{R} \tag{6.47}$$

$$[\boldsymbol{y}]_{r_n} = \sqrt{[\boldsymbol{y}]_{p_n}^2 + [\boldsymbol{y}]_{q_n}^2}, \quad \forall n \in \Omega, \tag{6.48}$$

where $\boldsymbol{x}$ and $\boldsymbol{y}$ denote integer and continuous variables, respectively, while $\boldsymbol{d}$, $\boldsymbol{A}$, $\boldsymbol{B}$ and $\boldsymbol{f}$ represent coefficient matrices corresponding to the objective function and constraint conditions. Equation (6.47) represents the linear part of the model, encompassing Eqs. (6.2)–(6.5), (6.8)–(6.17), (6.21)–(6.22), (6.23)–(6.33), (6.35)–(6.36), and (6.40)–(6.45). Equation (6.48) defines a set of quadratic equality constraints corresponding to constraints (6.38) and (6.46). For each $n \in \Omega$, a triplet (p_n, q_n, r_n) is introduced, where the set Ω constructs the index set for constraint (6.48).

It is important to emphasize that the objective of this section is not to directly solve problem $\langle \mathbf{P} \rangle$ for its optimal (final) solution, as this solution is valid only for the mildest fault scenarios. Instead, solving problem $\langle \mathbf{P} \rangle$ yields suboptimal (intermediate) solutions, which, if found to be feasible (or even optimal) under specific fault scenarios, enable further calculation of the load-shedding index R_{EENS}. The next section will provide a detailed discussion of the solution methodology for problem $\langle \mathbf{P} \rangle$.

6.3 Proposed Solution Approach

The objective of the proposed method is to compute the R_{EENS} index during the solution process of problem $\langle \mathbf{P} \rangle$. This method is based on a relaxation of problem $\langle \mathbf{P} \rangle$, in which the "=" in Eq. (6.48) are replaced with "≥", transforming them into SOC constraints. The convexified problem is denoted as $\langle \mathbf{P_{cvx}} \rangle$ in the following discussion. Since $\langle \mathbf{P_{cvx}} \rangle$ is a mixed-integer second-order cone programming (MISOCP) problem, it can be directly solved using commercial solvers with built-in B&C algorithms. However, the direct solution may violate the non-convex equality constraints in Eq. (6.48) and is only valid for a single fault scenario.

To address these issues, this section proposes two improved strategies based on the B&C framework. Sect. 6.3.1 introduces an enhanced spatial branching strategy that enables the B&C algorithm to handle SOC non-convexity. Section 6.3.2 presents an improved cutting-plane generation strategy, which extracts resilience indices for all fault scenarios within a single B&C search tree.

6.3.1 Cone-Specific Spatial Branching Strategy

The first improvement introduces an enhanced spatial branching strategy, designed for cases where the solution at a branching node is locally optimal but does not satisfy constraint (6.48). The core idea of the improved spatial branching strategy is to create two new nodes with more compact feasible regions at the current node, thereby eliminating the infeasible solution. Unlike traditional spatial branching strategy, which requires decomposing constraint (6.48) into three bilinear terms (as shown in Refs. [25–27]), the proposed method directly operates on the solution space of Eq. (6.48), forming a spatial branching strategy specifically for SOC.

Suppose that the k-th equality in Eq. (6.48) is violated beyond a predefined threshold at the current node. For simplicity, the variables used in the k-th equality of Eq. (6.48), originally denoted as $([\mathbf{y}]_{pk}, [\mathbf{y}]_{qk}, [\mathbf{y}]_{rk})$, are rewritten as (x, y, z), with their current values represented by $(\hat{x}, \hat{y}, \hat{z})$. Since this solution is non-convex and infeasible, it follows that:

$$\delta = \hat{z} - \sqrt{\hat{x}^2 + \hat{y}^2} > 0, \ l_x \leq \hat{x} \leq u_x, \ l_y \leq \hat{y} \leq u_y, \tag{6.49}$$

where l_x and u_x (l_y and u_y) represent the lower and upper bounds of variable x (variable y) at the current node, respectively. The green surface in Fig. 6.2a represents the feasible region, where P_0 corresponds to the infeasible solution $(\hat{x}, \hat{y}, \hat{z})$. Since the feasible region in the x–y plane may cover one, two, or all four quadrants, it is necessary to verify the following conditions to construct the spatial branching strategy accordingly:

$$(l_x \cdot u_x < 0) \wedge \left(l_y \cdot u_y < 0\right), \tag{6.50}$$

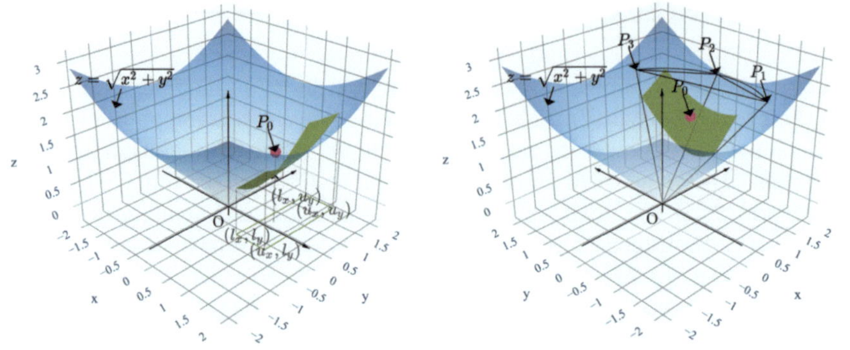

(a) illustration of feasible region and infeasible solution.

(b) triangle formed by O, P_1, P_2 and P_3 on the surface.

Fig. 6.2 3D representation of quadratic equality constraints and corresponding points/surfaces

where $\wedge$ represents the logical "and" operator. Based on the evaluation of Eq. (6.50), the SOC-specifc spatial branching strategy can be categorized into two cases.

Restriction to a Maximum of Two Quadrants: If Eq. (6.50) evaluates to false, it indicates that the feasible region in the x–y plane spans at most two quadrants. An example of this scenario is illustrated in Fig. 6.2a.

In this case, a subset of the blue conical surface is analyzed, with its vertices at P_1, P_3, and the origin O, as shown in Fig. 6.2b. This subset forms a surface resembling a triangle, bounded by the straight lines OP_1, OP_3, and the circular arc passing through points P_1, P_2, and P_3 on the blue surface. It can also be considered as the surface generated by rotating OP_1 around the z-axis until it reaches OP_3. It is important to note that the selection of points P_1 and P_3 follows the principle of identifying the tightest subset that fully covers the feasible region, as depicted in Fig. 6.2b. In the spherical coordinate system, the radial distance ρ, polar angle θ, and azimuthal angle φ of points P_1 and P_3 are denoted as $(\rho_1, \theta_1, \varphi_1)$ and $(\rho_3, \theta_3, \varphi_3)$, respectively, and are computed using the following formulas:

$$\rho_1 = \rho_3 = \sqrt{2\left[\max(l_x^2, u_x^2) + \max(l_y^2, u_y^2)\right]} \tag{6.51}$$

$$\theta_1 = \theta_3 = \arctan(1) = \pi/4 \tag{6.52}$$

$$\varphi_1 = \min_{w_y \in \{l_y, u_y\}, w_x \in \{l_x, u_x\}} \arctan \frac{w_y}{w_x} \tag{6.53}$$

$$\varphi_3 = \max_{w_y \in \{l_y, u_y\}, w_x \in \{l_x, u_x\}} \arctan \frac{w_y}{w_x}, \tag{6.54}$$

where ρ_1 and ρ_3 represent the maximum radial distances of all solutions within the feasible region, while θ_1 and θ_3 are fixed at $\pi/4$ due to the geometric characteristics of the cone surface. The values of φ_1 and φ_3 are determined by the minimum and maximum azimuthal angles of the feasible solutions.

The space formed by the straight lines OP_1, OP_3, and P_1P_3, along with the arc P_1P_3, represents the convex relaxation of the feasible region for the variables (x, y, z). Although this region provides a tighter approximation compared to the entire space above the green surface, the infeasible solution P_0 still resides within this feasible domain. To address this issue, the space branching strategy creates two new nodes with tighter feasible domains, ensuring that P_0 is eliminated in either the downward or upward child nodes. Another selection for point P_2, as shown in Fig. 6.2b, is configured as follows:

$$\rho_2 = \rho_1 = \rho_3, \quad \theta_2 = \pi/4, \quad \varphi_2 = \arctan(\hat{y}/\hat{x}), \tag{6.55}$$

where the values of ρ_2 and θ_2 are the same as those of P_1 and P_3, while the azimuthal angle φ_2 is set to the azimuthal angle of the infeasible solution P_0.

To convert any 3D point spherical coordinates (ρ, θ, φ) to corresponding Cartesian coordinates (x, y, z), we introduce the operator cart$(\cdot)$, defined as follows:

$$\begin{aligned}(x, y, z) &= \text{cart}\,(\rho, \theta, \varphi)\\ &= (\rho\cos\varphi\sin\theta,\ \rho\sin\varphi\sin\theta,\ \rho\cos\theta)\end{aligned} \tag{6.56}$$

so that the Cartesian coordinates of the points P_1 (x_1, y_1, z_1), P_2 (x_2, y_2, z_2), and P_3 (x_3, y_3, z_3) can be calculated using the following equation:

$$\begin{cases}(x_1, y_1, z_1) = \text{cart}\,(\rho_1, \theta_1, \varphi_1)\\ (x_2, y_2, z_2) = \text{cart}\,(\rho_2, \theta_2, \varphi_2)\\ (x_3, y_3, z_3) = \text{cart}\,(\rho_3, \theta_3, \varphi_3)\end{cases}. \tag{6.57}$$

Then, the cutting planes of ΔOP_1P_2 and ΔOP_2P_3, marked in Fig. 6.3a, b, can be expressed in the form of a determinant (linear constraints) as:

$$f_{\Delta OP_1P_2}(x, y, z) = \begin{vmatrix} x & y & z \\ x_1 - x_0 & y_1 - y_0 & z_1 - z_0 \\ x_2 - x_0 & y_2 - y_0 & z_2 - z_0 \end{vmatrix} \le 0 \tag{6.58}$$

$$f_{\Delta OP_2P_3}(x, y, z) = \begin{vmatrix} x & y & z \\ x_2 - x_0 & y_2 - y_0 & z_2 - z_0 \\ x_3 - x_0 & y_3 - y_0 & z_3 - z_0 \end{vmatrix} \le 0. \tag{6.59}$$

The first row of Eqs. (6.58) (6.59) represents the variables, while the second and third rows are constants calculated from Eq. (6.57). Additionally, (x_0, y_0, z_0) corresponds to the coordinates of the origin O, which are (0,0,0).

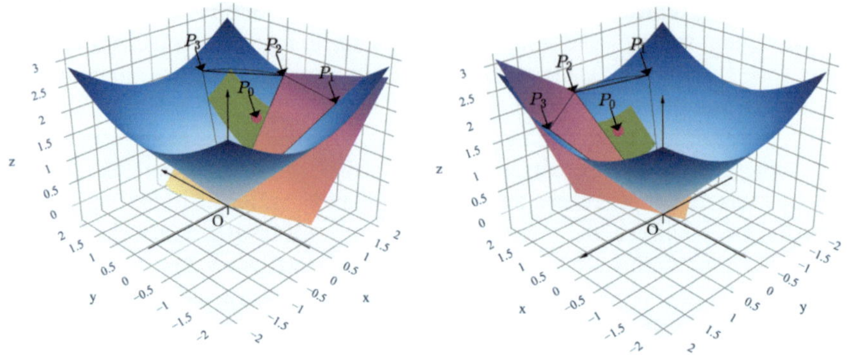

(a) downward branching constraint for ΔOP_1P_2. (b) upward branching constraint for ΔOP_2P_3

Fig. 6.3 SOC-specific spatial branching strategy when Eq. (6.50) is not satisfied

Based on Eqs. (6.58)–(6.59), the space branching rule for the infeasible solution P_0 can be expressed as:

$$f_{\Delta OP_1P_2}(x, y, z) \leq 0 \vee f_{\Delta OP_2P_3}(x, y, z) \leq 0, \tag{6.60}$$

where $\vee$ is the logical "or" operator. It should be emphasized that the branching constraints (6.58) and (6.59) are added to the two child nodes separately, not simultaneously.

Based on the following two propositions, the validity of the proposed space branching rule is proven.

Proposition 6.1 *The infeasible solution P_0 can be eliminated by applying the branching rule in Eq.* (6.60) *in both child nodes.*

Proof Since the proof process for downward and upward branching is similar, we use downward branching $f_{\Delta OP1P2} \leq 0$ as an example. For the infeasible solution $(\hat{x}, \hat{y}, \hat{z})$ that violates Eq. (6.48) in problem $\langle \mathbf{P} \rangle$, we assume it satisfies the following contradictory condition:

$$f_{\Delta OP_1P_2}(\hat{x}, \hat{y}, \hat{z}) \leq 0. \tag{6.61}$$

Substituting Eqs. (6.51)–(6.55) and (6.57) into Eq. (6.61), we obtain:

$$\begin{aligned} f_{\Delta OP_1P_2}(\hat{x}, \hat{y}, \hat{z}) = & \frac{1}{2}\rho_1\rho_2(\sin\varphi_1 - \sin\varphi_2)\hat{x} \\ & - \frac{1}{2}\rho_1\rho_2(\cos\varphi_1 - \cos\varphi_2)\hat{y} \\ & + \frac{1}{2}\rho_1\rho_2(\cos\varphi_1 \sin\varphi_2 - \sin\varphi_1 \cos\varphi_2)\hat{z} \leq 0. \end{aligned} \tag{6.62}$$

Since $\rho_1\rho_2 \neq 0$ and $\varphi_2 = \arctan(\hat{y}/\hat{x})$, Eq. (6.62) can be further simplified to:

$$\hat{x}\sin\varphi_1 - \hat{y}\cos\varphi_1 \leq \frac{\hat{z}}{\sqrt{\hat{x}^2+\hat{y}^2}}(\hat{x}\sin\varphi_1 - \hat{y}\cos\varphi_1). \tag{6.63}$$

Based on the definition in Eq. (6.53), $\varphi_1 \leq \arctan(\hat{y}/\hat{x})$; therefore $\hat{x}\sin\varphi_1 - \hat{y}\cos\varphi_1 \leq 0$. Thus, Eq. (6.63) simplifies to:

$$\sqrt{\hat{x}^2+\hat{y}^2} \geq \hat{z}. \tag{6.64}$$

Note that Eq. (6.64) violates the feasibility condition of P_0 in Eq. (6.49). Therefore, Eq. (6.61) is invalid.

Proposition 6.2 *The feasible region of each child node cannot be cut off by the cutting hyperplane in Eq.* (6.60)*.*

Proof Here, the proof for the downward branch is presented in detail, and the proof for the upward branch can be derived by analogy. Let (x^*, y^*, z^*) be a feasible solution for the downward branch, indicating that it is feasible for problem $\langle \mathbf{P} \rangle$. Suppose the following contradiction condition holds:

$$f_{\Delta OP_1P_2}(x^*, y^*, z^*) > 0, \tag{6.65}$$

where (x^*, y^*, z^*) subjects to:

$$z^* = \sqrt{(x^*)^2 + (y^*)^2},\ \varphi 1 \leq \arctan\frac{y^*}{x^*} \leq \varphi 2. \tag{6.66}$$

Substituting Eqs. (6.55) and (6.57) into Eq. (6.65) yields:

$$\begin{aligned} f_{\Delta OP_1P_2}(x^*, y^*, z^*) = &\frac{1}{2}\rho_1\rho_2(\sin\varphi_1 - \sin\varphi_2)\,x^* \\ &- \frac{1}{2}\rho_1\rho_2(\cos\varphi_1 - \cos\varphi_2)y^* \\ &+ \frac{1}{2}\rho_1\rho_2(\cos\varphi_1\sin\varphi_2 - \sin\varphi_1\cos\varphi_2)z^* > 0. \end{aligned} \tag{6.67}$$

The above equation can be further simplified. Considering $\rho_1\rho_2 \neq 0$ and $z^* \geq 0$, we get:

$$\frac{(\sin\varphi_2 - \sin\varphi_1)\,x^* - (\cos\varphi_2 - \cos\varphi_1)\,y^*}{\sin(\varphi_2 - \varphi_1)\,z^*} < 1. \tag{6.68}$$

Let $\varphi^* = \arctan(y^*/x^*)$, which gives $x^* = z^*\cos\varphi^*$ and $y^* = z^*\sin\varphi^*$. Substituting this into Eq. (6.68) leads to:

$$\sin(\varphi_2 - \varphi^*) + \sin(\varphi^* - \varphi_1) < \sin(\varphi_2 - \varphi_1). \tag{6.69}$$

Substituting $\varphi_2 - \varphi^*$ with A, and $\varphi_1 - \varphi^*$ with B, we get:

$$\begin{aligned} &\sin A + \sin B < \sin(A + B) \\ &= \sin A \cos B + \sin B \cos A. \end{aligned} \tag{6.70}$$

Since the domains of A, B, and $A + B$ are all within [0, π], the values of cos A and cos B are always less than 1, and sin A and sin B are always greater than 0. Therefore, Eq. (6.70) is invalid, completing the proof.

Four Quadrants Case: The following analysis is based on the premise that Eq. (6.50) is "true". Since the proof of Proposition 2 requires that the domains of A, B, and $A + B$ be within [0, π], the spatial branching strategy of Eq. (6.71) cannot be applied when the domain is [-π, π]. To address this issue, the following branching rule is proposed:

$$\begin{cases} x \geq 0 \vee x \leq 0, \text{ if } u_x - l_x \geq u_y - l_y \\ y \geq 0 \vee y \leq 0, \text{ if } u_y - l_y \geq u_x - l_x \end{cases}. \tag{6.71}$$

This branching rule simplifies the solution space into two subspaces defined by the planes $x = 0$ or $y = 0$. A schematic of the branching strategy is shown in Fig. 6.4. Unlike Eq. (6.60), this strategy eliminates the non-convex infeasible point P_0 from the feasible domain. The idea behind Eq. (6.71) is to create two subnodes with reduced feasible domains, each of which can only cover two quadrants of the x–y plane (the part violating Eq. (6.50), thus allowing further branching in each subnode using Eq. (6.60).

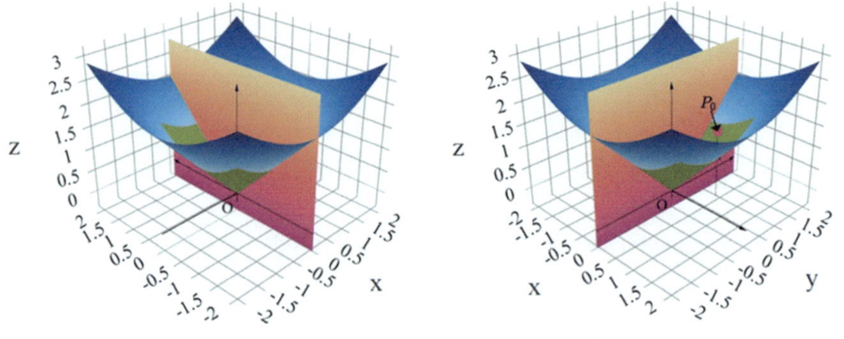

(a) downward branching constraint (Plane x = 0). (b) upward branching constraint (plane x = 0).

Fig. 6.4 Illustration of the SOC-specific spatial branching strategy when Eq. (6.50) is satisfied

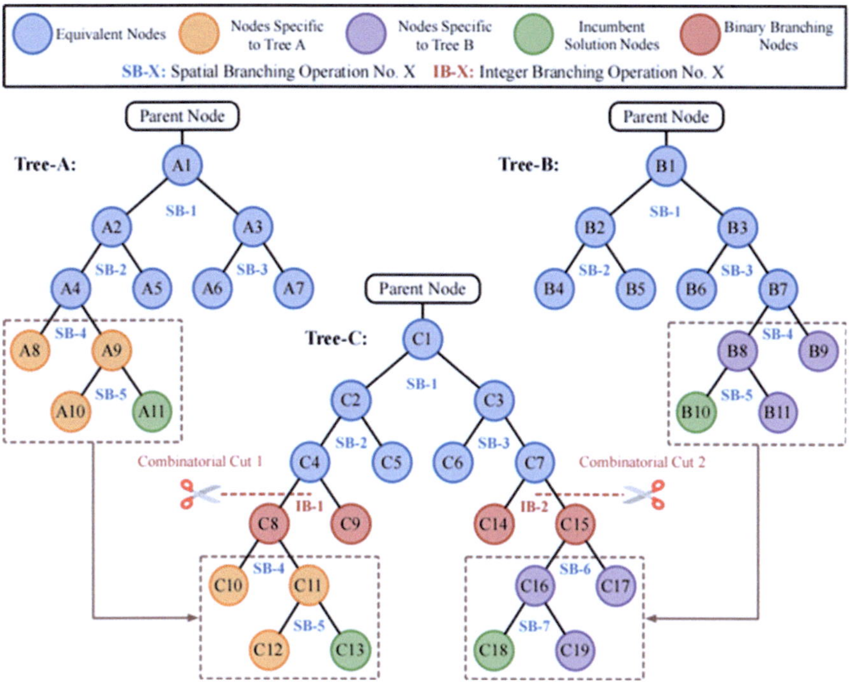

Fig. 6.5 Illustration of B&B tree simplification with branching guidance

6.3.2 Combinatorial Cut Derivation

The computation redundancy in the iterative resilience evaluation method [14, 29] mainly arises from repeatedly visiting the same branch nodes, as shown in Fig. 6.5. Assume that Tree A and Tree B represent two trees solving the load recovery model (i.e., for the same problem ⟨**P**⟩ but with different values of x_v in different scenarios). Using the above spatial branching strategy, the solution process begins at the root nodes (A1 and B1) and terminates when the incumbent solution is found.[1] Although different values of x_v may lead to different tree structures, certain nodes are equivalent, as indicated by the blue circles in Fig. 6.5. Since both Tree A and Tree B operate on the same problem ⟨**P**⟩, the space branching operations that generate nodes A2–A7 for Tree A may also apply to Tree B, and the new nodes B2–B7 may lead to the same current optimal solution B10. In other words, if leaf node A7 is further branched, B10 would only be a few steps away from Tree A. However, since A7 cannot yield a better solution than the current optimal solution A11, it is pruned. Thus, the traditional method requires regenerating Tree B from root node B1 and spends extra time handling infeasible solutions already pruned in Tree A.

[1] Incumbent solution refers to the best feasible solution found at any point during the search process, and the corresponding node is referred to as the current optimal node.

To address this redundancy, a second improvement to the spatial branch-and-bound (SB&B) algorithm is proposed to obtain solutions for all fault scenarios within a single branch-and-bound search tree. The schematic of the improved B&C algorithm is shown as Tree C in Fig. 6.5. Since it only applies to the problem ⟨P⟩, where the binary variables x_v and $x_{ij,t}$ are the decision variables, the optimal solution for each fault scenario is the current optimal solution of problem ⟨**P**⟩. It is assumed that A11 and B10 correspond to C13 and C18, respectively. The infeasible solutions removed during the solving process of Tree A (e.g., A1–A7 and B1–B7) are handled by C1–C7 in Tree C. Since problem ⟨P⟩ is an MINLP, both space branching and integer branching operations can be executed jointly. When binary variables x_v and $x_{ij,t}$ are fixed at C8 (or C15), it means the scenario is determined, and specific space branching for Tree A (or Tree B) will be performed. If C8 is the next node to branch, the algorithm usually terminates upon reaching C13. To force the algorithm to attempt another branching possibility at C7, the current solution at C7 must be discarded manually. Therefore, a combined cut-plane generation strategy is proposed.

Let $\boldsymbol{x}^*$ be the vector composed of the values of x_v and $x_{ij,t}$ at the current optimal solution node. The corresponding combined cut [33] is given.

$$\sum_{n\in\Gamma_0}[\boldsymbol{x}]_n + \sum_{n\in\Gamma_1}(1-[\boldsymbol{x}]_n) \geq 1, \tag{6.72}$$

where Γ_0 and Γ_1 are defined as:

$$\Gamma_0 = \{n : [\boldsymbol{x}^*]_n = 0\},\ \Gamma_1 = \{n : [\boldsymbol{x}^*]_n = 1\} \tag{6.73}$$

and the main idea is to force the change of at least one binary variable value to remove the current solution.

In the proposed method, when a current optimal solution is feasible under Eq. (6.48) and its objective value's relative error to the current lower bound of the problem is smaller than a preset threshold, the combined cut-plane (6.72) is added to the B&C search tree. The first condition ensures that the integrality and non-convexity constraints are satisfied, while the second condition is used to verify optimality.

Therefore, the obtained current optimal solution corresponds to the optimal load recovery strategy for the fault scenario determined by that solution. After adding the combined cut-plane (6.72) to remove the current optimal solutions, the B&C algorithm continues exploring the remaining nodes in the search tree until resilience metrics are calculated for all scenarios.

6.3.3 General Methodological Steps

The new resilience evaluation method is formed by embedding the two improvements into the B&C algorithm. Algorithm 1 provides the general procedure of the algorithm, which is applicable to most solvers that support SOC. As previously mentioned, $\langle \mathbf{P_{cvx}} \rangle$ is the relaxed version of problem $\langle \mathbf{P} \rangle$, obtained by replacing the equality symbol " =" in Eq. (6.48) with "$\geq$". If the integer constraints of $\langle \mathbf{P_{cvx}} \rangle$ are further removed, the resulting relaxed problem is $\langle \mathbf{P_{rel}} \rangle$, which is an SOCP that can be directly solved. If the result satisfies all the integer and non-convex constraints (even though they are not directly applied), this is a fortunate outcome. Otherwise, the standard branching process is applied to the selected variables. $\langle \mathbf{P_{cvx}} \rangle^{(n)}$ represents the problem at the nth node of the search tree, with its local objective and solution denoted as $\mathrm{obj}^{(n)}$ and $\mathrm{sol}^{(\mathrm{n})}$, respectively. For a node to be branched, the algorithm first checks whether it satisfies the integer constraints. If it does, the proposed improved spatial branching strategy is called, as shown in lines 7–12 of Algorithm 1. Otherwise, branching is done on the binary variables,[2] as described in lines 14–15. Both branching operations create child nodes with tighter feasible regions compared to their parent node. If a solution that is feasible for problem $\langle \mathbf{P} \rangle$ is found, meaning it satisfies the integer and non-convex constraints, the corresponding scenario index is recorded, and the node is removed from the search tree using the combination cutting-plane generation strategy, as shown in lines 19–21. Otherwise, the search tree is generated according to the standard branch-and-cut algorithm procedure.

[2] Available methods include pseudo reduced cost branching, pseudo shadow price branching, maximum infeasibility branching, and strong branching [25].

Algorithm 1: Common procedural work

[1] Available methods include pseudo reduced cost branching, pseudo shadow price branching, maximum infeasibility branching, and strong branching [25].

Input: Original problem $\langle \mathbf{P} \rangle$.

Output: Resilience metrics R_{EENS}

Remove the integer constraints of the variables and relaxing Eq. (6.48) into SOC constraints, the relaxed problem $\langle \mathbf{P} \rangle$ is created as $\langle \mathbf{P}_{\mathbf{rel}} \rangle$;

Let $\langle \mathbf{P}_{\mathbf{rel}} \rangle$ be $\langle \mathbf{P}_{\mathbf{rel}} \rangle^{(0)}$, solve $\langle \mathbf{P}_{\mathbf{rel}} \rangle^{(0)}$ to obtain $\mathrm{sol}^{(0)}$ and $\mathrm{obj}^{(0)}$, and set $\mathbb{N} \leftarrow \{\langle \mathbf{P}_{\mathbf{rel}} \rangle^{(0)}\}, \mathbb{V} \leftarrow \mathcal{V}$, $lb \leftarrow \mathrm{obj}^{(0)}$:

Initialize the resilience index $R_{EENS} \leftarrow 0$, counter $\mathrm{n} \leftarrow 0$, and set the optimal tolerance ε;

while $\mathbb{N}$ is non-empty or $\mathbb{V}$ is non-empty **do**

  Select the $\langle \mathbf{P}_{\mathbf{rel}} \rangle^{(k)}$ from $\mathbb{N}$ with the smallest objective value $\mathrm{obj}^{(k)}$, update $\mathbb{N} \leftarrow \mathbb{N} \setminus \langle \mathbf{P}_{\mathbf{rel}} \rangle^{(k)}$;

  if $\mathrm{sol}^{(k)}$ is an integer feasible solution **then**

    Calculate the error δ of all variable triplets in Eq. (6.49) within Eq. (6.48), and select the triplet $(\hat{x},\ \hat{y},\ \hat{z})$ with the maximum δ;

    Obtain the current node's lower and upper bounds: l_x, u_x, l_y and u_y;

    if Eq. (6.50) is violated **then**

      Create child nodes $\langle \mathbf{P}_{\mathbf{rel}} \rangle^{(n+1)}$ and $\langle \mathbf{P}_{\mathbf{rel}} \rangle^{(n+2)}$ using Eq. (6.60);

    else

      Create child nodes $\langle \mathbf{P}_{\mathbf{rel}} \rangle^{(n+1)}$ and $\langle \mathbf{P}_{\mathbf{rel}} \rangle^{(n+2)}$ using Eq. (6.71);

  else

    Select a variable that is restricted to binary but has a fractional value in the relaxed problem $\langle \mathbf{P}_{\mathbf{rel}} \rangle$;

    Impose the integer constraint on the selected variable, create child nodes $\langle \mathbf{P}_{\mathbf{rel}} \rangle^{(n+1)}$ and $\langle \mathbf{P}_{\mathbf{rel}} \rangle^{(n+2)}$;

  for each index $i \in \{n+1, n+2\}$ **do**

    if $\langle \mathbf{P}_{\mathbf{rel}} \rangle^{(i)}$ is neither infeasible nor an unbounded problem **then**

      Solve $\langle \mathbf{P}_{\mathbf{rel}} \rangle^{(i)}$ to obtain the optimal solution $\mathrm{sol}^{(i)}$ and objective value $\mathrm{obj}^{(i)}$;

      if $\mathrm{sol}^{(i)}$ is feasible for $\langle \mathbf{P} \rangle$ and $\frac{\mathrm{obj}^{(i)} - lb}{lb} \leq \varepsilon$ **then**

        Obtain the corresponding scenario index v, update $\mathbb{V} \leftarrow \mathbb{V} \setminus \{v\}$, and $R_{EENS} \leftarrow R_{EENS} + Kv \cdot \mathrm{obj}^{(i)}$;

        Apply constraint (6.72) to all nodes in $\mathbb{N}$ to remove this solution;

      else

        Update $\mathbb{N} \leftarrow \mathbb{N} \setminus \{\langle \mathbf{P}_{\mathbf{rel}} \rangle^{(i)}\}$;

  Update $\mathrm{n} \leftarrow \mathrm{n}+2$, and $lb \leftarrow \min_{i \in \{i:\ \langle \mathbf{Prel} \rangle^{(i)} \in \mathbb{N}\}} \mathrm{obj}^{(i)}$;

6.3.4 *Advanced Implementation for CPLEX*

Algorithm 1 is a highly customized B&C algorithm, where all processes—including both the traditional B&C components and the novel methods proposed in this chapter—require manual implementation. In practice, however, rewriting the entire B&C algorithm from scratch is unnecessary. Instead, a more modern and efficient approach is to customize an existing commercial solver (e.g., CPLEX[3]) using callbacks [27, 28, 35]. This approach allows handling only the spatial branching and combinatorial cut generation while delegating the integer branching process to the solver.

The modern implementation of Algorithm 1 can be summarized as follows. First, formulate the problem $\langle \mathbf{P}_{\mathbf{cvx}} \rangle$ (which is a MISOCP) and pass it to the CPLEX solver. Then, register the *branch* callback, *lazy constraint* callback, and *incumbent* callback with the solver so that they are periodically executed during the optimization process. The steps in the branch callback correspond to lines 7–11 of Algorithm 1, while the steps in the *lazy constraint* callback correspond to lines 17–23. If the *incumbent* callback is triggered, only lines 20–21 need to be executed to remove the current optimal solution. With this deep customization of CPLEX, the solver needs to be called only once, eliminating the need for multiple solver calls during the iterative process.

6.4 Numerical Performance Analysis

This section evaluates the effectiveness of the proposed resilience assessment method using three test systems of different scales. Case A represents a small-scale system, consisting of an IEEE 39-bus EPS and a 11-node NGS. Case B corresponds to a medium-scale system with 118-bus-24-node IEGS. Case C represents a large-scale system, comprising a 200-bus EPS in Illinois of United States, USA, and a 40-node NGS in the Rhine-Ruhr region of Germany. The topologies of these three cases are illustrated in Figs. 6.6, 6.7 and 6.8, based on synthetic geographic data. The resilience assessment in this section is conducted under a simulated six-hour hurricane scenario, where line failures are generated using the probabilistic failure model described in Refs. [14, 15]. A total of 50 representative failure scenarios (each specifying the time and location of failures) are selected to define the parameters $A_{ij,t,v}$. The occurrence frequency of each scenario is used to determine K_v, with specific case parameters detailed in Ref. [36]. The computations are performed on a personal computer running Ubuntu 18.04, equipped with an Intel(R) Xeon(R) Gold 6230R CPU and 256 GB RAM.

To evaluate the performance of the proposed fast resilience assessment algorithm (M8), this section compares it with seven existing methods. The seven different

[3] The version used in this chapter is 22.1.1, and we suggest that the version should be at least 12.8.0.

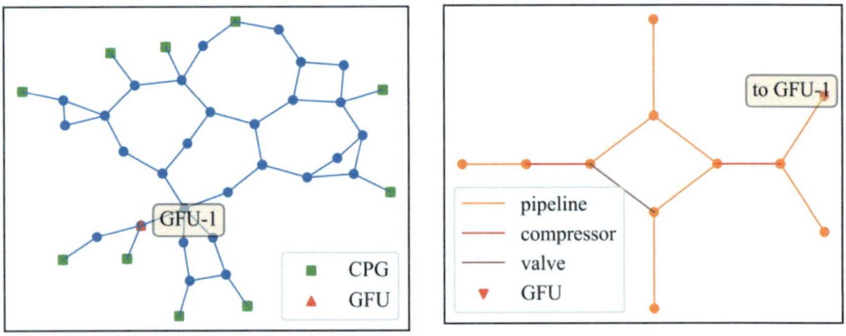

(a) 39-bus power system. (b) 11-node gas system.

Fig. 6.6 Topology of Case A

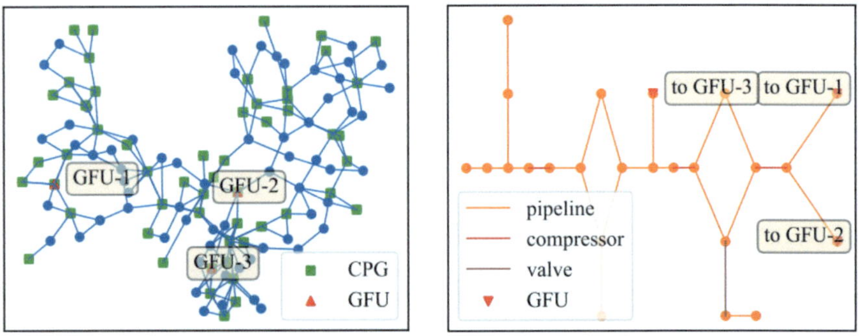

(a) 118-bus power system. (b) 24-node gas system.

Fig. 6.7 Topology of Case B

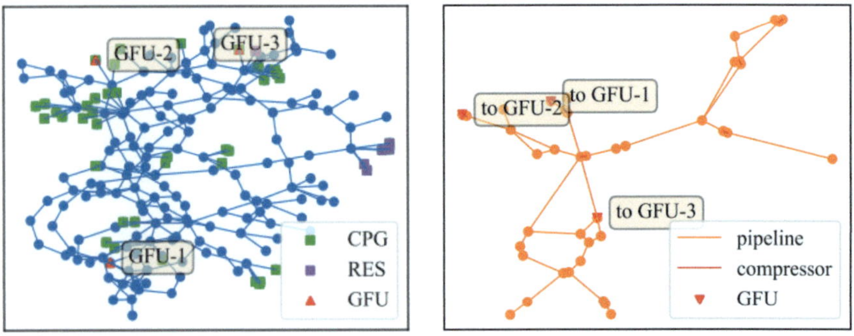

(a) 200-bus power system. (b) 40-node gas system.

Fig. 6.8 Topology of Case C

Table 6.2 Solver-specific parameter settings

Solver	Method	Parameter	Value	Description
Gurobi	M1–M4	Params.Non-convex	2	Enabled
		Params.NLPHeur	0	Disabled
SCIP	M5	Heuristics/subnlp/freq	− 1	Disabled
CPLEX	M6–M8	Preprocessing.reduce	0	Disabled
		Preprocessing.aggregator	0	Disabled
		Preprocessing.reformulations	0	Disabled
		Mip.strategy.heuristicfreq	− 1	Disabled
		Threads	8	Up to 8 threads

algorithms for solving non-convex optimization problems (M1–M7) are summarized as follows:

M1—Iteration of convex relaxation [19]
M2—Iteration of CCP [22]
M3—Iteration of modified SLP [23]
M4—Iteration of solver Gurobi v10.0.0 [25]
M5—Iteration of solver SCIP v8.0.3 [26]
M6—Iteration of bilinear-specific spatial B&B [27]
M7—Iteration of the proposed SOC-specific spatial B&B
M8—The proposed non-convex and non-iterative method.

The characteristics of M1–M8 are summarized as follows. M1–M3 solve the convexified problem $\langle \mathbf{P}_{cvx} \rangle$, while M4–M8 directly solve the original problem $\langle \mathbf{P} \rangle$. Among them, M1 does not guarantee solution feasibility or optimality, whereas M2 and M3 ensure feasibility but not optimality. In contrast, M4–M8 employ global optimization techniques, ensuring that $\langle \mathbf{P} \rangle$ is solved to an ε-optimal solution. For resilience assessment, M1–M7 evaluate each failure scenario independently by fixing $x_v = 1 (v \in \mathcal{V})$, solving the problem separately for each scenario. In contrast, M8 directly obtains resilience metrics for all failure scenarios within a single execution. The thresholds for non-convex constraint violations and optimality criteria are both set to 1.0%, and the time limit is set to 7200 s. Solver parameter configurations are detailed in Table 6.2, with unspecified parameters set to their default values.

6.4.1 Findings from Resilience Assessment

The evaluation results are presented in Tables 6.3 and 6.4. Here, Table 6.3 summarizes the model size and key evaluation metrics for each test case, while Table 6.4 compares the proposed algorithm with seven existing methods. Among them, M5 serves as the benchmark algorithm, and the resilience evaluation metrics for the other seven methods are computed based on their relative error with respect to the benchmark. Accuracy is defined as the success rate, with the left column of Table 6.4 listing the number of failure scenarios v where the LS_v obtained by each method matches the benchmark M5 result, while the right column indicates the total number of generated failure scenarios, which is 50.

The results indicate that M1–M3 fail to guarantee accuracy across all test cases, whereas M6–M8 produce identical evaluation results to the benchmark algorithm M5 in all three cases. Although M4 performs well in Cases A and B, it fails to obtain a valid solution within the given time limit for Case C. The primary issue with M1 is its inability to ensure solution feasibility; enforcing the non-convex constraint (6.48) makes it difficult for M1 to detect potential load shedding. The feasibility recovery strategies of local optimization methods M2 and M3 are highly dependent on the initial solution settings [22, 23]. After extreme events, the feasible region of the IEGS may shrink significantly. While the global optimal solution may avoid load shedding, M2 and M3 can still converge to suboptimal operating points that result in unavoidable load shedding. In contrast, M4–M8 leverage the spatial branch-and-bound algorithm in global optimization, making them more effective in handling

Table 6.3 Case study and evaluation results

	Term	Case A	Case B	Case C
Model	Number of binary variables	600	1320	2004
	Number of continuous variables	2310	7170	11,970
	Number of linear constraints	2694	8255	13,765
	Number of quadratic constraints	252	906	1470
	Number of non-convex constraints	348	1146	1938
Metrics	EENS (MWh)	175.68	70.31	24,079

Table 6.4 Comparative analysis of resilience assessment methods

	Case A		Case B		Case C	
	Metrics (%)	Accuracy	Metrics (%)	Accuracy	Metrics (%)	Accuracy
M1	↓9.36	47/50	↓8.54	49/50	↓5.34	48/50
M2	↓0.12	50/50	↑24.76	47/50	↑6.12	48/50
M3	↑1.15	49/50	↑23.07	48/50	↑31.01	45/50
M4	↓0.08	50/50	↓0.04	50/50	–	–
M5	Benchmark	50/50	Benchmark	50/50	Benchmark	50/50
M6	↑0.03	50/50	↑0.07	50/50	↑0.08	50/50
M7	↑0.02	50/50	↓0.02	50/50	↑0.06	50/50
M8	↑0.02	50/50	↓0.01	50/50	↑0.05	50/50

1 Symbol "–" means the method failed within the time limit
2 Symbol "↓ (↑) " denotes that the corresponding value is lower (higher) than the benchmark value, respectively

such challenges. It is important to note that the core objective of global optimization is to ensure solution quality (optimality) rather than strictly achieving the global optimum. In practical resilience evaluation applications, a feasible solution that is sufficiently close to the optimal one is generally adequate.

6.4.2 Effectiveness Assessment

This subsection proposes two improvements aimed at addressing non-convexity and redundancy issues in resilience analysis: a spatial branching strategy for SOC-specific and a combinatorial cutting-plane generation strategy. Their effectiveness is evaluated and validated through the following methods:

Performance in Managing Non-Convexity: To assess the performance of M1–M8 in handling non-convex constraints, the constraint violation values δ for all quadratic equality constraints in Eq. (6.48) were computed using Eq. (6.49) for each system and failure scenario, and the maximum violation was recorded. These violation data are presented in boxplots in Fig. 6.9, where the bottom and top of each box represent the first quartile and third quartile, respectively, and the orange line inside the box denotes the median. The whiskers extend 1.5 times the interquartile range from the box, with outliers marked as data points beyond the whisker ends. The blue dashed line represents the threshold for constraint violations. As shown in Fig. 6.9, algorithms incorporating the spatial branching strategy for SOC (adopted in M7 and M8) ensure that all violation values remain below the threshold, guaranteeing solution feasibility. Except for M4, which exhibits anomalies in Case C, the constraint violations for M2–M6 remain stable. In contrast, M1 performs the worst, with constraint violations exceeding the given threshold across all scenarios.

Efficiency in Redundancy Reduction: The proposed combinatorial cutting-plane generation strategy (adopted exclusively by M8) theoretically does not affect solution accuracy. Since the search tree generated by M8 can be regarded as a superset of

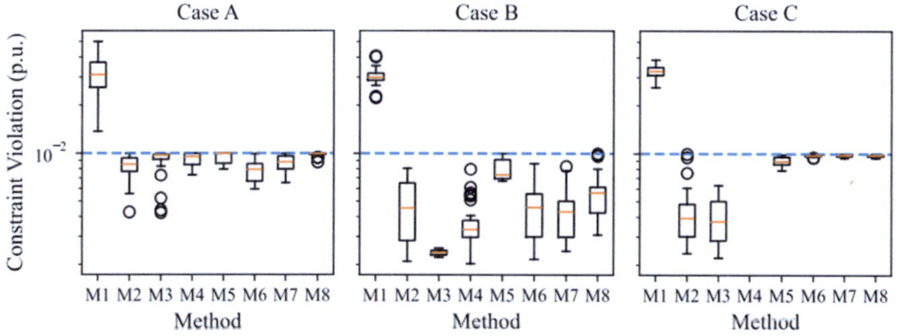

Fig. 6.9 Comparison of maximum non-convex constraint violations

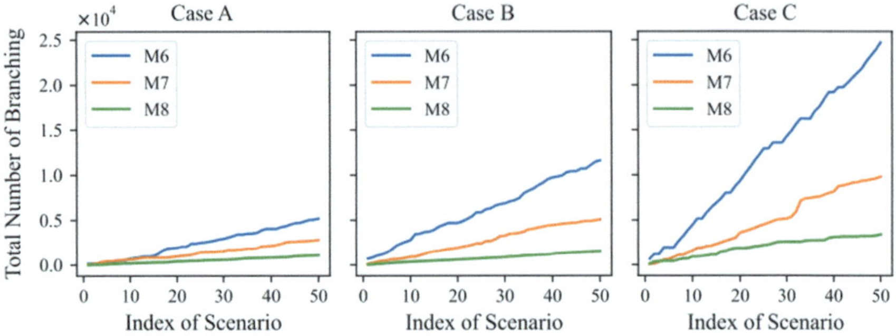

Fig. 6.10 Comparison of spatial branching counts for M6, M7, and M8 across three scenarios

all search trees generated by M7, its optimization process simplifies the problem losslessly by eliminating redundant nodes. The consistency of M8 with M7 is further validated in Table 6.4. Based on this accuracy verification, the efficiency of the proposed strategy is further assessed by statistically analyzing the total number of spatial branching operations. The combinatorial cutting-plane generation strategy significantly reduces the number of spatial branching operations by preventing the regeneration of previously generated branches. To provide an intuitive visualization of this advantage, the total number of spatial branching operations executed in each scenario for M6, M7, and M8 is recorded and plotted in Fig. 6.10. In this figure, each point (x, y) represents the total number of spatial branching operations performed after analyzing the first x failure scenarios. Figure 6.10 demonstrates that M8 requires significantly fewer spatial branching operations than other methods across all three test cases, with its advantage becoming more pronounced as system size increases.

6.4.3 Computational Efficiency

To verify the superiority of the proposed method in terms of computational efficiency, this section compares the CPU time required for resilience assessment across M1–M8 under three test cases. Table 6.5 presents the total CPU time consumed by each method, while the distribution of computation time for each test case and failure scenario is illustrated using violin plots in Fig. 6.11. The evaluation results indicate that M1 achieves the shortest runtime but at the cost of solution accuracy. The local optimization methods M2 and M3 demonstrate slightly better efficiency than the advanced solvers employed in M4 and M5, but they do not exhibit the significant advantages described in Refs. [22, 23]. This discrepancy arises primarily because the feasible region of the OEGF model proposed in this chapter's fault recovery stage is substantially smaller than that of the normal operation model in Refs. [22, 23], leading to a significant reduction in the number of near-optimal solutions and consequently increasing computation time. Additionally, M2 and M3 may converge

Table 6.5 Total CPU time comparison (seconds)

	M1	M2	M3	M4	M5	M6	M7	M8
Case A	19.3	172.9	79.4	898.3	354.8	305.1	223.4	92.9
Case B	39.0	687.5	751.6	10,659.8	2504.5	2490.4	719.6	358.5
Case C	105.6	2692.8	3944.9	–	5068.5	18,337.7	6418.1	1050.2

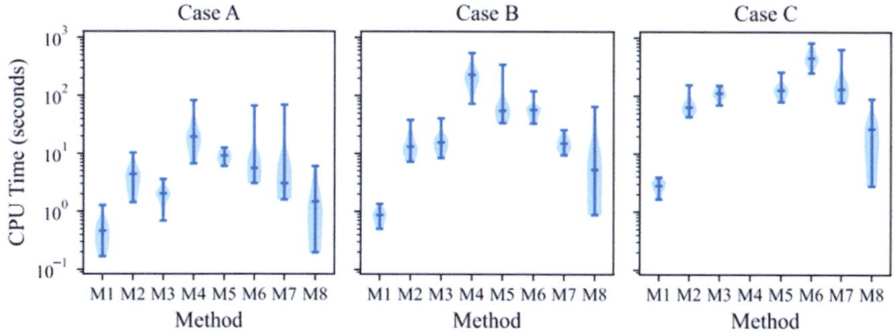

Fig. 6.11 Violin plot of CPU time distribution for M1–M8 across different cases and failure scenarios

to suboptimal solutions that deviate considerably from the global optimum, thereby compromising the accuracy of IEGS resilience assessment. In contrast, to ensure both feasibility and optimality, M4–M8 inevitably require more computational time. While directly employing existing solvers (such as M4 and M5) is straightforward, it results in excessive computational complexity. Experimental results demonstrate that, when the branching strategy in the spatial B&B algorithm is well-designed (as in M6 and M7), computational efficiency can be significantly improved. In summary, the proposed method M8 exhibits the best computational efficiency, outperforming M4–M8 in terms of CPU time.

6.4.4 Accuracy and Controllability

In practical engineering applications, the accuracy requirements for resilience assessment may vary depending on the specific application scenario. Therefore, this section conducts an in-depth exploration of the controllability of the proposed method's accuracy from both modeling and algorithmic perspectives.

First, this section compares the transient gas flow model adopted in this study with the conventional steady-state model. The steady-state model can be obtained by setting $\partial \psi / \partial t = 0$ and $\partial f / \partial t = 0$ in Eqs. (6.18)–(6.19), respectively. The resilience metrics obtained by the above two models are shown in Table 6.6. A comparison of resilience indicators under these two models reveals that using the steady-state model may lead to misleading conclusions. This is because NGS often struggle to maintain steady-state conditions, particularly when EPS failures occur, requiring GFUs to rapidly increase output to prevent load shedding, which in turn causes a surge in gas demand. Given that the gas velocity in pipelines is approximately 40 km/h [17], the additional gas energy flow from supply facilities may take several hours to reach GFUs. The conventional steady-state model fails to capture this transient process, thereby underestimating the risk of load shedding.

Furthermore, this section examines the precision adjustability of the proposed method in controlling the non-convexity threshold. Algorithm 1, Line 19 provides an interface for modifying this parameter to verify whether the solution of $\langle \mathbf{P}_{\mathbf{rel}} \rangle$ is applicable to problem $\langle \mathbf{P} \rangle$. Although a smaller threshold enhances solution accuracy, it significantly increases computational time. In the previous analysis, the non-convexity threshold was set to a default value of 1%. To further validate the controllability of accuracy, additional tests were conducted for M8 under different threshold settings (2.50, 1.50, 0.75, and 0.50%). Figure 6.12 illustrates the runtime of M8 under various threshold settings, where orange dots represent the default 1% threshold, and blue dots correspond to the newly adjusted thresholds. The results indicate that as the threshold decreases, the runtime exhibits an overall increasing trend. Given the trade-off between accuracy and efficiency, this controllability is crucial for practical applications. Additionally, cross marks in Fig. 6.12 indicate the runtime of M4 and M5. A comparison reveals that when the threshold is set to 0.50%, the efficiency of the proposed method M8 is comparable to that of M4 and M5, further verifying its superiority.

Table 6.6 Resilience metrics across different modeling approaches

	Non-convex	Gas flow	Case A	Case B	Case C
Proposed	Yes	Transient	175.68	70.31	24.79
[27]	Yes	Steady-state	167.94	65.89	21.64

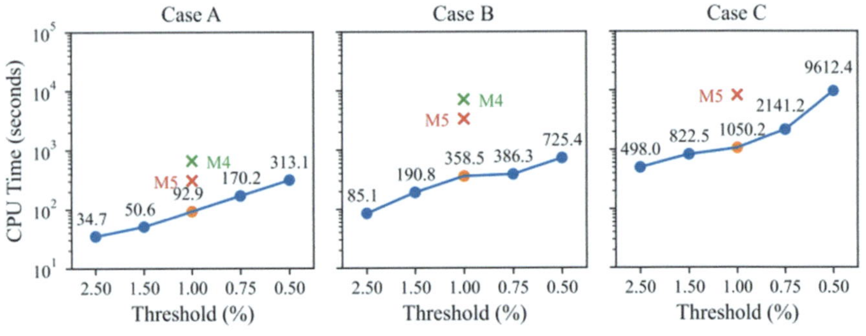

Fig. 6.12 CPU time analysis of M8 under varying non-convexity thresholds (including M4 and M5 for benchmarking)

6.5 Discussion

6.5.1 The SOC-Specific Spatial Branching Strategy

The proposed improved spatial branching strategy is designed to address non-convexity issues of the form $z = \sqrt{x^2 + y^2}$ (where $(x, y, z) \in \mathbb{R}^3$). In this chapter, the nonlinear terms in the power flow equations and transient gas flow equations can be appropriately restructured to fit this form. If non-convex constraints in other models can also be reformulated in this manner, the proposed method is similarly applicable. For example, in the alternating current optimal power flow (AC OPF) model, the branch power flow equation is $v_i l_{ij} = s_{ij}^2$, where v_i represents the square of the voltage magnitude, l_{ij} represents the square of the current, and s_{ij} represents the apparent power flow. By substituting z with $v_i + l_{ij}$, x with $v_i - l_{ij}$, and y with $2s_{ij}$, the branch power flow equation can be rewritten as $v_i + l_{ij} = \left(v_i - l_{ij}\right)^2 + \left(2s_{ij}\right)^2$. Therefore, the proposed SOC-specific spatial branching strategy can be directly applied to this model. However, the limitation of this strategy lies in its applicability only to non-convex constraint problems in three-dimensional space. Future research will focus on exploring the extension of this strategy to solve problems in higher-dimensional spaces.

6.5.2 The Combinatorial Cut Derivation Strategy

The proposed improved combination cutting-plane generation strategy allows for the non-iterative computation of resilience indicators, effectively avoiding redundant spatial branching operations. However, this strategy can only be enabled after the search tree has been generated, making it unsuitable for the initial phase of the spatial B&B algorithm (which is why heuristic methods are disabled in the solver settings of this section). Given the advantages of heuristic algorithms in enhancing solution efficiency, future work will focus on integrating heuristic algorithms into the proposed method via the CPLEX *heuristic* callback function. Additionally, the selection of resilience indicators is another important direction for future research. For simplicity, this chapter uses the load-shedding indicator R_{EENS}, but to achieve a more comprehensive resilience assessment for IEGSs, future research could incorporate other resilience indicators, such as critical load supply or the resilience triangle [7, 12].

6.6 Conclusions

This chapter presents a creative non-convex and non-iterative resilience assessment method for IEGS. The method effectively addresses the non-convexity in the dynamic characteristics of the gas network and the redundancy in scenario enumeration within the B&C algorithm framework through two key improvements. The first improvement is a SOC-specific spatial branching strategy, and the second improvement is a combined cutting-plane generation strategy. These two improvements are embedded into the CPLEX solver and dynamically invoked through *callback* functions.

The effectiveness of the proposed algorithm is validated through experimental results on three IEGSs of different scales, with resilience assessment outcomes consistent with the benchmark methods. Additionally, comparison with seven existing solution methods further demonstrates that the proposed approach offers significant advantages in both solution accuracy and computational efficiency.

References

1. Martínez Ceseña, E. A., Loukarakis, E., Good, N., Ochoa, L. F., Mancarella, P., & Papaefthymiou, G. (2020). Integrated electricity-heat-gas systems: Techno-economic modeling, optimization, and application to multi-energy districts. *Proceedings of the IEEE, 108*(9), 1392–1410.
2. Bao, M., Ding, Y., Singh, C., & Zhao, J. (2019). A multi-state model for reliability assessment of integrated gas and power systems utilizing universal generating function techniques. *IEEE Transactions on Smart Grid, 10*(6), 6271–6283.
3. Huang, W., Zhang, X., Li, K., Wang, W., & Sun, H. (2022). Resilience-oriented planning of urban multi-energy systems with generalized energy storage sources. *IEEE Transactions on Power Systems, 37*(4), 2906–2918.
4. Amirioun, M. H., Aminifar, F., & Shahidehpour, M. (2019). Resilience-promoting proactive scheduling against hurricanes in multiple energy carrier microgrids. *IEEE Transactions on Power Systems, 34*(3), 2160–2168.
5. He, C., Dai, C., Wu, L., & Wang, C. (2018). Robust network hardening strategy for enhancing resilience of integrated electricity and natural gas distribution systems against natural disasters. *IEEE Transactions on Power Systems, 33*(5), 5787–5798.
6. Stanković, A. M., Tomsovic, K. L., De Caro, F., Parisio, A., & Liu, C.-C. (2023). Methods for analysis and quantification of power system resilience. *IEEE Transactions on Power Systems, 38*(5), 4774–4787.
7. Dehghani, A., Sedighizadeh, M., & Haghjoo, F. (2021). An overview of the assessment metrics of the concept of resilience in electrical grids. *International Transactions on Electrical Energy Systems, 31*(12), e13159.
8. Li, S., Ding, T., Jia, W., Zhang, J., & Liu, H. (2022). A machine learning-based vulnerability analysis for cascading failures of integrated power-gas systems. *IEEE Transactions on Power Systems, 37*(3), 2259–2270.
9. Huang, B., & Wang, J. (2022). Applications of physics-informed neural networks in power systems—A review. *IEEE Transactions on Power Systems, early access.* Advance Online Publication.
10. Buldyrev, S. V., Parshani, R., Paul, G., Stanley, H. E., & Havlin, S. (2010). Catastrophic cascade of failures in interdependent networks. *Nature, 464*(7291), 1025–1028.

11. Fang, J., Su, C., Chen, Z., He, H., & Hu, S. (2018). Power system structural vulnerability assessment based on an improved maximum flow approach. *IEEE Transactions on Smart Grid, 9*(2), 777–785.
12. Jasiūnas, J., Lund, P. D., & Mikkola, J. (2021). Energy system resilience—A review. *Renewable and Sustainable Energy Reviews, 150*, 111476.
13. Li, G., Yan, K., Zhang, R., & Wang, Z. (2022). Resilience-oriented distributed load restoration method for integrated power distribution and natural gas systems. *IEEE Transactions on Sustainable Energy, 13*(1), 341–352.
14. Zhang, H., Wang, P., Yao, S., Zhang, J., & Zhang, Y. (2020). Resilience assessment of interdependent energy systems under hurricanes. *IEEE Transactions on Power Systems, 35*(5), 3682–3694.
15. Zhao, T., Zhang, H., Liu, X., Wu, W., & Li, Z. (2021). Resilient unit commitment for day-ahead market considering probabilistic impacts of hurricanes. *IEEE Transactions on Power Systems, 36*(2), 1082–1094.
16. Wang, J., Zuo, W., Rhode-Barbarigos, L., Hong, T., & Yan, D. (2019). Literature review on modeling and simulation of energy infrastructures from a resilience perspective. *Reliability Engineering & System Safety, 183*, 360–373.
17. Bayani, R., & Manshadi, S. D. (2022). Natural gas short-term operation problem with dynamics: A rank minimization approach. *IEEE Transactions on Smart Grid, 13*(4), 2761–2773.
18. Liu, W., Li, P., Yang, W., Zhang, S., & Zhao, Y. (2019). Optimal energy flow for integrated energy systems considering gas transients. *IEEE Transactions on Power Systems, 34*(6), 5076–5079.
19. Sun, G., Chen, S., Wei, Z., Liu, F., & Wang, Y. (2020). Corrective security-constrained optimal power and gas flow with binding contingency identification. *IEEE Transactions on Sustainable Energy, 11*(2), 1033–1042.
20. Chen, S., Conejo, A. J., Sioshansi, R., & Wang, X. (2020). Operational equilibria of electric and natural gas systems with limited information interchange. *IEEE Transactions on Power Systems, 35*(1), 662–671.
21. Wang, C., Ju, P., Wu, F., & Zhou, Z. (2022). Long-term voltage stability-constrained coordinated scheduling for gas and power grids with uncertain wind power. *IEEE Transactions on Sustainable Energy, 13*(1), 363–377.
22. Yang, L., Xu, Y., Sun, H., Zhou, Q., & He, C. (2020). Two-stage convexification-based optimal electricity-gas flow. *IEEE Transactions on Smart Grid, 11*(2), 1465–1475.
23. Mhanna, S., Saedi, I., & Mancarella, P. (2022). Iterative LP-based methods for the multiperiod optimal electricity and gas flow problem. *IEEE Transactions on Power Systems, 37*(1), 153–166.
24. Nagarajan, H., Lu, M., Wang, S., & Linderoth, J. T. (2019). An adaptive, multivariate partitioning algorithm for global optimization of non-convex programs. *Journal of Global Optimization, 74*(4), 639–675.
25. Gurobi Optimization, LLC. (2023). *Gurobi optimizer reference manual* [Online]. Retrieved from https://www.gurobi.com
26. Vigerske, S., & Gleixner, A. (2018). SCIP: Global optimization of mixed-integer nonlinear programs in a branch-and-cut framework. *Optimization Methods and Software, 33*(3), 563–593.
27. Liu, P., Wu, Z., Gu, W., Wang, Y., & Wang, J. (2022). An improved spatial branch-and-bound algorithm for non-convex optimal electricity-gas flow. *IEEE Transactions on Power Systems, 37*(2), 1326–1339.
28. Liu, P., Wu, Z., Gu, W., Zhang, H., & Wang, Y. (2023). Spatial branching for conic non-convexities in optimal electricity-gas flow. *IEEE Transactions on Power Systems, 38*(1), 972–975.
29. Shen, Y., Gu, C., Yang, X., Zhang, B., & Liu, Z. (2021). Impact analysis of seismic events on integrated electricity and natural gas systems. *IEEE Transactions on Power Delivery, 36*(4), 1923–1931.
30. Shi, Q., Li, F., Kuruganti, T., Liu, H., & Tang, Y. (2021). Resilience-oriented DG siting and sizing considering stochastic scenario reduction. *IEEE Transactions on Power Systems, 36*(4), 3715–3727.

31. Yang, Y., Guan, X., & Zhai, Q. (2017). Fast grid security assessment with N-k contingencies. *IEEE Transactions on Power Systems, 32*(3), 2193–2203.
32. Bragin, M. A., Luh, P. B., Yan, J. H., Wang, C., & Sun, Y. (2014). Surrogate Lagrangian relaxation and branch-and-cut for unit commitment with combined cycle units. *2014 IEEE PES General Meeting—Conference & Exposition*, pp. 1–5.
33. Michels, A. S., Lopes, T. C., Sikora, C. G. S., & Morabito, R. (2019). A Benders' decomposition algorithm with combinatorial cuts for the multi-manned assembly line balancing problem. *European Journal of Operational Research, 278*(3), 796–808.
34. Fang, J., Zeng, Q., Ai, X., Li, G., & Chen, Z. (2018). Dynamic optimal energy flow in the integrated natural gas and electrical power systems. *IEEE Transactions on Sustainable Energy, 9*(1), 188–198.
35. Liu, P., Wu, Z., Gu, W., Zhang, H., & Wang, Y. (2022). A non-convex and non-iterative approach for fast vulnerability analysis of smart grids. *IEEE PES ISGT-Europe, 2022*, 1–5.
36. Liu, P. Data [Online]. Retrieved from https://pan.seu.edu.cn:443/link/FB38BE747C1D278108A954CC56D54D02

Chapter 7
Resilient-Oriented IES Planning Considering Risk and Economic Trade-Offs

Abstract To address the disaster uncertainty of extreme events, this chapter proposes a resilience planning method for integrated energy systems (IES) that balances economic efficiency and risk. First, a hierarchical fault scenario generation strategy is developed by integrating stratified sampling, importance sampling, stochastic fault simulation, and average loss filtering, forming a set of representative fault scenarios covering multiple crisis levels and their probabilistic characteristics. Second, a two-stage resilience planning framework is established, incorporating pre-disaster protection and post-disaster multi-energy coordinated emergency scheduling. Conditional value-at-risk (CVaR) is introduced to formulate a tail risk analytical expression for low-probability, high-impact events, ensuring controllability between economic efficiency and risk balance. To enhance computational efficiency for large-scale models, a tailored Benders decomposition (TBD) algorithm accelerated by primal-dual cuts is proposed. The effectiveness of the proposed model and algorithm is validated through case studies. This work provides a systematic solution for disaster resilience planning in integrated energy systems, ensuring a balanced trade-off between cost and risk control.

Keywords Seismic events · Integrated energy system · Resilience enhancement · Risk-economic trade-offs

Nomenclature

IES	Integrated energy systems
PDS	Power distribution system
NGS	Natural gas system
DHS	District heating system
CVaR	Conditional value-at-risk
VaR	Value-at-Rrisk
C&CG	Column-and-Constraint Generation
PHA	Progressive hedging algorithm

Z. Wu et al., *Integrated Energy System Resilience*,
https://doi.org/10.1007/978-981-95-2363-4_7

HILP	High-impact low-probability
TBD	Tailored Benders decomposition
BD	Benders decomposition
R-SRIP	Risk-based seismic-resilient IES planning
MEG	Mobile emergency generator
GT	Gas turbine
CHP	Combined heat and power
HP	Heat pumps
HB	Electric boiler
UB	Upper bound
LB	Lower bound
MINLP	Mixed-integer nonlinear programming
MILP	Mixed-integer linear programming
MP	Master problem
EMP	Enhanced master problem
SP	Subproblem
SP-r	Relaxed subproblem
SP-e	Exact subproblem
β	Index for risk aversion
s	Index for fault scenario
t	Index for time interval
$i/j/n$	Index of power, gas and heat node
m	Index of MEG type
k	Index of CHP boundary operation point
ki/ij	Index of power line, gas pipeline and heat pipeline
ς	Index for uncertainty
$\mathcal{N}_{\mathrm{P/G/H}}$	Set of node in PDS/ NGS/ DHS
$\mathcal{N}_{\mathrm{Sub}}^{\mathrm{P}}$	Set of substatio in PDS
$\mathcal{N}_{\mathrm{GT}}^{\mathrm{P}}$	Set of GT node in PDS
$\mathcal{N}_{\mathrm{CHP}}^{\mathrm{P}}$	Set of CHP unit node in PDS
$\mathcal{N}_{\mathrm{EB}}^{\mathrm{P}}$	Set of electric boiler node in PDS
$\mathcal{N}_{\mathrm{MEG}}^{\mathrm{P}}$	Set of MEG node in PDS
$\mathcal{N}_{\mathrm{Sta/CHP/GT}}^{\mathrm{G}}$	Set of gas station node in NGS
$\mathcal{N}_{\mathrm{CHP}}^{\mathrm{G}}$	Set of CHP node in NGS
$\mathcal{N}_{\mathrm{GT}}^{\mathrm{G}}$	Set of GT node in NGS
$\mathcal{N}_{\mathrm{CHP}}^{\mathrm{H}}$	Set of CHP node in DHS
$\mathcal{N}_{\mathrm{EB}}^{\mathrm{H}}$	Set of EB node in DHS
$\mathcal{L}_{\mathrm{P/G/H}}$	Set of power line/gas pipeline/heat pipeline in PDS/NGS/DHS
$\mathcal{L}_{\mathrm{Loop}\ /\ \mathrm{Loop}\ ,n}$	Set of loop/power line in loop n in PDS
$\mathcal{L}_{\mathrm{G}}^{\mathrm{Com}}$	Set of gas pipeline with compressors
$\mathcal{S}$	Set of fault scenario
$\mathcal{T}$	Set of time interval
$\mathcal{T}_{i,m}^{\mathrm{Travel}}$	Set of dispatch time periods of the MEGs
$\mathcal{M}$	Set of MEG type

$\mathcal{K}$	Set of CHP boundary operation point
A_{ij}^{H}	Cross-sectional area of heat pipeline
L_{ij}^{H}	Cross-sectional area of length
$\lambda_{ij}^{\mathrm{H}}$	Cross-sectional area of heat transfer coefficient
ρ_w	Cross-sectional area of water density
f_{Inv}	IES investment cost
f_{Loss}	Expected value of IES emergency dispatch cost
C_{ij}^{Har}	Hardening costs for power line/gas pipeline/heat pipeline
C_{Cut}	Penalty cost for load curtailment
$C_{\mathrm{Rent},\ m}$	Renting cost coefficients of type m MEG
$C_{\mathrm{MEG},\ m}$	Operating cost coefficients of type m MEG
$E_{m,\ \max}^{\mathrm{MEG}}$	Maximum active power capacity of type m MEG
$G_{i,\ t}^{\mathrm{Load}}/H_{i,\ t}^{\mathrm{Load}}$	Gas/heat load
M	A large constant
$N_{\max}^{\mathrm{Har,\ P/G/H}}$	Maximum budget of power line/gas pipeline/heat pipeline to be hardened
$N_{m,\ \max}^{\mathrm{MEG}}$	Maximum number of type m MEG can be rented
$N_{i,\ \max}^{\mathrm{MEG}}$	MEG allowed to be connected to PDS node i
$P_{i,\ t}^{\mathrm{Load}}/Q_{i,\ t}^{\mathrm{Load}}$	Active/reactive power load
$P_{i,\ s,\ t}^{\mathrm{CHP},\ k}/H_{i,\ s,\ t}^{\mathrm{CHP},\ k}$	Boundary power/heat operation point of CHP
r_{ij}	Resistance of power line
x_{ij}	Reactance of power line
ϖ	Anticipated occurrence of seismic events during planning periods
ρ_s	Probability of fault scenario
γ_{g2e}	Gas to electricity constant
φ_{GT}	Power factor of GT
φ_{MEG}	Power factor of MEG
φ_{Com}	Inflow compressibility ratio of compressor
Γ_{Com}	Pressure boost ratio of compressor
Δt	Duration of time interval
ϕ_{ij}^{G}	Weymouth coefficient of gas pipeline
τ_t^{amb}	Ambient temperature of heat pipeline
η_{GT}	Efficiency of GT
η_{EB}	Efficiency of EB
$\eta_{\mathrm{CHP},\ p}$	Gas to power efficiency of CHP
$\eta_{\mathrm{CHP},\ h}$	Gas to heat efficiency of CHP
$\omega_i^{\mathrm{P}}/\omega_i^{\mathrm{G}}/\omega_i^{\mathrm{H}}$	Weight of power/gas/heat load in PDS/NGS/NGS
$H_{i,s,t}^{\mathrm{CHP}}$	Heat power of CHP
$H_{i,\ s,\ t}^{\mathrm{EB}}$	Heat power of EB
$H_{ij,s,t}^{\mathrm{Pipe,in}}$	Inlet heat power of heat pipeline
$H_{ij,\ s,\ t}^{\mathrm{Pipe,\ out}}$	Outlet heat power of heat pipeline
$H_{ij,\ s,\ t}^{\mathrm{Pipe,\ loss}}$	Heat loss of heat pipeline

$H_{i,s,t}^{\mathrm{Cut}}$	Curtailed heat load
$G_{i,s,t}^{\mathrm{Sta}}$	Gas injection at gas station
$G_{i,s,t}^{\mathrm{CHP}}$	Gas consumption of CHP
$G_{i,s,t}^{\mathrm{GT}}$	Gas consumption of GT
$G_{ij,s,t}^{\mathrm{Com,\,in}}$	Gas flow inlet of compressor
$G_{ij,s,t}^{\mathrm{Com,\,out}}$	Gas flow outlet of compressor
$G_{i,s,t}^{\mathrm{Cut}}$	Curtailed gas load
$G_{ij,s,t}^{\mathrm{Pipe}}$	Gas flow of pipeline
$P_{i,s,t}^{\mathrm{Sub}}/Q_{i,s,t}^{\mathrm{Sub}}$	Active/reactive power of substation
$P_{i,s,t}^{\mathrm{GT}}/Q_{i,s,t}^{\mathrm{GT}}$	Active/reactive power of GT
$P_{i,s,t}^{\mathrm{CHP}}$	Active power of CHP
$P_{i,s,t}^{\mathrm{EB}}$	Active power of EB
$P_{i,s,t}^{\mathrm{MEG}}/Q_{i,s,t}^{\mathrm{MEG}}$	Active/reactive power of total MEG injecting to PDS node i
$P_{i,m,s,t}^{\mathrm{MEG}}$	Active power of type m MEG injecting to PDS node i
$P_{ij,s,t}^{\mathrm{L}}/Q_{ij,s,t}^{\mathrm{L}}$	Active/reactive power flow on power line
$P_{i,s,t}^{\mathrm{Cut}}/Q_{i,s,t}^{\mathrm{Cut}}$	Active/reactive power load curtailment
$P_{i,\max}^{\mathrm{GT}}$	Maximum output active power of GT
$R_{i,\max}^{\mathrm{GT}}$	Maximum climbing power of GT
$R_{i,\max}^{\mathrm{EB}}$	Maximum creep power of EB
$R_{i,\max}^{\mathrm{CHP}}$	Maximum creep power of CHP unit
$P_{i,\max}^{\mathrm{Sub}}/Q_{i,\max}^{\mathrm{Sub}}$	Upper limit value of active/reactive power of substation
$V_{i,s,t}$	Squared node voltage
$V_{i,\min}/V_{i,\max}$	Lower/upper limit values of the node voltage
$h_{ij}^{\mathrm{P/G/H}}$	Binary, harden decision for power line/gas pipeline/heat pipeline
$\xi_{ij,s,t}^{\mathrm{P/G/H}}$	Damage status of power line/gas pipeline/heat pipeline
$\upsilon_{ij,s,t}^{\mathrm{P/G/H}}$	Binary, available status of power line/gas pipeline/heat pipeline
ι_{ij}	Binary, pre-event reconfiguration status of power line
$\alpha_{i,s,t}^{k}$	Operating level of CHP regarding boundary point k
$\mu_{i,m,s,t}$	Integer, the number of type m MEG connecting to PDS node i
$\tau_{ij,st}^{\mathrm{S,\,in}}/\tau_{ij,s,t}^{\mathrm{R,\,in}}$	Inflow temperature of heat supply/return pipeline
$\pi_{i,s,t}$	Pressure of gas node
$\pi_{ij,s,t}^{\mathrm{Com,\,in}}/\pi_{ij,s,t}^{\mathrm{Com,\,out}}$	Inlet/outlet pressure of gas compressor

7.1 Introduction

Resilience planning enhances system preparedness through preventive deployment and targeted reinforcement measures, making it a key approach for addressing multiple extreme disasters and climate change challenges.

Resilience planning measures against extreme natural disasters mainly include three categories: reinforcement of outdoor infrastructure such as power lines, natural gas pipelines, and thermal pipelines; allocation of emergency energy supply resources (e.g., backup diesel generators and reserved fuels for cogeneration units); and deployment of mobile energy vehicles (e.g., vehicle-mounted generators, mobile energy storage devices, and LNG storage trailers). In recent years, with the increasing demand for resilience enhancement, resilience-oriented IES planning has become a research hotspot in academia. As previously discussed, the multi-energy complementarity of IES offers significant potential advantages in enhancing system resilience against extreme disasters. However, without effective coordination, interdependence among multiple energy sources may increase system vulnerability. Existing studies indicate that implementing preventive resilience planning measures and leveraging multi-energy flow complementarity in post-disaster emergency dispatching can significantly reduce the impact of extreme natural disasters on IES [1].

Among various types of natural disasters, earthquakes are considered one of the most unpredictable and can severely damage IES infrastructure [2]. For instance, the 7.8-magnitude earthquake that struck Turkey in 2023 severely damaged critical infrastructure, leaving millions without electricity, natural gas, and heating for weeks, resulting in economic losses exceeding $34 billion. Due to the unpredictability and destructive nature of earthquakes, seismic resilience planning has become a key research focus in IES resilience planning. Ref. [3] proposed a grid restructuring strategy for electricity-gas coupled IES and simulated the energy supply recovery process after an earthquake. Ref. [4] introduced a comprehensive assessment framework to evaluate the earthquake resilience of electricity-gas coupled IES and identify system vulnerabilities. Ref. [5] developed a three-layer robust IES resilience planning model based on the "Defense-Attack-Defense" strategy to mitigate the impact of the most severe earthquake scenarios and adopted the Column-and-Constraint Generation (C&CG) algorithm for efficient model solving. Ref. [6] proposed a data-driven IES resilience planning approach that minimizes energy supply interruption losses under seismic disasters.

Although these studies have made progress, several challenges remain in the field. One of the core challenges in resilience planning is the high uncertainty of extreme disaster events. Given the current technological conditions, predicting earthquake magnitudes and occurrence probabilities with precision remains difficult. In existing literature, two primary methods—stochastic and robust approaches—are commonly used to model disaster uncertainty. The stochastic method generates a set of representative failure scenarios through clustering analysis to characterize earthquake-induced disaster uncertainty. However, traditional scenario clustering methods often overlook high-impact low-probability (HILP) extreme failure scenarios. The robust method typically assumes the worst-case of damaged IES components under a given disaster intensity and describes IES failure states based on the N-k criterion. The robust approach primarily focuses on worst-case extreme failure scenarios while inadequately considering more frequent medium- and small-scale failures [7, 8].

In resilience planning, decision-makers often face the challenge of balancing economic feasibility with resilience enhancement. The core goal of resilience planning is to minimize the overall impact of uncertain disaster scenarios and reduce tail risk under low-probability extreme events. Although increasing the allocation of disaster prevention resources can improve the emergency response capability of IES and enhance system resilience, resilience improvement generally exhibits significant diminishing marginal returns. Therefore, resilience planning must seek a balance between planning costs and resilience enhancement. In existing literature, stochastic methods typically optimize for minimizing average energy supply interruption losses across multiple failure scenarios. However, this approach often neglects the significant risks posed by low-probability extreme failure scenarios. In contrast, robust methods focus more on risk aversion, prioritizing energy supply interruption losses in extreme failure scenarios, which may lead to overly conservative planning schemes [9, 10].

At the mathematical level, IES resilience planning is typically a large-scale mixed-integer programming problem, posing significant computational challenges [11]. The commonly used approach is to directly solve the problem using commercial solvers. However, when handling large-scale problems, solvers often suffer from dimensionality issues. Efficiently solving resilience planning problems is crucial for practical engineering applications, as it enables decision-makers to conduct in-depth planning analyses and make effective decisions. To address computational challenges, various accelerated algorithms have been proposed in existing research. The progressive hedging algorithm (PHA) is an accelerated method based on Lagrangian dual decomposition. It decomposes the original large-scale model into multiple scenario subproblems, solving smaller-scale problems iteratively to reduce computational complexity [12]. However, as a heuristic method, the convergence speed of the PHA algorithm is highly dependent on parameter settings and does not guarantee global convergence [13].

Benders decomposition is another commonly used acceleration algorithm for solving large-scale planning problems, particularly those involving continuous recourse variables. Classical Benders decomposition gradually approximates the optimal solution by generating dual cutting planes [14, 15]. However, in IES resilience planning, the operational stage typically involves integer variables (e.g., deployment and scheduling of mobile emergency energy vehicles), making it difficult to generate effective Benders cuts during iterations. Some studies have proposed improved Benders decomposition methods that relax the operational subproblem to generate valid Benders cuts [16, 17]. However, the relaxed subproblem may weaken the feedback between the master and subproblems, leading to slow convergence [18, 19].

In summary, IES resilience planning methods remain in the exploratory stage, with several critical issues needing resolution. First, in disaster uncertainty modeling, existing methods fail to fully characterize the destructive impact and occurrence probabilities of different disaster severity levels, resulting in resilience planning schemes that perform poorly in both effectiveness and economic feasibility when facing uncertain disasters. Second, current IES resilience planning approaches cannot effectively resolve the inherent conflict between risk reduction and cost increase. Determining how to deploy resilience planning schemes from a risk-economic coordination perspective remains a key research challenge. Lastly, resilience planning models constructed for multiple failure scenarios face significant computational challenges, necessitating the development of more efficient solving algorithms.

7.2 Conceptual Framework of the Resilience Planning Model

Extreme natural disasters are characterized by their sudden onset, low probability, and high destructiveness, posing significant challenges to IES resilience planning [20, 21]. The key focus of this chapter is to address the trade-off between extreme risk reduction and increasing marginal costs in resilience planning. This chapter focuses on distribution-level IES, with its structure illustrated in Fig. 7.1.

Among various types of extreme natural disasters, earthquakes are a primary concern in IES resilience planning due to their unpredictability and destructive power [22–25]. The economic-risk trade-off in seismic resilience planning for IES presents a critical challenge. Therefore, this chapter selects earthquakes as the research subject. As shown in Fig. 7.1, when an earthquake occurs, seismic waves cause severe shaking of IES infrastructure, leading to structural damage to key components such as power lines and pipelines, which may result in large-scale energy supply disruptions [26–30]. Given the wide spatial coverage of the system, this study considers the uncertainty of seismic disasters across different spatial regions. It should be noted that this setting is also applicable to transmission-level IES, while for campus-level IES

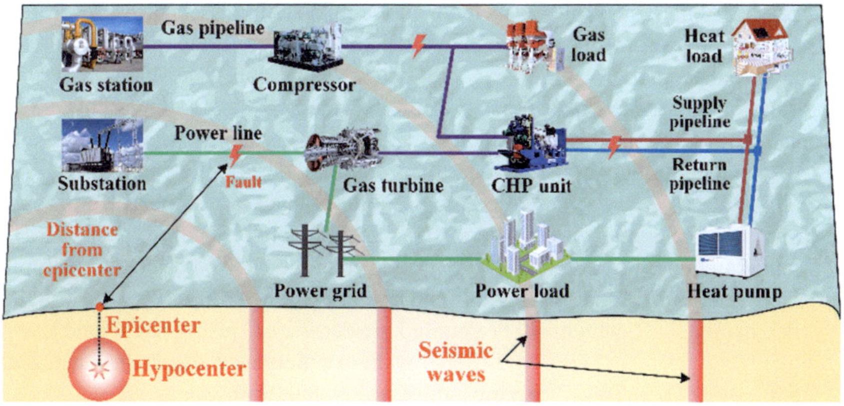

Fig. 7.1 Seismic impact on urban integrated energy system

with limited spatial coverage, earthquake disasters can be assumed to be uniformly distributed within the spatial range.

In earthquake engineering, seismic events are primarily characterized by magnitude, commonly measured using the Richter scale. The disaster impact of different magnitudes on IES exhibits an exponential difference. Low-magnitude earthquakes may cause only localized energy supply disruptions, whereas high-magnitude earthquakes can lead to system-wide outages [31–36].

To address the strong uncertainty associated with low-probability extreme failure events, this chapter proposes an IES resilience planning method based on an economic-risk-balanced perspective. The overall framework is illustrated in Fig. 7.2. A risk-based seismic-resilient IES planning (R-SRIP) model adopts a centralized framework, assuming the presence of a central coordinating body (e.g., a disaster prevention committee) that efficiently manages the resilience resource scheduling and distribution of the power distribution system (PDS), natural gas system (NGS), district heating system (DHS) [37]. For simplicity, the model primarily focuses on the steady-state operation of IES and does not consider the dynamic energy flow conversion process on short time scales. The resilience planning process consists of three main steps:

(1) **Data Collection and Model Preparation**: The required data includes IES system parameters, multi-energy load parameters, and historical seismic data. To characterize the uncertainty of earthquake disasters, this study proposes a probabilistic hierarchical fault scenario generation method to generate representative failure scenarios of different magnitudes along with their probabilistic characteristics.
(2) **Development of R-SRIP Model**: The first stage considers the pre-disaster planning phase, focusing on reinforcing power lines, natural gas pipelines, and heating pipelines to mitigate disaster impacts on the IES. The second stage considers post-disaster operations, conducting coordinated electricity-gas-heat emergency dispatch under hierarchical failure scenarios. Each fault scenario represents the failure state of IES components under different disaster levels. The model coordinates pre-disaster deployable resources and post-disaster emergency measures to minimize incident losses, including energy supply disruption costs and emergency resource dispatch costs. The emergency dispatch equipment considered in this study includes mobile emergency generator (MEG), gas turbines (GT), combined heat and power (CHP) units, heat pumps (HP), and electric boilers (HB). The objective function of the R-SRIP model includes investment costs, expected emergency dispatch costs under hierarchical failure scenarios, and tail-end risk in extreme failure scenarios (quantified using CVaR, which will be detailed in Sect. 7.3). Constraints include investment budget constraints, multi-energy flow emergency dispatch constraints, equipment operation constraints, and CVaR-related constraints.
(3) **Analysis of Risk-Economic Balance in Planning Schemes**: Sensitivity analysis of the risk control parameters in the R-SRIP model is conducted to determine the optimal planning scheme that achieves a balance between risk and economic

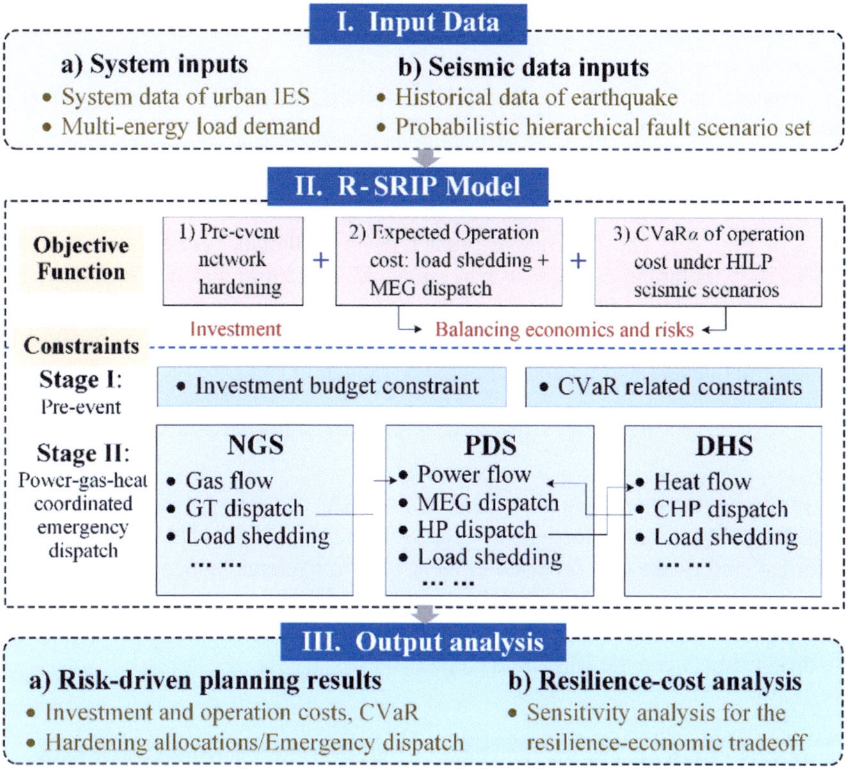

Fig. 7.2 Overall framework of the proposed R-SRIP model

efficiency. This ensures that the formulated IES resilience planning scheme meets budget constraints while achieving the desired level of risk control.

7.3 Generation of Hierarchical Failure Scenarios and Extreme Risk Modeling Under Disaster Uncertainty

In this section, Sect. 7.3.1 introduces the hierarchical failure scenario generation method, while Sect. 7.3.2 presents the tail risk quantification method based on CVaR.

7.3.1 Probabilistic Hierarchical Failure Scenario Generation

The overall framework of the hierarchical failure scenario generation method is illustrated in Fig. 7.3.

(1) **Characterization of Seismic Features**: First, the disaster potential and probabilistic characteristics of earthquakes are characterized, as shown in Module I of Fig. 7.3. The destructive impact of an earthquake depends on multiple factors, including earthquake magnitude M, epicentral distance R, soil type S, and uncertainty parameters ς related to seismic wave propagation. Based on [38], the general analytical expression for the relationship between earthquake destructiveness and related parameters is given in (7.1).

$$\ln(G) = c + F_1(M) + F_2(R) + F_3(S) + \varsigma, \tag{7.1}$$

where G represents peak ground acceleration (PGA). Referring to existing literature on determining the distance between the epicenter and IES locations [30, 38–40], historical earthquake data are used to identify active seismic zones within the planning area and determine the primary occurrence directions of past seismic events. The epicenter location is then simulated using the Monte Carlo method to establish the distance between the IES and the epicenter.

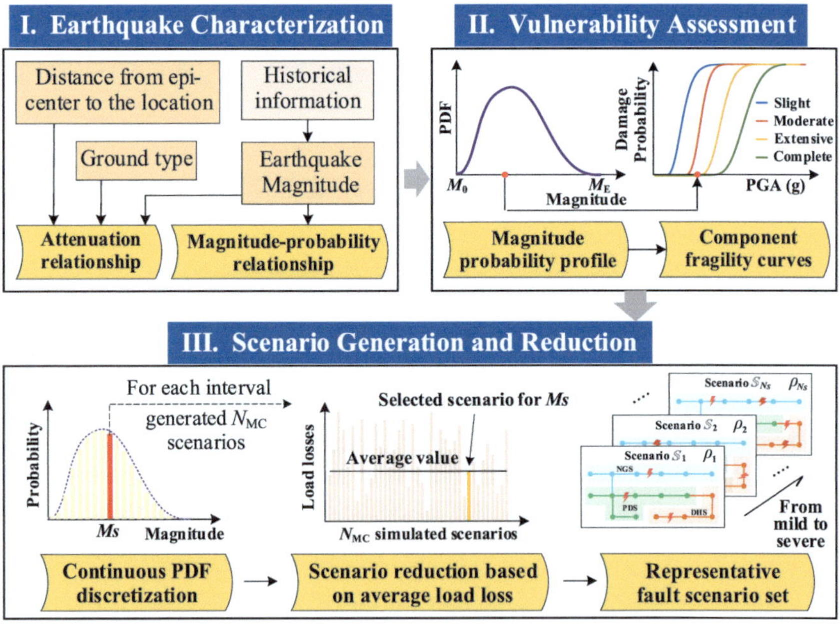

Fig. 7.3 Hierarchical failure scenario generation of IES under seismic disasters

In the seismic resilience planning of IES, the target magnitude range $[M_0, M_E]$ is defined, where the lower bound M_0 represents the minimum magnitude capable of causing damage to IES equipment, while the upper bound M_E corresponds to the maximum magnitude considered in disaster planning. Based on historical statistical data, a lognormal distribution is used to fit the probability distribution curve of the target magnitude range [41, 42].

(2) **IES Vulnerability Assessment:** The model primarily focuses on the damage to power lines, natural gas pipelines, and heating pipelines, utilizing fragility curves to model the probability of failure of these components under seismic events [41]. As shown in Module II of Fig. 7.3, fragility curves map the probability of IES reaching different post-disaster states under a given peak ground acceleration G (which characterizes different magnitudes). Following [30, 38], the "Connection Loss" approach is used to model the post-disaster states of IES.

Taking the PDS as an example, the post-disaster state due to earthquakes is represented by the percentage of damaged power lines. The post-disaster states include minor damage, moderate damage, severe damage, and complete failure, denoted as $d \in D = \{D_1, D_2, D_3, D_4\}$.. The fragility curves corresponding to each post-disaster state are described by (7.2).

$$P(d|G) = \phi[(1/\omega_d)\ln(G/\overline{G}_d)], \tag{7.2}$$

where $\phi[\cdot]$ denotes the standard normal cumulative distribution function, and ω_d represents the lognormal standard deviation. Furthermore, the expected number of damaged power lines under different earthquake magnitudes is calculated using (7.3). The expected number of damaged pipelines in the NGS and DHS is also determined using the "Connection Loss" method.

$$N_{ED} = [P(d \geq D_n|G) - P(d \geq D_{n+1}|G)]N_{D_n}. \tag{7.3}$$

(3) ***Hierarchical Failure Scenario Generation:*** Existing studies often use distance-based scenario clustering algorithms to form representative failure scenario sets [22, 43]. However, extreme failure scenarios with low probability and high impact are usually far from the clustering centers, making it difficult for typical scenario sets to encompass such extreme cases. To address this issue, this study proposes a probabilistic hierarchical failure scenario generation method combining stratified sampling and importance sampling, incorporating stochastic failure simulation and mean loss filtering. The specific steps are as follows:

Step 1. Discretize the target magnitude range $[M_0, M_E]$ into N_s equal intervals, with the center value of each interval representing the classified magnitude M_s. Calculate the probability ρ_s of each classified interval occurring.

Step 2. For each classified magnitude M_s, determine the possible peak ground acceleration (PGA) range based on (7.1) and perform random sampling within this range. Based on the sampled PGA values, compute the expected number of damaged components in the PDS, NGS, and DHS using (7.2) and (7.3).

Step 3. Use the Monte Carlo method to randomly generate N_{MC} failure scenarios for each classified magnitude. The number of failed components in each scenario is set as the expected value obtained in Step 2, and failure components are randomly selected based on the length attributes of power lines and pipelines [43]. The repair time of each failure component is randomly generated based on the Weibull distribution. A total of $N_s \times N_{\text{MC}}$ random failure scenarios are generated across all classified magnitudes.

Step 4. Simulate the cascading failures and energy supply restoration process of the IES for each random failure scenario, recording the total system load loss.

Step 5. For each classified magnitude M_s, calculate the mean load loss L_s across the N_{MC} random failure scenarios and determine the scenario with the smallest absolute deviation from the mean loss L_s as the representative failure scenario for that magnitude classification.

The resulting hierarchical failure scenario set is denoted as $\mathcal{S} = [\mathbb{S}_1, \mathbb{S}_2, \ldots, \mathbb{S}_{NS}]$, with corresponding scenario probabilities $\mathcal{P} = [\rho_1, \rho_2, \ldots, \rho_{NS}]$. This scenario set captures a range of failures from high-probability minor damage to low-probability severe damage, effectively representing the destructive impact of seismic events on the IES. Additionally, it reflects the probabilistic characteristics of different disaster levels, enabling the quantification of tail risk associated with low-probability extreme events.

7.3.2 Tail Risk Quantification Based on CVaR

To mitigate the severe energy supply disruptions caused by high-magnitude earthquakes, the model incorporates the CVaR metric [33] to quantify and control tail risk in extreme scenarios. Figure 7.4 provides an illustration of CVaR.

In Fig. 7.4, VaR_α represents the maximum possible loss at a given confidence level. CVaR_α extends by representing the expected loss beyond VaR_α. The calculation formula is given in (7.4):

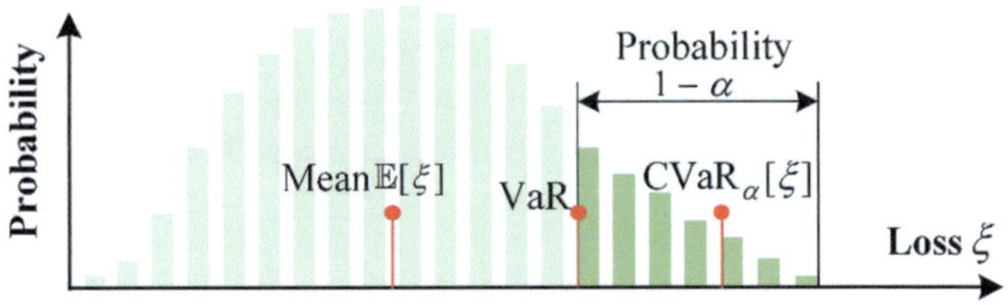

Fig. 7.4 CVaR with 1-α probability of occurrence

$$\mathrm{CVaR}_\alpha = \mathrm{VaR}_\alpha + \frac{1}{1-\alpha}\sum_{s\in\mathcal{S}} \rho_s [f_{\mathrm{Loss},\,s} - \mathrm{VaR}_\alpha]^+, \tag{7.4}$$

where $f_{\mathrm{Loss},\,s}$ denotes the loss function. Compared to value-at-risk (VaR), CVaR more effectively captures potential risks in extreme events. By comprehensively considering both expected costs and CVaR under uncertain failure scenarios, the model ensures a balanced trade-off between resilience and economic feasibility. The nonlinear component of (7.4) can be transformed using linearization techniques, which will be detailed in Sect. 7.4.

7.4 The Risk-Based Seismic-Resilient IES Planning Model

This section introduces the mathematical formulation of the established IES resilience planning model.

7.4.1 *Objective Function*

The objective function of the planning model is presented in (7.5), which includes investment costs, emergency dispatch costs, and the tail risk loss CVaR under extreme scenarios. The parameter $\beta \in [0, 1]$ is a risk aversion parameter used to balance economic efficiency and risk control effectiveness. A larger β value strengthens the risk control effect of resilience planning. Equation (7.6) represents the IES investment cost, including the reinforcement costs of power lines, natural gas pipelines, and heating pipelines. Equation (7.7) represents the expected value of IES emergency dispatch costs under graded fault scenarios, covering energy supply interruption costs in the PDS, NGS and DHS, as well as the rental and dispatch costs of MEGs.

$$\min : f_{\mathrm{Inv}} + (1-\beta) f_{\mathrm{Loss}} + \beta\ \mathrm{CVaR}_\alpha[f_{\mathrm{Loss}}] \tag{7.5}$$

$$f_{\mathrm{Inv}} = \sum_{ij\in\mathcal{L}_{\mathrm{P}}} C_{ij}^{\mathrm{Har}} h_{ij}^{\mathrm{P}} + \sum_{ij\in\mathcal{L}_{\mathrm{G}}} C_{ij}^{\mathrm{Har}} h_{ij}^{\mathrm{G}} + \sum_{ij\in\mathcal{L}_{\mathrm{H}}} C_{ij}^{\mathrm{Har}} h_{ij}^{\mathrm{H}} \tag{7.6}$$

$$\begin{aligned} f_{\mathrm{Loss}} = \varpi \sum_{s\in\mathcal{S}} \rho_s \sum_{t\in\mathcal{T}} \Delta t \Big[& \sum_{i\in\mathcal{N}_{\mathrm{P}}} C_{\mathrm{Cut}} \omega_i^{\mathrm{P}} P_{i,\,s,\,t}^{\mathrm{Cut}} + \sum_{i\in\mathcal{N}_{\mathrm{G}}} C_{\mathrm{Cut}} \omega_i^{\mathrm{G}} G_{i,\,s,\,t}^{\mathrm{Cut}} \gamma_{g2e} \\ & + \sum_{i\in\mathcal{N}_{\mathrm{H}}} C_{\mathrm{Cut}} \omega_i^{\mathrm{H}} H_{i,\,s,\,t}^{\mathrm{Cut}} + \sum_{i\in\mathcal{N}_{\mathrm{MEG}}^{\mathrm{P}}} \sum_{m\in\mathcal{M}} (C_{\mathrm{Rent},m} \mu_{i,\,m,\,s,\,t} + C_{\mathrm{MEG},m} P_{i,\,m,\,s,\,t}^{\mathrm{MEG}}) \Big]. \end{aligned} \tag{7.7}$$

Here, subscript s represents the fault scenario index. t denotes the scheduling period index. i and j are node indices. m is the type index of MEGs, and ij refers to power lines, natural gas pipelines, and heating pipelines. $\mathcal{L}_{\mathrm{P}}/\mathcal{L}_{\mathrm{G}}/\mathcal{L}_{\mathrm{H}}$ represent the sets of power lines, natural gas pipelines, and heating pipelines, respectively. $\mathcal{S}$ is the fault scenario set. $\mathcal{T}$ is the scheduling period set. $\mathcal{N}_{\mathrm{P}}/\mathcal{N}_{\mathrm{G}}/\mathcal{N}_{\mathrm{H}}$ are the sets of power, natural gas, and heating nodes, respectively. $\mathcal{N}_{\mathrm{MEG}}^{\mathrm{P}}$ is the set of candidate access nodes for MEGs. $\mathcal{M}$ is the set of MEG types. f_{Inv} represents the IES investment cost. f_{Loss} denotes the expected value of IES emergency dispatch costs. $\mathrm{CVaR}_{\alpha}\left[f_{\mathrm{Loss}}\right]$ is the CVaR value of emergency dispatch costs at confidence level. C_{ij}^{Har} is the reinforcement cost of line or pipeline. C_{Cut} is the unit load reduction cost. $C_{\mathrm{Rent},\,m}$ and $C_{\mathrm{MEG},\,m}$ are the rental cost and unit operating cost of type m MEGs, respectively. γ_{g2e} is the gas-to-electricity conversion coefficient. ϖ represents the expected number of earthquakes during the planning period. ρ_s is the occurrence probability of fault scenario. Δt is the scheduling period duration. ω_i^{P}, ω_i^{G}, and ω_i^{H} are the weighting factors of electricity, gas, and heat loads, respectively. h_{ij}^{P}, h_{ij}^{G}, and h_{ij}^{H} are 0–1 binary variables indicating the reinforcement state of power lines, gas pipelines, and heating pipelines (1 for reinforced, 0 for unreinforced). $\mu_{i,m,s,t}$ is the number of type m MEGs connected to power node i. $P_{i,m,s,t}^{\mathrm{MEG}}$ is the active power injected by type m MEGs into power node i. $P_{i,s,t}^{\mathrm{Cut}}$, $G_{i,s,t}^{\mathrm{Cut}}$ and $H_{i,s,t}^{\mathrm{Cut}}$ denote the curtailed active power, curtailed gas flow, and curtailed heating power, respectively.

7.4.2 Pre-disaster Investment Budget Constraints

The number of reinforced elements in the PDS, NGS, and DHS is limited by the investment budget, as expressed in (7.8). This constraint ensures the rational allocation of emergency resources within the budget.

$$\sum_{ij\in\mathcal{L}_{\psi}} h_{ij,s}^{\psi} \le N_{\max}^{\mathrm{Har},\,\psi};\ \forall\psi\in\{\mathrm{P,\ G,\ H}\},\ \ ij\in\mathcal{L}_{\psi},\ \ s\in\mathcal{S}. \tag{7.8}$$

7.4.3 Post-disaster Emergency Dispatch Constraints

(1) ***Power Distribution System Operation Constraints***: The PDS constraints are defined in (7.9)–(7.24). To balance model accuracy and computational complexity, a linearized DistFlow model is adopted for power flow modeling [44, 45]. Equations (7.9)–(7.10) describe active and reactive power balance constraints at nodes. Equation (7.11) defines the voltage drop constraint along power lines, using the big-M method to relax voltage drop for non-operating power lines. Here, ι_{ij} is a 0-1 binary variable representing the pre-disaster operational status of power lines (0 for out of service, 1 for in service), and $\upsilon_{ij,s,t}^{\mathrm{P}}$

indicates post-disaster damage status (0 for damaged, 1 for undamaged). Equations (7.12)–(7.13) enforce power flow constraints based on the operational and damage states of power lines. The branch power flow constraint is linearized using two sets of polygonal constraints [46, 47], as shown in (7.14).

$$\sum_{k:k\to i} P^{\mathrm{L}}_{ki,s,t} - \sum_{j:i\to j} P^{\mathrm{L}}_{ij,s,t} + P^{\mathrm{Sub}}_{i,s,t} + P^{\mathrm{GT}}_{i,s,t} + P^{\mathrm{MEG}}_{i,s,t} + P^{\mathrm{CHP}}_{i,s,t} - P^{\mathrm{EB}}_{i,s,t} = P^{\mathrm{D}}_{i,t} - P^{\mathrm{Cut}}_{i,s,t}; \quad \forall i \in \mathcal{N}_{\mathrm{P}}, \quad s \in \mathcal{S}, \quad t \in \mathcal{T} \tag{7.9}$$

$$\sum_{k:k\to i} Q^{\mathrm{L}}_{ki,s,t} - \sum_{j:i\to j} Q^{\mathrm{L}}_{ij,s,t} + Q^{\mathrm{Sub}}_{i,s,t} + Q^{\mathrm{GT}}_{i,s,t} + Q^{\mathrm{MEG}}_{i,s,t} = Q^{\mathrm{D}}_{i,t} - Q^{\mathrm{Cut}}_{i,s,t}; \ \forall i \in \mathcal{N}_{\mathrm{P}}, \ s \in \mathcal{S}, \ t \in \mathcal{T} \tag{7.10}$$

$$\left| V_{i,s,t} - V_{j,s,t} - 2P^{\mathrm{L}}_{ij,s,t} R_{ij} - 2Q^{\mathrm{L}}_{ij,s,t} X_{ij} \right| \le (2 - \iota_{ij} - \upsilon^{\mathrm{P}}_{ij,s,t}) M; \quad \forall ij \in \mathcal{L}_{\mathrm{P}}, \quad s \in \mathcal{S}, \quad t \in \mathcal{T} \tag{7.11}$$

$$-S^{\mathrm{L}}_{ij,\max} \upsilon^{\mathrm{P}}_{ij,s,t} \le P^{\mathrm{L}}_{ij,s,t}, Q^{\mathrm{L}}_{ij,s,t} \le S^{\mathrm{L}}_{ij,\max} \upsilon^{\mathrm{P}}_{ij,s,t}; \ \forall ij \in \mathcal{L}_{\mathrm{P}}, \quad s \in \mathcal{S}, \quad t \in \mathcal{T} \tag{7.12}$$

$$-S^{\mathrm{L}}_{ij,\max} \iota_{ij} \le P^{\mathrm{L}}_{ij,s,t}, Q^{\mathrm{L}}_{ij,s,t} \le S^{\mathrm{L}}_{ij,\max} \iota_{ij}; \forall ij \in \mathcal{L}_{P}, \quad s \in \mathcal{S}, \quad t \in \mathcal{T} \tag{7.13}$$

$$-\sqrt{2} S^{\mathrm{L}}_{ij,\max} \le P^{\mathrm{L}}_{ij,s,t} \pm Q^{\mathrm{L}}_{ij,s,t} \le \sqrt{2} S^{\mathrm{L}}_{ij,\max}; \tag{7.14}$$

where $P^{\mathrm{L}}_{ij,s,t}$ and $P^{\mathrm{L}}_{ki,s,t}$ indicate active power flow, and $P^{\mathrm{Sub}}_{i,s,t}$ indicates substation active power output. $P^{\mathrm{GT}}_{i,s,t}$ is GT active power output, and $P^{\mathrm{MEG}}_{i,s,t}$ is MEG active power output. $P^{\mathrm{CHP}}_{i,s,t}$ *is* CHP units active power output, and $P^{\mathrm{EB}}_{i,s,t}$ is EB active power input. $Q^{\mathrm{L}}_{ij,s,t}$ and $Q^{\mathrm{L}}_{ki,s,t}$ indicate reactive power flow. $Q^{\mathrm{Sub}}_{i,s,t}$ and $Q^{\mathrm{GT}}_{i,s,t}$ are substation and GT reactive power output, and $Q^{\mathrm{MEG}}_{i,s,t}$ is MEG reactive power output. $V_{i,s,t}$ and $V_{j,s,t}$ are squared nodal voltages.

Active and reactive load-shedding correlation constraints are shown in (7.15)–(7.16). Radial topology constraints for pre-disaster PDS are shown in (7.17)–(7.18), where $|\mathcal{N}_P|$ indicates the number of nodes in the PDS. Equation (7.19) indicates the logical relationship between the damage state $\upsilon^{\mathrm{P}}_{ij,s,t}$, reinforcement state h^{P}_{ij}, and destruction state $\xi^{\mathrm{P}}_{ij,s,t}$ [30]. Equations (7.20)–(7.22) indicate GT power and ramping limits, and Eq. (7.23) indicates substation power limits. Equation (7.24) sets voltage limits.

$$[0, 0] \le [P^{\mathrm{Cut}}_{i,s,t}, Q^{\mathrm{Cut}}_{i,s,t}] \le [P^{\mathrm{D}}_{i,t}, Q^{\mathrm{D}}_{i,t}]; \forall i \in \mathcal{N}_{\mathrm{P}}, \ s \in \mathcal{S}, \ t \in \mathcal{T} \tag{7.15}$$

$$P^{\mathrm{Cut}}_{i,s,t} Q^{\mathrm{D}}_{i,t} - Q^{\mathrm{Cut}}_{i,s,t} P^{\mathrm{D}}_{i,t} = 0; \forall i \in \mathcal{N}_{\mathrm{P}}, \ s \in \mathcal{S}, \ t \in \mathcal{T} \tag{7.16}$$

$$\sum\nolimits_{ij\in\mathcal{L}_{\mathrm{P}}} \iota_{ij} = |\mathcal{N}_{\mathrm{P}}| - 1; \forall ij \in \mathcal{L}_{\mathrm{P}} \tag{7.17}$$

$$\sum\nolimits_{ij\in\mathcal{L}_{\mathrm{Loop},n}} \iota_{ij} = \left|\mathcal{L}_{\mathrm{Loop},n}\right| - 1; \forall ij \in \mathcal{L}_{\mathrm{Loop},n}, \quad \mathcal{L}_{\mathrm{Loop},n} \in \mathcal{L}_{\mathrm{Loop}} \tag{7.18}$$

$$\upsilon^{\mathrm{P}}_{ij,s,t} = h^{\mathrm{P}}_{ij} + \zeta^{\mathrm{P}}_{ij,s,t} - h^{\mathrm{P}}_{ij}\zeta^{\mathrm{P}}_{ij,s,t}; \forall ij \in \mathcal{L}_{\mathrm{P}}, \quad s \in \mathcal{S}, \quad t \in \mathcal{T} \tag{7.19}$$

$$0 \le P^{\mathrm{GT}}_{i,s,t} \le P^{\mathrm{GT}}_{i,\max}; \forall i \in \mathcal{N}^{\mathrm{P}}_{\mathrm{GT}}, \ s \in \mathcal{S}, \ t \in \mathcal{T} \tag{7.20}$$

$$-R^{\mathrm{GT}}_{i,\max}\Delta t \le P^{\mathrm{GT}}_{i,s,t} - P^{\mathrm{GT}}_{i,s,t-1} \le R^{\mathrm{GT}}_{i,\max}\Delta t; \forall i \in \mathcal{N}^{\mathrm{P}}_{\mathrm{GT}}, \quad s \in \mathcal{S}, \quad t \in \mathcal{T} \tag{7.21}$$

$$\left|Q^{\mathrm{GT}}_{i,s,t}\right| \le P^{\mathrm{GT}}_{i,s,t}\tan(\cos^{-1}\varphi_{\mathrm{GT}}); \forall i \in \mathcal{N}^{\mathrm{P}}_{\mathrm{GT}}, \ s \in \mathcal{S}, \ t \in \mathcal{T} \tag{7.22}$$

$$[0, 0] \le [P^{\mathrm{Sub}}_{i,s,t}, \ Q^{\mathrm{Sub}}_{i,s,t}] \le [P^{\mathrm{Sub}}_{i,\max}, \ Q^{\mathrm{Sub}}_{i,\max}]; \forall i \in \mathcal{N}^{\mathrm{P}}_{\mathrm{Sub}}, \quad s \in \mathcal{S}, \ t \in \mathcal{T} \tag{7.23}$$

$$V^2_{i,\min} \le V_{i,s,t} \le V^2_{i,\max}; \forall i \in \mathcal{N}_{\mathrm{P}}, \ s \in \mathcal{S}, \ t \in \mathcal{T}, \tag{7.24}$$

where $\mathcal{L}_{\mathrm{Loop}}$ is the set of power loops and $\mathcal{L}_{\mathrm{Loop},n}$ is the set of power lines contained in loop n. $\mathcal{N}^{\mathrm{P}}_{\mathrm{GT}}$ is the set of GT nodes, and $\mathcal{N}^{\mathrm{P}}_{\mathrm{Sub}}$ is the set of substation nodes. $P^{\mathrm{GT}}_{i,\max}$ is the GT's maximum output active power of the GT. $R^{\mathrm{GT}}_{i,\max}$ is the maximum climbing power of the GT. φ_{GT} is the power factor of the GT. $P^{\mathrm{Sub}}_{i,\max}$ and $Q^{\mathrm{Sub}}_{i,\max}$ are the upper limit values of active and reactive power of the substation, respectively. $V_{i,\min}$ and $V_{i,\max}$ are the lower and upper limit values of the node voltages, respectively.

Emergency dispatching of MEGs after a disaster, with specific constraints as shown in (7.25)–(7.31) [48]. Equation (7.25) indicates the maximum number of MEGs at candidate nodes, and Eq. (7.26) indicates the maximum deployable number per type. Equations (7.27)-(7.28) indicate rated power and capacity limits. Equation (7.29) indicates active power injection limits per node, and Eq. (7.30) indicates travel time and deployment constraints. Equation (7.31) is reactive power limit.

$$\sum\nolimits_{m\in\mathcal{M}} \mu_{i,m,s,t} \le N^{\mathrm{MEG}}_{i,\max}; \forall i \in \mathcal{N}^{\mathrm{P}}_{\mathrm{MEG}}, \quad s \in \mathcal{S}, \ t \in \mathcal{T} \tag{7.25}$$

$$\sum\nolimits_{i\in\mathcal{N}_{\mathrm{P}}} \mu_{i,m,s,t} \le N^{\mathrm{MEG}}_{m,\max}; \forall m \in \mathcal{M}, \quad s \in \mathcal{S}, \ t \in \mathcal{T} \tag{7.26}$$

$$0 \le P^{\mathrm{MEG}}_{i,m,s,t} \le P^{\mathrm{MEG}}_{m,\max}\,\mu_{i,m,s,t}; \forall i \in \mathcal{N}^{\mathrm{P}}_{\mathrm{MEG}}, \quad s \in \mathcal{S}, \ t \in \mathcal{T} \tag{7.27}$$

$$\sum\nolimits_{t\in\mathcal{T}} P^{\mathrm{MEG}}_{i,m,s,t}\Delta t \le E^{\mathrm{MEG}}_{m,\max}\mu_{i,m,s,t}; \forall i \in \mathcal{N}^{\mathrm{P}}_{\mathrm{MEG}}, \ m \in \mathcal{M}, \ s \in \mathcal{S} \tag{7.28}$$

$$P^{\mathrm{MEG}}_{i,s,t} = \sum\nolimits_{m\in\mathcal{M}} P^{\mathrm{MEG}}_{i,m,s,t}; \forall i \in \mathcal{N}^{\mathrm{P}}_{\mathrm{MEG}}, \quad s \in \mathcal{S}, \ t \in \mathcal{T} \tag{7.29}$$

$$P^{\text{MEG}}_{i,m,s,t} = 0; \forall i \in \mathcal{N}^{\text{P}}_{\text{MEG}},\ m \in \mathcal{M},\ s \in \mathcal{S},\ t \in \mathcal{T}^{\text{Travel}}_{i,m} \tag{7.30}$$

$$\left|Q^{\text{MEG}}_{i,s,t}\right| \le P^{\text{MEG}}_{i,s,t} \tan(\cos^{-1} \varphi_{\text{MEG}}); \forall i \in \mathcal{N}^{\text{P}}_{\text{MEG}},\ s \in \mathcal{S},\ t \in \mathcal{T}, \tag{7.31}$$

where $\mathcal{T}^{\text{Travel}}_{i,m}$ is the set of dispatch time periods of the MEGs. $N^{\text{MEG}}_{i,\max}$ is the maximum number of MEGs that can be accessed by the nodes, and $N^{\text{MEG}}_{m,\max}$ is the maximum number of MEGs of type m that can be leased. $P^{\text{MEG}}_{m,\max}$ is the maximum output active power of the MEGs of type m. $E^{\text{MEG}}_{m,\max}$ is the maximum capacity of type m MEGs, and φ_{MEG} is the MEG power factor.

(2) ***Natural Gas System Constraints***: The NGS constraints are expressed in (7.32)–(7.42). Equation (7.32) uses a steady-state gas flow model for simplification [30]. Equation (7.33) represents the Weymouth constraint for the natural gas pipeline, characterizing the numerical relationship between the pipeline flow rate and the nodal air pressure. Equation (7.34) represents the numerical relationship constraint between the compressor inlet and outlet gas volumes. Equation (7.35) represents the inlet and outlet pressure difference constraint for the compressor pipeline.

$$\sum_{k:k\to i} G^{\text{Pipe}}_{ki,s,t} - \sum_{j:i\to j} G^{\text{Pipe}}_{ij,s,t} + G^{\text{Sta}}_{i,s,t} - G^{\text{GT}}_{i,s,t} - G^{\text{CHP}}_{i,s,t} = G^{\text{Load}}_{i,t} - G^{\text{Cut}}_{i,s,t};\quad \forall i \in \mathcal{N}_{\text{G}},\ s \in \mathcal{S},\ t \in \mathcal{T} \tag{7.32}$$

$$G^{\text{Pipe}^2}_{ij,s,t} = (\pi^2_{i,s,t} - \pi^2_{j,s,t})\phi^{\text{GL}}_{ij} \upsilon^{\text{G}}_{ij,s,t}; \forall ij \in \mathcal{L}_{\text{G}},\ s \in \mathcal{S},\ t \in \mathcal{T} \tag{7.33}$$

$$G^{\text{Com, out}}_{ij,s,t} = G^{\text{Com, in}}_{ij,s,t}(1 - \varphi_{\text{Com}})\upsilon^{\text{G}}_{ij,s,t}; \forall ij \in \mathcal{L}^{\text{Com}}_{\text{G}},\ s \in \mathcal{S},\ t \in \mathcal{T} \tag{7.34}$$

$$\pi^{\text{Com, in}}_{ij,s,t} \upsilon^{\text{G}}_{ij,s,t} \le \pi^{\text{Com,out}}_{ij,s,t} \le \Gamma_{\text{Com}} \pi^{\text{Com, in}}_{ij,s,t} \upsilon^{\text{G}}_{ij,s,t}; \quad \forall ij \in \mathcal{L}^{\text{Com}}_{\text{G}},\ s \in \mathcal{S},\ t \in \mathcal{T}, \tag{7.35}$$

where $\mathcal{L}^{\text{Com}}_{\text{G}}$ is the set of compressor pipelines. $G^{\text{Load}}_{i,t}$ is the gas loaded gas flow rate. and φ_{Com} is the compressor loss coefficient. Γ_{Com} is the compressor pressurization coefficient. ϕ^{G}_{ij} is the natural gas Weymouth coefficient of the pipeline. $G^{\text{Pipe}}_{ki,s,t}$ and $G^{\text{Pipe}}_{ij,s,t}$ are the gas pipeline gas flow rates, and $G^{\text{Sta}}_{i,s,t}$ is the gas gate station gas flow rate. $G^{\text{CHP}}_{i,s,t}$ is the cogeneration unit gas flow rate, and $G^{\text{GT}}_{i,s,t}$ is the GT gas flow rate. $G^{\text{Com, in}}_{ij,s,t}$ and $G^{\text{Com, out}}_{ij,s,t}$ are the compressor pipeline inlet and output gas flow rate, respectively. $\upsilon^{\text{G}}_{ij,s,t}$ is the damage status of the natural gas pipeline after the disaster. $\pi_{i,s,t}$ and $\pi_{j,s,t}$ are the nodal air pressures. $\pi^{\text{Com, in}}_{ij,s,t}$ and $\pi^{\text{Com,out}}_{ij,s,t}$ are the inlet and outlet air pressures of the compressor pipeline, respectively.

Equation (7.36) represents the upper and lower constraints on gas load reductions. Equation (7.37) represents the upper and lower gas flow constraints for gas pipelines.

Equation (7.38) represents the upper and lower constraints on the gas flow rate for natural gas gate stations. Equation (7.39) represents the upper and lower bound constraints on the nodal gas pressure. Here, $\mathcal{N}_{\text{Sta}}^{\text{G}}$ is the set of natural gas gate station nodes. $G_{i,\max}^{\text{Sta}}$ is the maximum gas flow rate of the natural gas gate station. $\pi_{i,\min}$ and $\pi_{i,\max}$ are the minimum and maximum values of the node air pressure, respectively.

$$0 \le G_{i,s,t}^{\text{Cut}} \le G_{i,t}^{\text{Load}}; \forall i \in \mathcal{N}_{\text{G}},\ \ s \in \mathcal{S},\ \ t \in \mathcal{T} \tag{7.36}$$

$$-\overline{G}_{ij}^{\text{Pipe}} \upsilon_{ij,s,t}^{\text{G}} \le G_{ij,s,t}^{\text{Pipe}} \le \overline{G}_{ij}^{\text{Pipe}} \upsilon_{ij,s,t}^{\text{G}}; \forall ij \in \mathcal{L}_{\text{G}},\ \ s \in \mathcal{S},\ \ t \in \mathcal{T} \tag{7.37}$$

$$0 \le G_{i,s,t}^{\text{Sta}} \le G_{i,\max}^{\text{Sta}}; \forall ij \in \mathcal{N}_{\text{Sta}}^{\text{G}},\ \ s \in \mathcal{S},\ \ t \in \mathcal{T} \tag{7.38}$$

$$\pi_{i,\min} \le \pi_{i,s,t} \le \pi_{i,\max}; \forall i \in \mathcal{N}_{\text{G}},\ \ s \in \mathcal{S},\ \ t \in \mathcal{T}, \tag{7.39}$$

Equation (7.40) has the same meaning as (7.19) in the NGS and represents the logical relationship between the damaged state $\upsilon_{ij,s,t}^{\text{G}}$ of the gas pipeline and the pipeline reinforcement state h_{ij}^{G} and the pipeline suffered damage state $\zeta_{ij,s,t}^{\text{G}}$. Equation (7.41) represents the conversion constraint between the input gas flow rate and the output active power of the GT. Equation (7.42) represents the energy conversion constraint for the CHP unit.

$$\upsilon_{ij,s,t}^{\text{G}} = h_{ij}^{\text{G}} + \zeta_{ij,s,t}^{\text{G}} - h_{ij}^{\text{G}} \zeta_{ij,s,t}^{\text{G}}; \forall ij \in \mathcal{L}_{\text{G}},\ \ s \in \mathcal{S},\ \ t \in \mathcal{T} \tag{7.40}$$

$$G_{i,s,t}^{\text{GT}} \gamma_{g2e} = P_{n,s,t}^{\text{GT}} / \eta_{GT}; \forall i \in \mathcal{N}_{\text{GT}}^{\text{G}},\ \ n \in \mathcal{N}_{\text{GT}}^{\text{P}},\ \ s \in \mathcal{S},\ \ t \in \mathcal{T} \tag{7.41}$$

$$\begin{aligned} &G_{i,s,t}^{\text{CHP}} \gamma_{g2e} = P_{n,s,t}^{\text{CHP}} / \eta_{CHP,p} + H_{k,s,t}^{\text{CHP}} / \eta_{\text{CHP},h}; \\ &\forall i \in \mathcal{N}_{\text{CHP}}^{\text{G}},\ \ n \in \mathcal{N}_{\text{CHP}}^{\text{P}},\ \ k \in \mathcal{N}_{\text{CHP}}^{\text{H}},\ \ s \in \mathcal{S},\ \ t \in \mathcal{T}, \end{aligned} \tag{7.42}$$

where $\mathcal{N}_{\text{GT}}^{\text{G}}$ and $\mathcal{N}_{\text{CHP}}^{\text{G}}$ are the set of GT and CHP unit nodes for the NGS. $\mathcal{N}_{\text{GT}}^{\text{P}}$ and $\mathcal{N}_{\text{CHP}}^{\text{P}}$ are the set of GT and CHP unit nodes for the PDS, respectively. $\mathcal{N}_{\text{CHP}}^{\text{H}}$ is the set of nodes of CHP units for the DHS. η_{GT} is the GT conversion efficiency. $\eta_{\text{CHP},p}$ and $\eta_{\text{CHP},h}$ are the gas-to-electricity and gas-to-heat efficiencies of the CHP units, respectively. $H_{i,s,t}^{\text{CHP}}$ is the output thermal power of the CHP units.

(3) ***District Heating System Constraints***: The DHS operation constraints are shown in (7.43)–(7.55). Equation (7.43) represents the node thermal power balance constraints, which ensure that the input thermal power of each thermal node is equal to the output thermal power. Equation (7.44) represents the heat transfer constraints for the HPs. Equation (7.45) represents the transmission heat loss of the heating pipeline.

$$\sum_{k:k\to i} H_{ki,s,t}^{\text{Pipe, out}} - \sum_{j:i\to j} H_{ki,s,t}^{\text{Pipe, in}} + H_{i,s,t}^{\text{CHP}} + H_{i,s,t}^{\text{EB}} = H_{i,t}^{\text{D}} - H_{i,s,t}^{\text{Cut}}; \quad \forall i \in \mathcal{N}_{\text{H}}, \ \ s \in \mathcal{S}, \ \ t \in \mathcal{T} \tag{7.43}$$

$$H_{ij,s,t}^{\text{Pipe, out}} = H_{ij,s,t}^{\text{Pipe, in}} - H_{ij,s,t}^{\text{Pipe,loss}} \upsilon_{ij,s,t}^{\text{H}}; \forall ij \in \mathcal{L}_{\text{H}}, \ \ s \in \mathcal{S}, \ \ t \in \mathcal{T} \tag{7.44}$$

$$H_{ij,s,t}^{\text{Pipe, loss}} = \frac{\lambda_{ij}^{H} L_{ij}^{H}}{\rho_w A_{ij}^{H}} (\tau_{ij,s,t}^{S,\text{in}} + \tau_{ij,s,t}^{R,\text{in}} - 2\tau_t^{\text{Amb}}); \forall ij \in \mathcal{L}_H, \ \ s \in \mathcal{S}, \ \ t \in \mathcal{T}, \tag{7.45}$$

where λ_{ij}^{H} is the heat transfer coefficient of the heat pipe, and A_{ij}^{H} is the cross-sectional area of the heat pipe. L_{ij}^{H} is the length of the heat pipe. ρ_w is the density of the heat medium, and τ_t^{amb} is the ambient temperature around the heat pipe. $\tau_{ij,s,t}^{\text{S, in}}$ and $\tau_{ij,s,t}^{\text{R, in}}$ are the temperatures of the beginning node of the water supply pipe and return pipe, respectively. $H_{ij,s,t}^{\text{Pipe, in}}$ and $H_{ij,s,t}^{\text{Pipe, out}}$ are the thermal power of the beginning and end node of the heating pipe, respectively. $H_{ij,s,t}^{\text{Pipe, loss}}$ is the heat loss of the heating pipe. $H_{i,s,t}^{\text{CHP}}$ is the output thermal power of the CHP unit, and $H_{i,s,t}^{\text{EB}}$ is the output thermal power of the EB, and $\upsilon_{ij,s,t}^{\text{G}}$ is the damaged state of the heat supply pipe after the disaster.

Equation (7.46) represents the upper and lower constraints on the thermal power of the beginning and end nodes of the heating pipeline, where $H_{ij,\max}^{\text{Pipe}}$ is the maximum thermal power flow of the heating pipe. Equation (7.47) represents the logical relationship between the damaged state of the heating pipeline, the reinforced state of the pipeline, and the damaged state of the pipeline. Equation (7.48) represents the upper and lower bound constraints on the heat load reduction.

$$-H_{ij,\max}^{\text{Pipe}} \upsilon_{ij,s,t}^{\text{H}} \le H_{ij,s,t}^{\text{Pipe, in/out}} \le H_{ij,\max}^{\text{Pipe}} \upsilon_{ij,s,t}^{\text{H}}; \forall ij \in \mathcal{L}_{\text{H}}, \ \ s \in \mathcal{S}, \ \ t \in \mathcal{T} \tag{7.46}$$

$$\upsilon_{ij,s,t}^{\text{H}} = h_{ij}^{\text{H}} + \zeta_{ij,s,t}^{\text{H}} - h_{ij}^{\text{H}} \zeta_{ij,s,t}^{\text{H}}; \forall ij \in \mathcal{L}_{\text{H}}, \ \ s \in \mathcal{S}, \ \ t \in \mathcal{T} \tag{7.47}$$

$$0 \le H_{i,s,t}^{\text{Cut}} \le H_{i,t}^{\text{Load}}; \forall i \in \mathcal{N}_{\text{H}}, \ \ s \in \mathcal{S}, \ \ t \in \mathcal{T}. \tag{7.48}$$

Equations (7.49)–(7.52) represent the constraints on the correlation between the output thermal power and the output electrical power of the CHP unit, which are modeled using a linearized combination of the unit's feasible domain boundary points [49]. Equation (7.53) represents the upper and lower bound constraints on the electrical creep power of the CHP unit. Equation (7.54) represents the upper and lower bound constraints on the thermal power output of the EB. Equation (7.55) represents the conversion constraint between the output thermal power of the EB and the input electric power. Equation (7.56) represents the upper and lower limit constraints on the climbing power of the EB.

$$P_{i,s,t}^{\mathrm{CHP}} = \sum\nolimits_{k \in \mathcal{K}} \alpha_{i,s,t}^{k} P_{i,s,t}^{\mathrm{CHP},k}; \forall i \in \mathcal{N}_{\mathrm{CHP}}^{\mathrm{H}},\ s \in \mathcal{S},\ t \in \mathcal{T} \tag{7.49}$$

$$H_{i,s,t}^{\mathrm{CHP}} = \sum\nolimits_{k \in \mathcal{K}} \alpha_{i,s,t}^{k} H_{i,s,t}^{\mathrm{CHP},k}; \forall i \in \mathcal{N}_{\mathrm{CHP}}^{\mathrm{H}},\ s \in \mathcal{S},\ t \in \mathcal{T} \tag{7.50}$$

$$0 \le \alpha_{i,s,t}^{k} \le 1; \forall i \in \mathcal{N}_{\mathrm{CHP}}^{\mathrm{H}}, k \in \mathcal{K},\ s \in \mathcal{S},\ t \in \mathcal{T} \tag{7.51}$$

$$\sum\nolimits_{k \in \mathcal{K}} \alpha_{i,s,t}^{k} = 1; \forall i \in \mathcal{N}_{\mathrm{CHP}}^{\mathrm{H}},\ s \in \mathcal{S},\ t \in \mathcal{T} \tag{7.52}$$

$$-R_{i,\max}^{\mathrm{CHP}} \Delta t \le P_{i,s,t}^{\mathrm{CHP}} - P_{i,s,t-1}^{\mathrm{CHP}} \le R_{i,\max}^{\mathrm{CHP}} \Delta t; \forall i \in \mathcal{N}_{\mathrm{CHP}}^{\mathrm{H}},\ s \in \mathcal{S},\ t \in \mathcal{T} \tag{7.53}$$

$$0 \le H_{i,s,t}^{\mathrm{EB}} \le H_{i,\max}^{\mathrm{EB}}; \forall i \in \mathcal{N}_{\mathrm{EB}}^{\mathrm{H}},\ s \in \mathcal{S},\ t \in \mathcal{T} \tag{7.54}$$

$$H_{i,s,t}^{\mathrm{EB}} = P_{n,s,t}^{\mathrm{EH}} \eta_{\mathrm{EB}}; \forall i \in \mathcal{N}_{\mathrm{EB}}^{\mathrm{H}},\ n \in \mathcal{N}_{\mathrm{EB}}^{\mathrm{P}},\ s \in \mathcal{S},\ t \in \mathcal{T} \tag{7.55}$$

$$-R_{i,\max}^{\mathrm{EB}} \Delta t \le P_{i,s,t}^{\mathrm{EB}} - P_{i,s,t-1}^{\mathrm{EB}} \le R_{i,\max}^{\mathrm{EB}} \Delta t; \forall i \in \mathcal{N}_{\mathrm{EB}}^{\mathrm{H}},\ s \in \mathcal{S},\ t \in \mathcal{T}, \tag{7.56}$$

where subscript k is the boundary point of the operational domain of the CHP unit. $\mathcal{N}_{\mathrm{EB}}^{\mathrm{P}}$ and $\mathcal{N}_{\mathrm{EB}}^{\mathrm{H}}$ is the set of EB nodes of the DHS and PDS, respectively. $\mathcal{K}$ is the set of boundary points of the operational domain of the CHP unit. $P_{i,s,t}^{\mathrm{CHP},k}$ and $H_{i,s,t}^{\mathrm{CHP},k}$ are the boundary points of the feasible domain of electric power and thermal power of the CHP unit. $R_{i,\max}^{\mathrm{CHP}}$ is the maximum creep power of the CHP unit, and $R_{i,\max}^{\mathrm{EB}}$ is the maximum creep power of the EB. η_{EB} is the conversion efficiency of the EB. $\alpha_{i,s,t}^{k}$ is the occupancy coefficient of the power of the boundary points of the CHP unit.

(4) ***CVaR Constraints***: To simplify the complexity of the nonlinear CVaR constraints, the CVaR expression (7.4) in the hierarchical fault scenario is linearized by introducing an auxiliary variable κ_s for an equivalent transformation [31], as shown in (7.57)–(7.58).

$$\mathrm{CVaR}_{\alpha} = \mathrm{VaR}_{\alpha} + \frac{1}{1-\alpha} \sum_{s \in \mathcal{S}} \rho_s \kappa_s; \tag{7.57}$$

$$\kappa_s \ge f_{\mathrm{Loss},s} - \mathrm{VaR}_{\alpha},\ \ \kappa_s \ge 0; \forall i \in \mathcal{N}_{\mathrm{EB}}^{\mathrm{H}},\ n \in \mathcal{N}_{\mathrm{EB}}^{\mathrm{P}},\ s \in \mathcal{S},\ t \in \mathcal{T}. \tag{7.58}$$

The scheduling constraints (7.25)–(7.28) for MEGs in the second stage of the constructed toughness planning model contain integer variables. In mathematical form, the constructed model is a two-stage stochastic planning model containing integer recourse variables.

7.5 Solution Algorithm for the Planning Model

In this section, a TBD algorithm based on primal-dual cuts acceleration is introduced to decouple and iteratively solve the R-SRIP model established in the previous section, thereby enhancing the computational efficiency.

7.5.1 Model Linearization

The R-SRIP model contains bilinear terms resulting from the Weymouth constraint (7.33) and the multiplication of continuous and binary variables in (7.34), (7.35) and (7.44). Mathematically, the model is formulated as a mixed-integer nonlinear programming (MINLP) problem, whose non-convex and nonlinear characteristics pose significant challenges for the solution process [50]. To address this, the original model is reformulated as a mixed-integer linear programming (MILP) model. Specifically, the Weymouth constraint (7.33) is linearized using a Taylor expansion-based method, whose approximation accuracy has been validated in existing literature [51]. Additionally, the bilinear terms are equivalently linearized using the Big-M method. After linearization, the compact form of the planning model is as follows:

$$\min_{\boldsymbol{x},\boldsymbol{y}_s,VaR_\alpha,\kappa_s} \quad \boldsymbol{c}^{\mathrm{T}}\boldsymbol{x} + (1-\beta)\sum_{s\in\mathcal{S}} \rho_s \boldsymbol{d}^{\mathrm{T}}\boldsymbol{y}_s + \beta(\mathrm{VaR}_\alpha + \sum_{s\in\mathcal{S}} \frac{1}{1-\alpha}\rho_s\kappa_s) \tag{7.59}$$

$$\text{s.t.} \quad \boldsymbol{A}\boldsymbol{x} \le \boldsymbol{b}, \ \boldsymbol{x} \in \{0, 1\} \tag{7.60}$$

$$\kappa_s \ge \boldsymbol{d}^{\mathrm{T}}\boldsymbol{y}_s - \mathrm{VaR}_\alpha, \ \kappa_s \in \mathbb{R}_+; \quad \forall s \in \mathcal{S} \tag{7.61}$$

$$\boldsymbol{D}\boldsymbol{x} + \boldsymbol{E}\boldsymbol{u}_s \le \boldsymbol{k}_s; \quad \forall s \in \mathcal{S} \tag{7.62}$$

$$\boldsymbol{M}\boldsymbol{u}_s + \boldsymbol{N}\boldsymbol{z}_s \le \boldsymbol{e}_s; \quad \forall s \in \mathcal{S} \tag{7.63}$$

$$\boldsymbol{y}_s = \{\boldsymbol{u}_s, \boldsymbol{z}_s\}, \ \boldsymbol{u}_s \in \mathbb{R}, \boldsymbol{z}_s \in \mathbb{Z}_+. \tag{7.64}$$

Equation (7.59) represents the model's objective function, corresponding to (7.5) in the original model. Equations (7.60)–(7.61) define the first-stage constraints. The vector x represents equipment reinforcement decisions, pre-disaster power line reconfiguration variables, and post-disaster damaged line and pipeline variables. Equation (7.60) corresponds to (7.8), (7.17)–(7.19), (7.40), and (7.47) in the original model. Equation (7.61) represents constraints related to the CVaR metric, corresponding to (7.58). The term $\boldsymbol{c}^{\mathrm{T}}\boldsymbol{x}$ denotes investment costs, while $\boldsymbol{d}^{\mathrm{T}}\boldsymbol{y}_s$ represents the second-stage recourse function, which accounts for IES emergency dispatch costs under hierarchical disaster scenarios. Equations (7.62)–(7.63) define the second-stage constraints. The vector represents continuous decision variables in the second stage, while z represents integer decision variables related to MEG leasing and dispatching, as defined in (7.64). Equation (7.62) represents constraints on continuous decision variables, corresponding to (7.9)–(7.16), (7.20)–(7.24), (7.29)–(7.39), (7.41)–(7.46), and (7.48)–(7.56) in the original model. Equation (7.63) represents logical constraints between continuous and integer variables, corresponding to (7.25)–(7.28). The parameters A, D, E, M, N, b, c, d, k_s, e_s represent coefficient matrices associated with various model parameters.

7.5.2 *Benders Decomposition Algorithm with Integer Recourse Variables*

BD is an effective approach for solving large-scale MILP problems by decomposing the original large-scale model into a master problem (**MP**) and subproblems (**SP**), corresponding to the first and second stages of the model, respectively. The algorithm iteratively solves the **MP** and **SP**, where the subproblem generates Benders cuts based on duality theory and feeds them back to the **MP** to iteratively approximate the recourse function until convergence to the optimal solution. However, the second stage of the established planning model contains integer recourse variables, preventing direct application of duality theory to generate effective Benders cuts. To address this, the integer recourse variables in the second stage must be relaxed to generate valid Benders cuts. The specific steps of the BD algorithm with relaxed integer recourse variables [52] are as follows.

(1) ***Generating Benders Cuts from the Relaxed Subproblem:*** Given a solution $\hat{\boldsymbol{x}}$ obtained from the master problem, a relaxed subproblem (**SP-r**) is constructed, as shown in (7.65)–(7.70). By relaxing z_s to a continuous variable and computing the relaxed variable $\boldsymbol{\upsilon}$ based on (7.67), effective Benders cuts can be generated and fed back to the master problem [50].

$$\mathbf{SP-r}:\min_{y_s}\ \Psi_s=\boldsymbol{d}^{\mathrm{T}}\boldsymbol{y}_s \tag{7.65}$$

$$\text{s.t.}\ \ \boldsymbol{Dx}+\boldsymbol{Eu}_s+\boldsymbol{F}z_s\le\boldsymbol{d}_s;\quad\forall s\in\mathcal{S} \tag{7.66}$$

$$\boldsymbol{x}=\hat{\boldsymbol{x}}:\quad\boldsymbol{\upsilon} \tag{7.67}$$

$$\boldsymbol{Mu}_s+\boldsymbol{N}z_s\le\boldsymbol{e}_s;\quad\forall s\in\mathcal{S} \tag{7.68}$$

$$0\le z_s\le z_{\max};\quad\forall s\in\mathcal{S} \tag{7.69}$$

$$\boldsymbol{y}_s = \{\boldsymbol{u}_s, \boldsymbol{z}_s\}, \ \ \boldsymbol{u}_s \in \mathbb{R}, \ \ \boldsymbol{z}_s \in \mathbb{R}_+. \tag{7.70}$$

(2) ***Generating Upper Bound via Exact Subproblem***: Eqs. (7.71)–(7.73) define an exact subproblem (**SP-e**) without relaxation of integer recourse variables. In each iteration, solving the **SP-e** for each failure scenario provides an upper bound on the model's optimal solution. Once the **MP** is solved to obtain $\hat{\boldsymbol{x}}$, the **SP-e** for all failure scenarios are solved, and the optimal solution $\hat{\boldsymbol{y}}_s$ is recorded to obtain an upper bound (UB). It is worth noting that the solution spaces of **SP-r** and **SP-e** are independent for different failure scenarios, allowing these subproblems to be solved in parallel.

$$\mathbf{SP-e}: \min_{\boldsymbol{x}, \psi_s, VaR_\alpha, \kappa_s} \ \boldsymbol{c}^{\mathrm{T}}\boldsymbol{x} + (1-\beta)\sum_{s\in\mathcal{S}} \rho_s\psi_s + \beta(\mathrm{VaR}_\alpha + \sum_{s\in\mathcal{S}} \frac{1}{1-\alpha}\rho_s\kappa_s) \tag{7.71}$$

$$\text{s.t. } (7.66), \ (7.68) \tag{7.72}$$

$$\boldsymbol{y}_s = \{\boldsymbol{u}_s, \boldsymbol{z}_s\}, \ \ \boldsymbol{u}_s \in \mathbb{R}, \ \ \boldsymbol{z}_s \in \mathbb{R}_+. \tag{7.73}$$

(3) ***Generating Lower Bound via the Master Problem:*** The master problem is constructed using the Benders cuts generated from the **SP-r**, as shown in (7.74)–(7.77). Since **SP-r** contains load-shedding constraints, the planning model possesses a "complete recourse" property, ensuring that all Benders cuts generated in the iterative process are "optimal cuts" [53]. The objective function of **MP** (7.74) provides a lower bound (LB) for the model's optimal solution.

$$\mathbf{MP}: \min_{\boldsymbol{x}, \psi_s, VaR_\alpha, \kappa_s} \ \boldsymbol{c}^{\mathrm{T}}\boldsymbol{x} + (1-\beta)\sum_{s\in\mathcal{S}} \rho_s\psi_s + \beta(\mathrm{VaR}_\alpha + \sum_{s\in\mathcal{S}} \frac{1}{1-\alpha}\rho_s\kappa_s) \tag{7.74}$$

$$\text{s.t. } \ \boldsymbol{A}\boldsymbol{x} \le \boldsymbol{b}, \ \boldsymbol{x} \in \{0, 1\}, \ \ VaR_\alpha, \ \kappa_s \in \mathbb{R}_+ \tag{7.75}$$

$$\psi_s \ge \hat{\Psi}_s^k + (\boldsymbol{x} - \hat{\boldsymbol{x}}_s^k)^{\mathrm{T}}\boldsymbol{v}^k; \quad \forall k = 1, \ldots, K-1, \ s \in \mathcal{S} \tag{7.76}$$

$$\kappa_s \ge \psi_s - VaR_\alpha, \ \ \kappa_s \ge 0; \quad \forall s \in \mathcal{S}. \tag{7.77}$$

The algorithm terminates when the gap between the LB and UB falls below a predefined threshold.

7.5.3 Improved Benders Decomposition Algorithm Based on Primal-Dual Cuts Acceleration

In the BD algorithm with relaxed integer recourse variables, the dual variables are derived from the relaxed subproblem, resulting in weak Benders cuts (7.76) that provide limited approximation of the recourse function $\boldsymbol{d}^{\mathrm{T}}\boldsymbol{y}_s$. Additionally, the CVaR metric plays a crucial role in the objective function, but the above algorithm approximates the recourse function using weak duality, leading to unsatisfactory computational efficiency. Therefore, to accelerate the convergence of the master and subproblems, a TBD algorithm based on primal-dual cuts acceleration is proposed.

(1) ***Enhanced Benders Cuts for Recourse Function*:** During the iterative process, the current optimal solution $\hat{z}_s$ from the **SP-e** is substituted back into **SP-e**, forming a linear programming model (**SP-e***), as shown in (7.78)–(7.81). Solving **SP-e*** yields dual variableγ, which is then used to generate enhanced Benders cuts and feed them back to the master problem. Since γ is obtained from the non-relaxed subproblem, the enhanced Benders cuts provide more effective feedback to the master problem [54], enabling a more accurate approximation of the recourse function and accelerating the convergence process.

$$\mathbf{SP-e*}: \min_{\boldsymbol{y}_s} \quad \Gamma_s = \boldsymbol{d}^{\mathrm{T}}\boldsymbol{y}_s \tag{7.78}$$

$$\text{s.t.}(7.72) - (7.73) \tag{7.79}$$

$$\boldsymbol{x} = \hat{\boldsymbol{x}}: \quad \gamma \tag{7.80}$$

$$z_s = \hat{z}_s. \tag{7.81}$$

(2) ***Enhanced Primal Cuts for the CVaR Metric***: The CVaR metric in the second-stage recourse function is primarily determined by a few extreme scenarios, known as "binding scenarios" [55], whose corresponding objective values are greater than or equal to the VaR indicator of the recourse function. By identifying these binding scenarios, enhanced primal cuts can be generated to further accelerate convergence. Specifically, after solving the exact subproblems **SP-e** for all failure scenarios $s \in \mathcal{S}$,, the scenarios are sorted by their objective values. By locating the scenario that satisfies a predefined condition, the binding scenario is identified. Subsequently, enhanced primal cuts are constructed as shown in (7.82)–(7.83) and fed back to the **MP**.

$$\nu_{\sigma(s)} = \begin{cases} 0; & s = 1, \ldots, \hat{s} - 1 \\ \Pi_s - \alpha; & s = \hat{s} \\ \rho_s; & s = \hat{s} + 1, \ldots, |\mathcal{S}| \end{cases} \tag{7.82}$$

$$\mathrm{CVaR}_\alpha \geq \frac{1}{1-\alpha} \sum\nolimits_{s \in \mathcal{S}} \nu_s \boldsymbol{d}^{\mathrm{T}} \boldsymbol{y}_s. \tag{7.83}$$

Compared to Benders cuts generated based on duality, primal cuts provide stronger cutting information for the CVaR metric, improving the lower bound of the **MP**'s optimal solution and accelerating convergence [54, 55]. Notably, only primal cuts associated with binding scenarios are fed back to **MP** to avoid excessive computational burden during iteration. By incorporating enhanced Benders cuts (7.86)–(7.87) and enhanced primal cuts (7.88)–(7.89), the enhanced master problem (**EMP**) is obtained, as shown in (7.84)–(7.89). These two types of cuts enhance the iterative feedback between the master and subproblems, effectively refining the feasible region of the original model and accelerating convergence.

$$\textbf{EMP}: \min_{\boldsymbol{x},\, \psi_s,\, VaR_\alpha,\, \kappa_s} \boldsymbol{c}^{\mathrm{T}} \boldsymbol{x} + (1-\beta) \sum_{s \in \mathcal{S}} \rho_s \chi_s + \beta \mathrm{CVaR}_\alpha \tag{7.84}$$

$$\text{s.t. } (7.75) - (7.76) \tag{7.85}$$

$$\chi_s \geq \Gamma_s^k + (\boldsymbol{x} - \hat{\boldsymbol{x}}_s^k)^{\mathrm{T}} \boldsymbol{\gamma}^k; \quad \forall k = 1, \ldots, K-1, \; s \in \mathcal{S} \tag{7.86}$$

$$\chi_s \geq \psi_s; \quad \forall s \in \mathcal{S} \tag{7.87}$$

$$\mathrm{CVaR}_\alpha \geq \sum\nolimits_{s \in \mathcal{S}} \nu_s \boldsymbol{d}^{\mathrm{T}} \boldsymbol{y}_s / (1-\alpha); \quad \forall k = 1, \ldots, K-1. \tag{7.88}$$

$$(7.62) - (7.64) \tag{7.89}$$

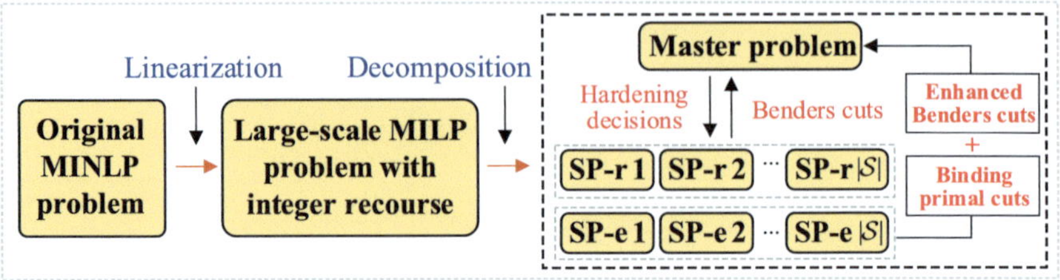

Fig. 7.5 Model linearization and decomposition process

The linearization and decomposition process of the TBD algorithm is illustrated in Fig. 7.5, with the detailed algorithmic steps provided as follows (Fig. 7.6).

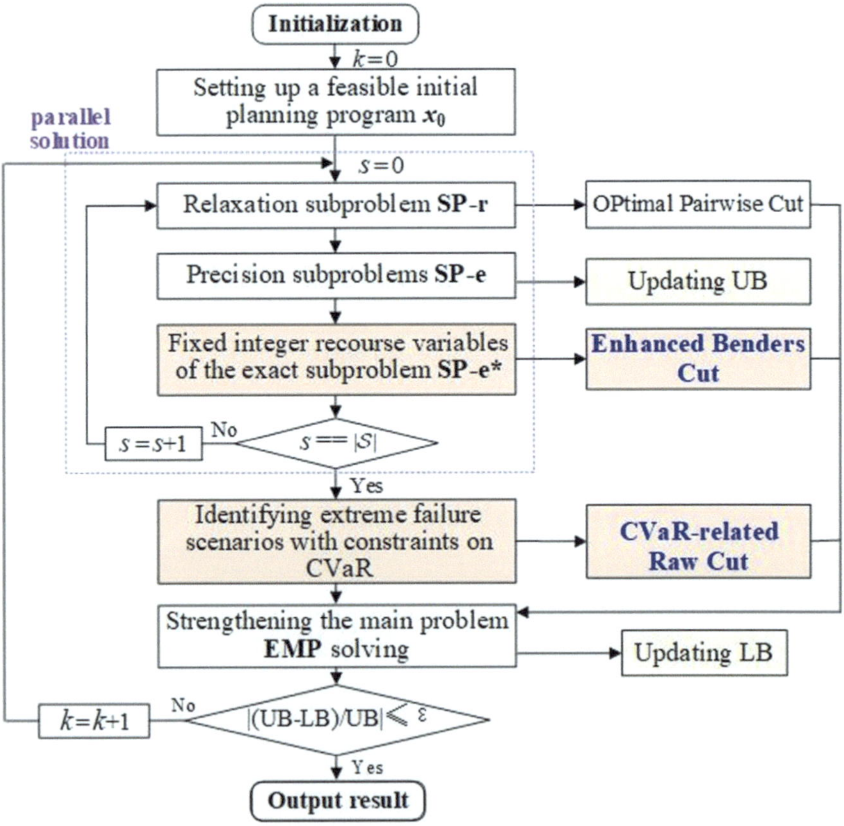

Fig. 7.6 Improving the Benders decomposition algorithm process

Algorithm 1: Tailored Benders decomposition algorithm

1: **Initialize**, set $k = 0$, model upper bound $UB = +\infty$, lower bound $LB = -\infty$
2: **While** $k \leq k_{\max}$ **do**
3: Set $\boldsymbol{x}$ as a set of feasible initial planning scenarios
4: **parallel for** $s \in S$
5: Solve the exact subproblem **SP-e**, record $\hat{z}_s^k$ and $\hat{\Gamma}_s^k$, and update the upper bound based on $UB_k = \min\{UB_{k-1},\ c^{\mathrm{T}}\hat{x}^k + (1-\beta)\sum_{s\in S}\rho_s \hat{\Gamma}_s^k + \beta CVaR_\alpha(\hat{\Gamma}_s^k)\}$
6: Solve the relaxation subproblem **SP-r** to obtain the relaxation Benders cut
7: **End**
8: Identify the corresponding "binding scenario" of CVaR, and obtain strengthened primal cut (7.88)–(7.89)
9: **parallel for** $s \in S$
10: Solve the **SP-e*** with fixed integer recourse variables to obtain the strengthened Benders cut (7.86)–(7.87)
11: **End**
12: Solve EMP by updating the lower bound value based on $LB_k = \min\{LB_{k-1},\ c^{\mathrm{T}}\hat{x}^k + (1-\beta)\sum_{s\in S}\rho_s \hat{\chi}_s^k + \beta\hat{\Lambda}_s^k\}$
13: Calculate the gap between the bounds $\Delta = |(UB_k - LB_k)/LB_k|$
14: If $\Delta < \varepsilon$, return the optimal solution and end the iterative process
15: Otherwise, set $k = k + 1$ and return to step 4
16: **End**

7.6 Case Studies

To verify the effectiveness of the proposed R-SRIP method for IES with an economic-risk balance, this section selects the IES E33-G21-H27 system shown in Fig. 7.7 for case study testing. As illustrated in Fig. 7.7, the PDS and NGS are coupled through two GTs, while the PDS and DHS are coupled via a CHP unit and an EB. The total power, natural gas, and heating loads are 3.915MW+2.400MVar, 7.266 MW, and 2.900 MW, respectively. Detailed system parameters can be found in [56]. In the baseline case, the confidence level α and risk aversion parameter β are set to 0.8 and 0.5, respectively, and the target earthquake magnitude range is set to [4.0, 6.5]. All simulations were conducted on a laptop with a 3.60-GHz CPU and 32GB RAM using MATLAB programming and solved with Gurobi.

7.6.1 Planning Results Under Baseline Settings

This subsection verifies the effectiveness of the hierarchical fault scenario generation method and the R-SRIP model.

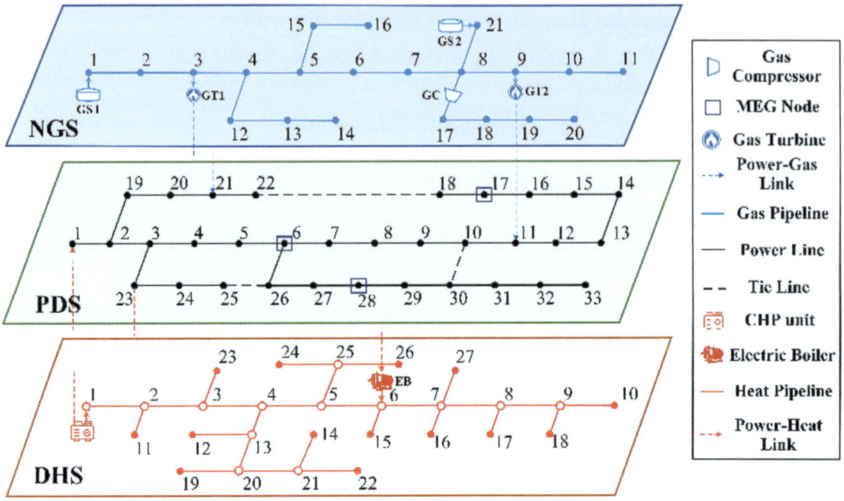

Fig. 7.7 IES E33-G21-H27 test case

First, the hierarchical fault scenario set is generated using the method described in Sect. 7.3. In the baseline case, the target earthquake magnitude range is evenly divided into 20 levels in ascending order, and 300 random fault scenarios are generated at each hierarchical magnitude using the Monte Carlo method. Figure 7.8 shows the trend of average lost load in IES under different disaster levels within the random scenarios. It can be observed that as the number of scenarios increases, the average lost load under different disaster levels gradually converges, and 300 random scenarios can ensure that the average lost load for each disaster level reaches near convergence.

In Fig. 7.9a, the magenta curve represents the fitted probability distribution of earthquake intensity, while the purple bar chart represents the occurrence probability of each classified magnitude. The proposed scenario generation method preserves the probability characteristics of different levels of disaster events, which enables

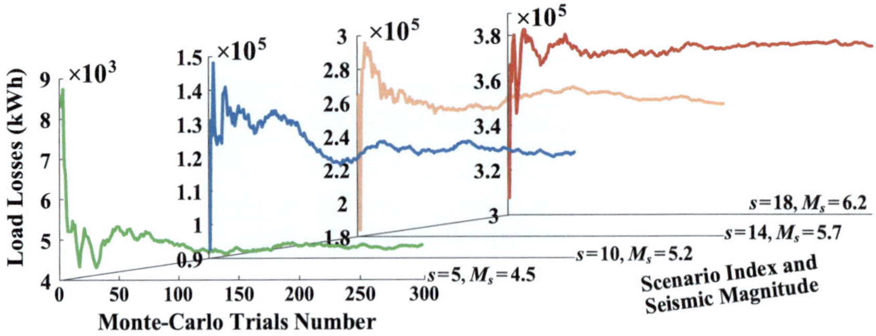

Fig. 7.8 Convergence curve of average lost load under different disaster levels

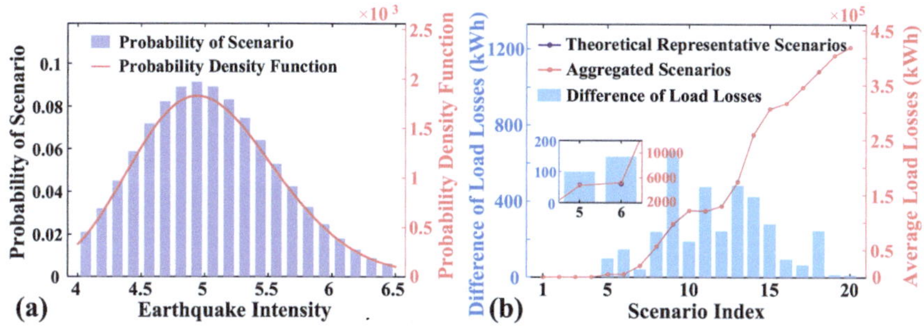

Fig. 7.9 **a** Seismic magnitude profile and probability of hierarchical seismic scenario; **b** Load losses differences between aggregated scenarios and theoretical representative scenarios

quantification of tail risk associated with low-probability, high-damage disasters. Figure 7.9b compares the lost load of selected typical scenarios with the average lost load of random scenarios for different disaster levels. The results show that the lost load differences for the typical scenario set are within an acceptable error range, demonstrating that the proposed hierarchical fault scenario generation method effectively reflects the differentiated impact of disasters on IES.

Fig. 7.10 illustrates the damage status of power lines, gas pipelines, and heating pipelines in the hierarchical fault scenarios, as well as the resilience planning scheme for IES based on the hierarchical fault scenarios.

Figure 7.10a–c shows that as the disaster level increases, the number of damaged components in the corresponding fault scenarios increases, which is consistent with the results in Fig. 7.9. Under a limited planning budget, components closer to energy supply sources (substations, natural gas stations, CHP units, and EBs) and critical loads are prioritized for reinforcement, helping to reduce the total lost load. Figure 7.10d presents the MEG dispatch results under hierarchical fault scenarios. Considering the deployment and scheduling costs of MEGs, no MEGs are deployed in low-level fault scenarios. However, in high-level fault scenarios, multiple MEGs are deployed in the planning scheme, indicating that the benefits of reducing lost load in high-level fault scenarios outweigh the deployment and scheduling costs.

To further validate the effectiveness of the R-SRIP model in enhancing IES resistance, Fig. 7.11 compares the emergency dispatch costs of IES before and after resilience planning. As shown in Fig. 7.11, without resilience planning, IES suffers significant economic losses under uncertain disaster scenarios. In contrast, after implementing resilience planning, the emergency dispatch costs for all disaster scenarios are significantly reduced. The CVaR indicator corresponding to extreme low-probability scenarios decreases from 8.31×10^6 \$ to 2.81×10^6 \$, a reduction of 66.1%, indicating that resilience planning can significantly reduce the extreme lost load risk caused by low-probability disaster events.

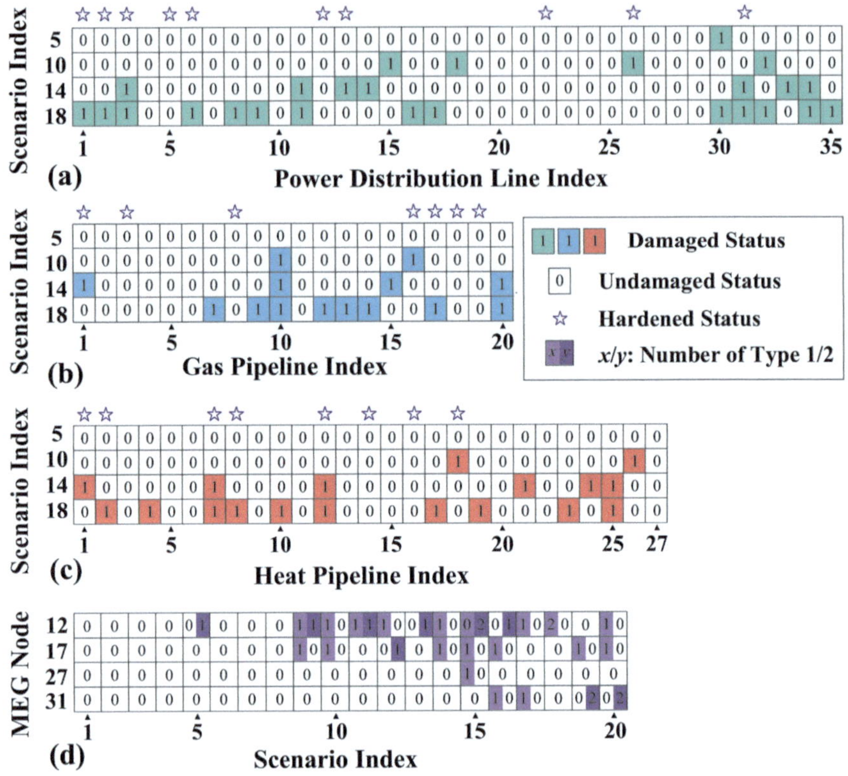

Fig. 7.10 Hierarchical fault scenarios and IES resilience planning scheme

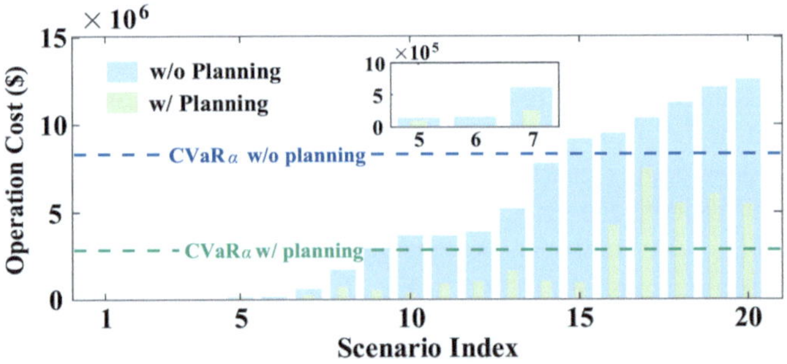

Fig. 7.11 Cost comparison before and after planning

7.6.2 Comparison with Existing Planning Methods

The confidence level α is set to 0.8, considering 20% of extreme scenarios to mitigate tail risk. This subsection compares the proposed method with three existing planning methods, with specific case settings as follows:

Case 1: Risk-neutral stochastic planning, focusing on the average performance of the plan under multiple fault scenarios.

Case 2: Robust planning, focusing on the worst-case fault scenario with the most severe lost load loss.

Case 3: Independent planning of electricity, gas, and heat systems. This method is based on the proposed resilience planning model but conducts resilience planning separately for the PDS, NGS and DHS.

Case 4: The proposed resilience planning method.

(1) ***Comparison of Planning Cost and Risk Control Effectiveness:*** The planning results of *Case 1* to *Case 4* are shown in Table 7.1.

Comparing *Case 1* and *Case 4*, although the total planning cost of *Case 4* is slightly higher, its CVaR value and emergency dispatch cost under the worst-case scenario are lower than those of *Case 1*, demonstrating better performance in reducing tail risk in extreme scenarios. *Case 2* has the lowest emergency dispatch cost under the worst-case scenario but incurs higher total planning costs due to its exclusive focus on extreme fault scenarios. In *Case 3*, where the PDS, NGS, and DHS are planned independently, the total planning cost is higher than that of *Case 4*, while its CVaR value and emergency dispatch cost in the worst-case scenario also increase. Compared with *Case 1* to *Case 3*, the R-SRIP method balances cost-effectiveness and resilience enhancement by effectively reducing tail risk while maintaining reasonable planning costs.

(2) ***Comprehensive Effectiveness of Planning Schemes under Random Fault Scenarios***: To further validate the performance of the R-SRIP method in future potential disaster events, this subsection generates 100 sets of random fault scenario collections and conducts out-of-sample testing for *Case 1* to *Case 4*. As shown in Fig. 7.12, the R-SRIP planning method outperforms other methods in both total planning cost and tail risk control effectiveness, demonstrating significant advantages in mitigating energy supply interruptions caused by potential

Table 7.1 Cost comparison of planning results among four cases (10^4\$)

	Investment	Operation	Total	$CVaR_\alpha$	Worst-case
Case 1	37.26	83.55	120.82	289.58	770.72
Case 2	38.71	153.22	191.93	–	434.85
Case 3	36.54	91.85	128.40	291.24	766.89
Case 4	37.98	86.60	124.59	281.77	749.04

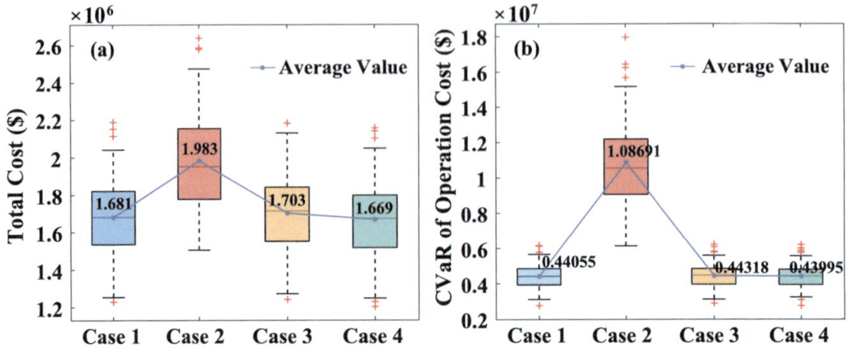

Fig. 7.12 Comparison of total planning cost and CVaR under out-of-sample scenarios

disasters. The above results confirm the superiority of the proposed planning method in addressing the high uncertainty of disasters.

7.6.3 Sensitivity Analysis of Planning Parameters

This section conducts a sensitivity analysis of key parameters in the planning model.

(1) ***Sensitivity Analysis of Risk Control Parameters*:** This subsection analyzes the impact of the risk control parameters α and β on the planning results. Figure 7.13 presents the efficient frontier curve of total planning cost and CVaR values under different confidence levels α. It can be observed that at the same confidence level α, as the risk control parameter β increases, the model's ability to control tail risk is enhanced. For example, when $\alpha = 0.95$, as β increases, the CVaR value decreases, indicating that the resilience planning scheme has improved risk control effectiveness. However, strengthening risk control requires more emergency resources, leading to an increase in the total cost of the resilience planning scheme. At the same β value, as α increases, the CVaR focuses more on rare but highly destructive extreme failure scenarios, which in turn increases the total planning cost. Therefore, efficient frontier curves corresponding to higher confidence levels α are positioned in the upper right section of Fig. 7.13.

As shown in Fig. 7.13, strengthening tail risk control in IES resilience planning requires a higher total planning cost. To achieve a balanced resilience planning scheme that accounts for both risk control and economic feasibility, the risk control parameters α and β must be appropriately set when constructing the model. Taking $\alpha = 0.8$ as an example, Fig. 7.14 illustrates the trend of planning cost and CVaR values as β varies. It can be observed that when $\beta \geq 0.4$, the marginal improvement in risk control effectiveness stabilizes. Therefore, in this case, setting $\beta = 0.4$ achieves an optimal balance between enhancing IES resilience and maintaining economic

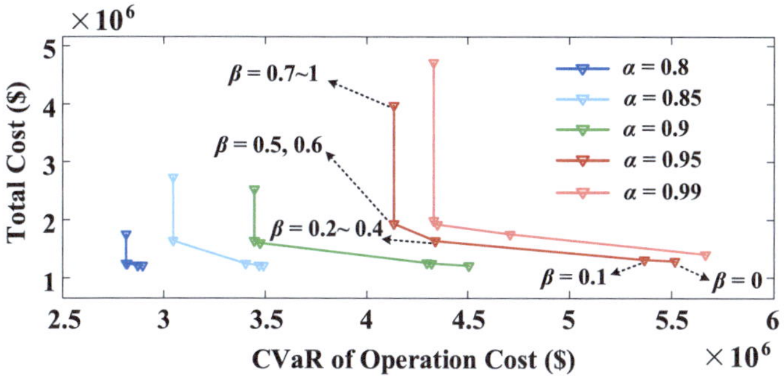

Fig. 7.13 Efficient frontier curves of total planning cost and CVaR value

feasibility. The above analysis provides a reference for setting risk parameters and making planning decisions in IES resilience planning.

Based on the above analysis, the R-SRIP method provides an effective framework for balancing risk control and economic feasibility in extreme disaster scenarios characterized by strong uncertainties. In practical applications, planners can reasonably set the values of α and β according to specific needs for resilience enhancement and risk control, ensuring that the formulated IES resilience planning scheme achieves the expected risk control targets within the budget constraints.

(2) ***Sensitivity Analysis of the Number of Tiered Failure Scenarios***: This subsection analyzes the impact of the number of tiered failure scenarios N_s on the planning results. Six sets of cases were designed, where N_s was set to 10, 15, 20, 30, 40, and 60, respectively. Using the method proposed in Sect. 7.3, 10 sets of tiered typical failure scenario sets were generated for each case, based on which the IES resilience planning scheme was formulated. Figure 7.15 presents the average and error bars of total planning costs corresponding to 10 scenario sets under six different cases.

The results indicate that within each case, the variation in total planning cost across different scenario sets is minimal, demonstrating the stability of the proposed tiered

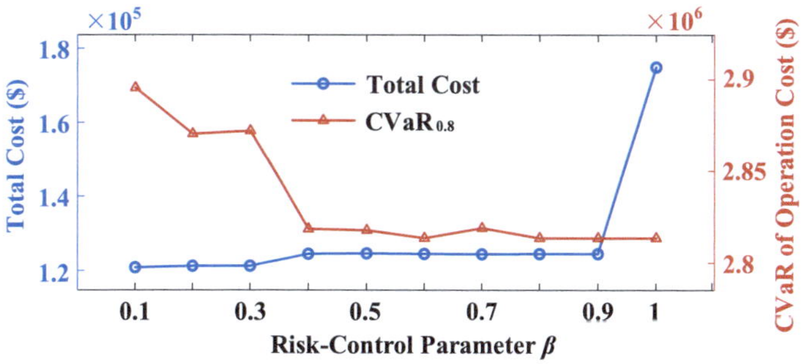

Fig. 7.14 Trend of planning cost and CVaR values with Changing β

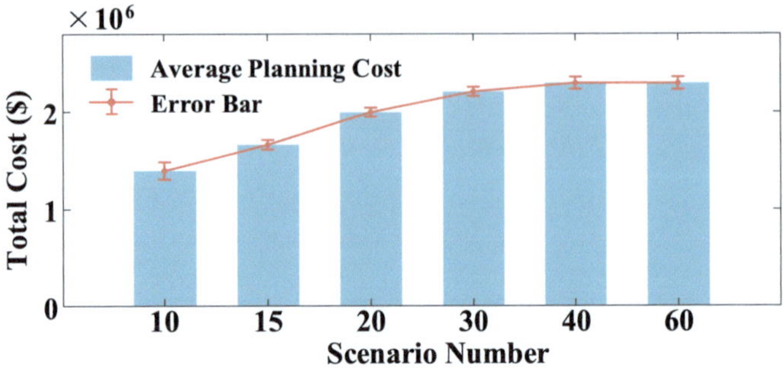

Fig. 7.15 Comparison of planning costs under different failure scenario numbers

failure scenario generation method. As N_s increases, the total planning cost rises. When $N_s \geq 30$, the increase in total cost slows down significantly, indicating that the proposed scenario generation method can effectively capture the destructiveness of disasters of different severity levels and identify tail risks caused by low-probability extreme events.

7.6.4 Computational Performance Testing of the Solution Algorithm

This subsection tests the computational performance of the proposed TBD algorithm. Table 7.2 compares the computational performance of the TBD algorithm with BD algorithm and Gurobi direct solving under different failure scenario settings. The columns "Obj. " "Iter. " "Com. " and "Gap" represent the objective function value, the number of iterations, the solution time, and the relative gap between the solution and the optimal value, respectively. The acceptable optimization gap and maximum solution time are set to 0.5% and 720 minutes, respectively.

As shown in Table 7.2, the TBD algorithm significantly outperforms the other two methods in solution time. Compared to the BD algorithm, the proposed algorithm introduces two types of enhanced cuts, which significantly accelerate the convergence of both the master and subproblems, thereby reducing the total solution time. Under different failure scenario settings, the proposed algorithm reduces solution time by 38.8% to 50.6% compared to the BD algorithm. The results validate the proposed algorithm's advantages in improving the computational efficiency of the R-SRIP model.

Table 7.2 Computational performance comparisons

Sce. Num.	TBD				BD[a]				Gurobi		
	Obj. (10^5 \$)	Iter.	Com. (min)	Gap[b](%)	Obj. (10^5 \$)	Iter.	Com. (min)	Gap[b](%)	Obj. (10^5 \$)	Com. (min)	Gap[c](%)
10	10.56	13	2.64	0.46	10.56	22	5.32	0.47	10.59	5.35	0.50
15	14.45	13	5.07	0.45	14.45	21	7.79	0.45	14.47	22.49	0.49
20	22.22	14	12.84	0.29	22.24	20	22.82	0.33	22.28	41.24	0.39
30	24.10	15	31.08	0.30	24.10	23	62.94	0.31	24.11	141.25	0.32
40	24.63	18	56.57	0.21	24.64	27	92.44	0.23	24.63	261.66	0.24
60	24.52	17	86.15	0.36	24.53	29	170.26	0.38	24.55	600.08	0.39

[a]The Benders Algorithm in [38] cannot converge within the time limit. Here, the enhanced Benders cuts corresponding to (7.86)–(7.87) are incorporated
[b]Gap here denotes the relative gap between UB and LB
[c]Gap here denotes the relative MIP optimality gap, derived from Gurobi software

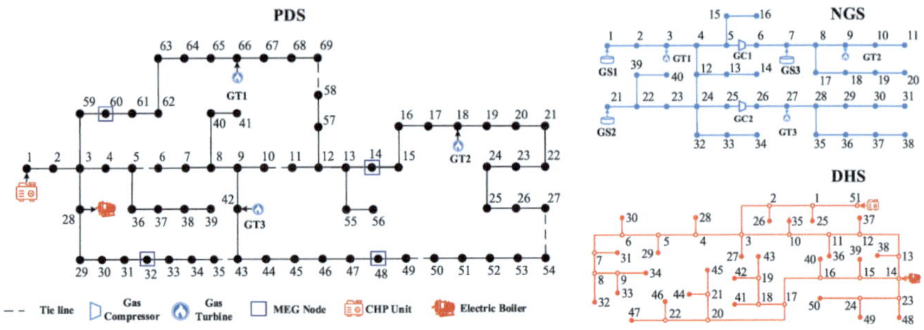

Fig. 7.16 IES E69-G40-H54 test case

7.6.5 *Scalability Testing*

To verify the scalability of the R-SRIP method, this subsection conducts scalability tests on the IES E69-G40-H54 system. The system topology is shown in Fig. 7.16. Based on the four case settings in Sect. 7.6.2, the planning results for the IES E69-G40-H54 system were tested and compared under *Case 1* to *Case 4*. Table 7.3 presents the planning cost comparison of different methods under these cases. The test results are consistent with those in Sect. 7.6.2, confirming that the proposed R-SRIP method effectively balances economic feasibility and risk control.

Subsequently, a scalability test was conducted on the solution algorithm in the IES E69-G40-H54 system. Considering the complexity of solving large-scale systems, the acceptable optimization gap was set to 0.05. Similar to the settings in Sect. 7.6.4, Table 7.4 presents the computational performance comparison of the TBD algorithm with two other solution methods under different failure scenario settings. The results show that compared to the BD algorithm, the proposed algorithm requires fewer iterations and reduces solution time by 50.5–68.3%. When the number of scenarios exceeds 30, Gurobi fails to converge to an acceptable optimization gap within the time limit. In contrast, the TBD algorithm can efficiently solve the planning model under all scenario settings. Overall, the R-SRIP model and solution algorithm demonstrate excellent scalability in large-scale systems.

Table 7.3 Cost comparison of planning results among four cases in scalability tests (10^4\$)

	Investment	Operation	Total	$CVaR_\alpha$	Worst-Case
Case 1	49.11	104.36	153.48	371.58	770.72
Case 2	49.87	193.06	242.93	–	434.85
Case 3	48.51	129.24	177.75	376.87	766.89
Case 4	49.33	121.41	170.74	322.32	749.04

Table 7.4 Computational performance comparisons in scalability tests

Sce. Num.	TBD				BD[a]				Gurobi		
	Obj. (10^5 $)	Iter.	Com. (min)	Gap (%)	Obj. (10^5 $)	Iter.	Com. (min)	Gap (%)	Obj. (10^5 $)	Com. (min)	Gap (%)
10	24.93	18	23.38	4.92	24.97	29	52.85	4.94	24.93	63.46	4.93
15	25.94	21	45.83	4.97	25.97	31	109.23	4.98	25.99	153.56	5.00
20	27.12	22	80.94	4.91	27.11	33	163.57	4.91	27.12	264.90	4.92
30	33.41	23	121.66	4.85	33.44	35	293.79	4.86	33.47	475.16	4.90
40	34.78	25	180.63	4.99	34.80	36	480.55	5.00	35.25	T[a]	6.25
60	34.94	26	262.38	4.97	34.96	35	827.64	4.98	37.17	T[a]	10.60

[a]T indicates exceeding the maximum solution time limit.

7.7 Conclusions

This chapter proposes a R-SRIP method that balances economic efficiency and risk under extreme events, considering the high uncertainty of extreme disasters and the diminishing marginal effect of disaster prevention planning. First, to account for the differentiated impacts of disasters of varying severity on the IES, a probabilistic hierarchical fault scenario generation method is developed based on stochastic fault simulation and average loss filtering. Then, an analytical formulation of tail-end risk for IES load loss under hierarchical disaster scenarios is established, based on which an IES resilience planning model with controllable economic-risk balance is constructed. Through a sensitivity analysis of planning costs and extreme load loss risks, a planning scheme that achieves an optimal trade-off between economic efficiency and risk resilience is determined. Finally, to address the computational challenges of large-scale models, a TBD algorithm is proposed, incorporating primary and dual cuts to accelerate convergence. Based on the case study simulation results, the following conclusions can be drawn:

1. The proposed hierarchical fault scenario generation method effectively captures the probabilistic characteristics of disasters of different severity levels and their varied impacts on the IES, providing a robust scenario set for resilience planning to mitigate extreme risks caused by low-probability events.
2. The R-SRIP model coordinates emergency resources across the PDS, NGS and DHS, enhancing the IES's capability to withstand disasters of different magnitudes. Compared with existing methods, the proposed approach can analyze the trade-off between planning costs and risk control effectiveness, enabling the formulation of a resilience planning strategy that balances resilience enhancement and cost control.
3. The proposed TBD algorithm can converge to the optimal solution within a limited number of iterations. Compared with Gurobi's direct solution and BD, the proposed algorithm significantly reduces the computation time of large-scale models and improves solving efficiency.

This chapter introduces an IES resilience planning method from the perspective of economic-risk balance, providing a solution to mitigate energy supply disruptions under extreme disaster events with high uncertainty. Future research will further explore information privacy protection and incentive compatibility in multi-entity planning of PDS, NGS, and DHS, as well as investigate the coupling relationship between extreme event uncertainty and resilience planning measures across different time scales.

References

1. Jasiūnas, J., Lund, P. D., & Mikkola, J. (2021). Energy system resilience—A review. *Renewable and Sustainable Energy Reviews, 150*, Article 111487.
2. Lead, F. A. T. (2023). Türkiye earthquake sequence on February 6, 2023.
3. Shen, Y. C., Gu, C. H., Xiang, Y., Fu, Y., Huo, D., Li, J. L., & Zhao, P. F. (2022). Dynamic service restoration for integrated energy systems under seismic stress. *IEEE Transactions on Sustainable Energy, 13*(1), 527–536. https://doi.org/10.1109/TSTE.2021.3097623
4. Wang, H., Hou, K., Zhao, J. B., Yu, X. D., Jia, H. J., & Mu, Y. F. (2022). Planning-oriented resilience assessment and enhancement of integrated electricity-gas system considering multi-type natural disasters. *Applied Energy, 315*, Article 118824. https://doi.org/10.1016/j.apenergy.2022.118824
5. Hasanzad, F., & Rastegar, H. (2022). Application of optimal hardening for improving resilience of integrated power and natural gas system in case of earthquake. *Reliability Engineering and System Safety, 223*, Article 108476. https://doi.org/10.1016/j.ress.2022.108476
6. Zhao, P., Gu, C., Cao, Z., Shen, Y., Teng, F., Chen, X., Wu, C., Huo, D., Xu, X., & Li, S. (2021). Data-driven multi-energy investment and management under earthquakes. *IEEE Transactions on Industrial Informatics, 17*(10), 6939–6950. https://doi.org/10.1109/TII.2021.3067895
7. Rajaei, F., Latify, M. A., & Ebrahimi, A. (2023). Incorporating DSO's situational awareness in resilience-oriented distribution system planning. *IEEE Transactions on Smart Grid, 14*(3), 1985–1997. https://doi.org/10.1109/TSG.2023.3249876
8. Espinoza, S., Poulos, A., Rudnick, H., de la Llera, J. C., Panteli, M., & Mancarella, P. (2020). Risk and resilience assessment with component criticality ranking of electric power systems subject to earthquakes. *IEEE Systems Journal, 14*(2), 2837–2848. https://doi.org/10.1109/JSYST.2019.2957440
9. Wang, Y., Rousis, A. O., & Strbac, G. (2021). A three-level planning model for optimal sizing of networked microgrids considering a trade-off between resilience and cost. *IEEE Transactions on Power Systems, 36*(6), 5657–5669. https://doi.org/10.1109/TPWRS.2021.3089085
10. Sun, Q. R., Wu, Z., Gu, W., Zhang, X. P., Liu, P. X., Pan, G. S., & Qiu, H. F. (2023). Tri-level multi-energy system planning method for zero energy buildings considering long- and short-term uncertainties. *IEEE Transactions on Sustainable Energy, 14*(1), 339–355. https://doi.org/10.1109/TSTE.2022.3211346
11. Liu, P., Wu, Z., Gu, W., Lu, Y., Sun, Q., Ye, Y., & Strbac, G. (2024). Fast resilience assessment of integrated electricity-gas systems considering non-convex gas flow dynamics. *IEEE Transactions on Smart Grid, 15*(4), 3514–3527. https://doi.org/10.1109/TSG.2024.3286789
12. Zou, Y., Xu, Y., & Zhang, C. (2023). A risk-averse adaptive stochastic optimization method for transactive energy management of a multi-energy microgrid. *IEEE Transactions on Sustainable Energy, 14*(3), 1599–1611. https://doi.org/10.1109/TSTE.2023.3267981
13. Cao, X., Sun, X., Xu, Z., Zeng, B., & Guan, X. (2022). Hydrogen-based networked microgrids planning through two-stage stochastic programming with mixed-integer conic recourse. *IEEE Transactions on Automation Science and Engineering, 19*(4), 3672–3685. https://doi.org/10.1109/TASE.2022.3144987

14. Du, Y., Xue, Y., Wu, W., Shahidehpour, M., Shen, X., Wang, B., & Sun, H. (2024). Coordinated planning of integrated electric and heating system considering the optimal reconfiguration of district heating network. *IEEE Transactions on Power Systems, 39*(1), 794–808. https://doi.org/10.1109/TPWRS.2024.3332156
15. Hu, B., & Wu, L. (2016). Robust SCUC considering continuous/discrete uncertainties and quick-start units: A two-stage robust optimization with mixed-integer recourse. *IEEE Transactions on Power Systems, 31*(2), 1407–1419. https://doi.org/10.1109/TPWRS.2015.2425148
16. Liu, X., Wang, Y., Liu, X., Zhao, T., & Wang, P. (2024). More efficient energy management for hybrid AC/DC microgrids with hard constrained neural network-based conversion loss surrogate models. *IEEE Transactions on Smart Grid, 15*(3), 2534–2552. https://doi.org/10.1109/TSG.2024.3321789
17. Zhao, T., Liu, X., Wang, P., & Blaabjerg, F. (2023). More efficient energy management for networked hybrid AC/DC microgrids with multivariable nonlinear conversion losses. *IEEE Systems Journal, 17*(2), 3212–3223. https://doi.org/10.1109/JSYST.2023.3267874
18. Dumas, J., Cointe, C., Wehenkel, A., Sutera, A., Fettweis, X., & Cornelusse, B. (2022). A probabilistic forecast-driven strategy for a risk-aware participation in the capacity firming market. *IEEE Transactions on Sustainable Energy, 13*(2), 1234–1243. https://doi.org/10.1109/TSTE.2021.3134567
19. Van Beesten, E. R., Romeijnders, W., & Roodbergen, K. J. (2024). Convex approximations of two-stage risk-averse mixed-integer recourse models. *Computational Optimization and Applications, 88*(1), 313–347. https://doi.org/10.1007/s10589-024-00423-5
20. Pribadi, K., Abduh, M., Wirahadikusumah, R., et al. (2021). Learning from past earthquake disasters: The need for a knowledge management system to enhance infrastructure resilience in Indonesia. *International Journal of Disaster Risk Reduction, 64*, Article 102424.
21. Panza, G., Kossobokov, V. G., Laor, E., et al. (2021). Earthquakes and sustainable infrastructure: Neodeterministic (NDSHA) approach guarantees prevention rather than cure. Elsevier.
22. Huang, W. J., Zhang, X., Li, K. P., et al. (2022). Resilience-oriented planning of urban multi-energy systems with generalized energy storage sources. *IEEE Transactions on Power Systems, 37*(4), 2906–2918.
23. Sharma, S., Verma, A., Xu, Y., & Panigrahi, B. K. (2019). Robustly coordinated bi-level energy management of a multi-energy building under multiple uncertainties. *IEEE Transactions on Sustainable Energy, 12*(1), 3–13.
24. Li, T., Han, X., Wu, W., et al. (2023). Robust expansion planning and hardening strategy of meshed multi-energy distribution networks for resilience enhancement. *Applied Energy, 341*, Article 121066.
25. Sun, Q., Wu, Z., Gu, W., et al. (2024). Resilience assessment for integrated energy system considering gas-thermal inertia and system interdependency. *IEEE Transactions on Smart Grid, 15*(2), 1509–1524.
26. Shen, Y. C., Gu, C. H., Yang, X. H., et al. (2021). Impact analysis of seismic events on integrated electricity and natural gas systems. *IEEE Transactions on Power Delivery, 36*(4), 1923–1931.
27. Shen, Y. C., Gu, C. H., Xiang, Y., et al. (2022). Dynamic service restoration for integrated energy systems under seismic stress. *IEEE Transactions on Sustainable Energy, 13*(1), 527–536.
28. Wang, H., Hou, K., Zhao, J. B., et al. (2022). Planning-oriented resilience assessment and enhancement of integrated electricity-gas system considering multi-type natural disasters. *Applied Energy, 315*, 118824.
29. Hasanzad, F., & Rastegar, H. (2022). Application of optimal hardening for improving resilience of integrated power and natural gas system in case of earthquake. *Reliability Engineering & System Safety, 223*, 108476.
30. Zhao, P., Gu, C., Cao, Z., et al. (2021). Data-driven multi-energy investment and management under earthquakes. *IEEE Transactions on Industrial Informatics, 17*(10), 6939–6950.
31. Poudyal, A., Poudel, S., & Dubey, A. (2023). Risk-based active distribution system planning for resilience against extreme weather events. *IEEE Transactions on Sustainable Energy, 14*(2), 1178–1192.

32. Rajaei, F., Latify, M. A., & Ebrahimi, A. (2023). Incorporating DSO's situational awareness in resilience-oriented distribution system planning. *IEEE Transactions on Smart Grid, 14*(3), 1985–1997.
33. Espinoza, S., Poulos, A., Rudnick, H., et al. (2020). Risk and resilience assessment with component criticality ranking of electric power systems subject to earthquakes. *IEEE Systems Journal, 14*(2), 2837–2848.
34. Bie, Z. H., Lin, Y. L., Li, G. F., et al. (2017). Battling the extreme: A study on the power system resilience. *Proceedings of the IEEE, 105*(7), 1253–1266.
35. Wang, Y., Rousis, A. O., & Strbac, G. (2021). A three-level planning model for optimal sizing of networked microgrids considering a trade-off between resilience and cost. *IEEE Transactions on Power Systems, 36*(6), 5657–5669.
36. Sun, Q. R., Wu, Z., Gu, W., et al. (2023). Tri-level multi-energy system planning method for zero energy buildings considering long- and short-term uncertainties. *IEEE Transactions on Sustainable Energy, 14*(1), 339–355.
37. Glasserman, P., Heidelberger, P., & Shahabuddin, P. (1999). Importance sampling and stratification for value-at-risk. Citeseer.
38. Nazemi, M., & Dehghanian, P. (2019). Seismic-resilient bulk power grids: Hazard characterization, modeling, and mitigation. *IEEE Transactions on Engineering Management, 67*(3), 614–630.
39. Venkateswaran, B., Saini, D. K., & Sharma, M. (2021). Techno-economic hardening strategies to enhance distribution system resilience against earthquake. *Reliability Engineering and System Safety, 213*, Article 107745.
40. Erenoglu, A. K., Sancar, S., Terzi, I. S., et al. (2022). Resiliency-driven multi-step critical load restoration strategy integrating on-call electric vehicle fleet management services. *IEEE Transactions on Smart Grid, 13*(4), 3118–3132.
41. Pasari, S. (2019). Inverse Gaussian versus lognormal distribution in earthquake forecasting: Keys and clues. *Journal of Seismology, 23*(3), 537–559.
42. Khan, M. Y., Shah, M. A., & Khanam, F. (2021). Earthquake stochastic modeling and estimating the probabilities of earthquake occurrences in the Hindu Kush region. *Arabian Journal of Geosciences, 14*(3), 1–16.
43. Shi, W., Liang, H., & Bittner, M. (2024). Stochastic planning for power distribution system resilience enhancement against earthquakes considering mobile energy resources. *IEEE Transactions on Sustainable Energy, 15*(1), 414–428.
44. Cao, X. Y., Cao, T. X., Gao, F., et al. (2021). Risk-averse storage planning for improving RES hosting capacity under uncertain siting choices. *IEEE Transactions on Sustainable Energy, 12*(4), 1984–1995.
45. Qiu, H. F., Wang, L., Gu, W., et al. (2022). Multistage scheduling of regional power grids against sequential outage and power uncertainties. *IEEE Transactions on Smart Grid, 13*(6), 4624–4637.
46. Zheng, W. Y., Huang, W. J., Hill, D. J., et al. (2021). An adaptive distributionally robust model for three-phase distribution network reconfiguration. *IEEE Transactions on Smart Grid, 12*(2), 1224–1237.
47. Zou, Y., Xu, Y., Feng, X., et al. (2023). Peer-to-peer transactive energy trading of a reconfigurable multi-energy network. *IEEE Transactions on Smart Grid, 14*(3), 2236–2249.
48. Zhang, G., Zhang, F., Zhang, X., et al. (2020). Mobile emergency generator planning in resilient distribution systems: A three-stage stochastic model with nonanticipativity constraints. *IEEE Transactions on Smart Grid, 11*(6), 4847–4859.
49. Du, Y., Xue, Y., Wu, W., et al. (2024). Coordinated planning of integrated electric and heating system considering the optimal reconfiguration of district heating network. *IEEE Transactions on Power Systems, 39*(1), 794–808.
50. Sun, Q. (n.d.). Supplementary material. Retrieved from https://drive.google.com/file/d/1Vomez LHGQVkDm20mvSVH0o0Ps1VsoDPe/view?usp=sharing
51. Liu, R. P., Hou, Y. H., Li, Y. J., et al. (2021). Sample robust scheduling of electricity-gas systems under wind power uncertainty. *IEEE Transactions on Power Systems, 36*(6), 5889–5900.

52. Hu, B., & Wu, L. (2016). Robust SCUC considering continuous/discrete uncertainties and quick-start units: A two-stage robust optimization with mixed-integer recourse. *IEEE Transactions on Power Systems, 31*(2), 1407–1419.
53. Taşkin, Z. C. (2010). Benders decomposition. *Wiley Encyclopedia of Operations Research and Management Science.*
54. Cao, X., Sun, X., Xu, Z., et al. (2022). Hydrogen-based networked microgrids planning through two-stage stochastic programming with mixed-integer conic recourse. *IEEE Transactions on Automation Science and Engineering, 19*(4), 3672–3685.
55. Mínguez, R., van Ackooij, W., & García-Bertrand, R. (2021). Constraint generation for risk-averse two-stage stochastic programs. *European Journal of Operational Research, 288*(1), 194–206.
56. Sun, Q. (n.d.). Test data. Retrieved from https://docs.google.com/spreadsheets/d/15DAQEg5K19QrhUsew_CbyPrJ3aw450vM/edit?pli=1#gid=679163298.

Chapter 8
Network-Based Modeling and Operation Optimization for Power-Hydrogen IESs under Multiple Uncertainties

Abstract Hydrogen-enriched compressed natural gas (HCNG) holds significant potential for renewable energy integration and hydrogen utilization. However, injecting hydrogen into the natural gas network alters the original fluid dynamics and complicates the physical properties of compressed gas, posing operational safety challenges to the electricity-HCNG-integrated energy system (E-HCNG-IES). To address this issue, this chapter examines the impact of HCNG on the dynamic characteristics of the gas network and proposes an improved HCNG network model, incorporating the influence of hydrogen blending into the pressure drop equation and linepack equation. Additionally, an optimal dispatch model for the E-HCNG-IES is developed, considering the full supply chain of HCNG, including production, storage, blending, transportation, and utilization. The dispatch problem is reformulated as a mixed-integer second-order cone programming (MISOCP) problem using second-order cone (SOC) relaxation and piecewise linearization techniques. To obtain a tighter solution, an iterative algorithm is introduced, combining the convex-concave procedure with a bound-tightening method. Finally, the proposed methodology is validated using two numerical test cases of E-HCNG-IES with varying hydrogen volume fractions. A detailed operational analysis indicates that although energy delivery capacity and pipeline linepack flexibility decrease with increasing hydrogen volume fraction, the E-HCNG-IES benefits economically and environmentally from higher hydrogen integration.

Keywords Hydrogen-enriched compressed natural gas · Improved HCNG network model · Electricity-HCNG-integrated energy system · Optimal dispatch

Nomenclature

WT, *wt*	Wind turbine
PV, *pv*	Photovoltaic cell
ELZ, *elz*	Electrolyzer
MR, *mr*	Methanation reactor

Z. Wu et al., *Integrated Energy System Resilience*,
https://doi.org/10.1007/978-981-95-2363-4_8

HST, *hst*	Hydrogen storage tank
GT, *gt*	Gas turbine
BZ, *bz*	Blending zone
CP, *cp*	Compressor
x	Pipeline length coordinate, m
t	Time coordinate, s
ρ	Gas density,kg/m^3
v	Gas velocity, m/s
p	Gas pressure, Pa
D	Pipeline diameter, m
θ	Pipeline inclination angle, °
λ	Pipeline friction coefficient
M	Gas molar mass, g/mol
Z	Gas compressibility factor
T	Gas temperature, K
m	Gas mass flow rate, kg/s
A	Pipeline cross-sectional area, m^2
q	Gas volume flow rate
p_i^t, p_j^t	Gas pressure at the inlet and outlet of pipeline ij, Pa
$q_{ij,\text{in}}^t, q_{ij,\text{out}}^t$	Gas flow rate at the inlet and outlet of pipeline ij, m^3/s
$gs \in \text{up}(i)$	Index of gas source node gs connected to node i
$j \in \text{up}(i)$	Index of upstream nodes relative to node i
$j \in \text{down}(i)$	Index of downstream nodes relative to node i
$gt \in \text{down}(i)$	Index of gas turbines connected to node i
q_{gs}^t	Gas flow rate injected into node i from gas source node gs, m^3/s
q_{gt}^t	Gas flow rate injected from node i into gas turbine, m^3/s
$L_{i,\text{gas}}^t$	Gas energy demand at node i, MW
$\tilde{S}_{ij}^t$	Linepack of pipeline ij, m^3
$\tilde{p}_{ij}^t$	Average gas pressure at the inlet and outlet of pipeline ij, Pa
p_n, T_n	Standard pressure and temperature, Pa, K
N_t	Index set of dispatch period t
Ω_{pipe}	Index set of pipeline ij in HCNG network
Ω_{node}	Index set of node i/j in HCNG network
N_{gs}	Index set of gas source node
N_{gt}	Index set of gas turbines
$k_{\text{hcng,buy}}^t$	Price of HCNG, ¥/m^3
$C_{ng,\text{buy}}^t$	Natural gas purchase cost, ¥
$C_{\text{res,curt}}^t$	Penalty cost for curtailed wind and solar power, ¥
$C_{\text{carb,tax}}^t$	Carbon tax cost, in ¥
C_{wheel}^t	Wheeling cost paid to public power grid, ¥
C_{om}^t	Operation and maintenance cost, ¥
$V_{ng,\text{buy}}^t$	Natural gas purchase volume of HCNG station, m^3/s
$k_{ng,\text{buy}}^t$	Natural gas price, ¥/m^3
$\varepsilon_{ng,\text{carb}}$	Carbon emission factor of natural gas, ton/m^3

$k_{\text{carb,tax}}$	Carbon tax price, ¥/ton
k_{wheel}	Unit electricity wheeling price, ¥/MWh
P^t_{grid}	Electric power transmitted through public grid, MW
$V^t_{\text{hst,ch/dis}}$	Charging/discharging flow rate of HST, m^3/s
$V^t_{\text{mr,elz/hst,ng}}$	Natural gas flow rate produced in MR from ELZ and HST, m^3/s
$V^t_{\text{bz,mix,ng/h}}$	Natural gas flow rate and hydrogen flow rate injected into BZ, m^3/s
$k_{\text{om.wt}}$	Unit maintenance cost of WT, \$/MWh, \$/m^3
$k_{\text{om.pv}}$	Unit maintenance cost of PV, \$/MWh, \$/m^3
$k_{\text{om.elz}}$	Unit maintenance cost of ELZ, \$/MWh, \$/m^3
$k_{\text{om.mr}}$	Unit maintenance cost of MR, \$/MWh, \$/m^3
$k_{\text{om.hst}}$	Unit maintenance cost of HST, \$/MWh, \$/m^3
$k_{\text{om.gt}}$	Unit maintenance cost of GT, \$/MWh, \$/m^3
$k_{\text{om.bz}}$	Unit maintenance cost of BZ, \$/MWh, \$/m^3
$P^t_{\text{gt/wt/pv}}$	Output electric power of GT/WT/PV, MW
P^t_{elz}	Input electric power of ELZ, MW
F^t_{nl}	Power flow of line *nl*, MW
$L^t_{n,\text{elec}}$	Electric load at node *n*, MW
X_{nl}	Line reactance of line *nl*, Ω
$\theta_{n/l,t}$	Phase angle of bus *n*/*l*
$V^t_{\text{elz},h}$	Hydrogen flow rate produced by ELZ, m^3/s
η_{elz}	Energy conversion efficiency of ELZ
$V^t_{\text{bz,elz},h}$	Hydrogen flow rate injected into BZ, m^3/s
$V^t_{\text{hst,ch}}$	Hydrogen flow rate injected into HST, m^3/s
$V^t_{\text{mr,elz},h}$	Hydrogen flow rate injected into MR, m^3/s
Cap_{elz}	Installed capacity of ELZ, MW
$\mu_{\text{elz,min / max}}$	Minimum and maximum output power coefficients of ELZ
η_{mr}	Energy conversion efficiency of MR
$V^t_{\text{mr,elz/hst},h}$	Hydrogen flow rate injected into MR from ELZ and HST, m^3/s
γ_{mr}	Electricity consumption coefficient of MR, kWh/m^3
Cap_{mr}	Installed capacity of MR, MW
S^t_{hst}	Stored hydrogen state of HST
Cap_{hst}	Installed capacity of HST, MW
$\mu_{\text{hst,min / max}}$	Minimum and maximum storage state of HST
σ_{hst}	Self-loss coefficient of HST
$\eta_{\text{hst,ch/dis}}$	Charging/discharging efficiency of HST
$V^t_{\text{bz,hst},h}$	Hydrogen flow rate withdrawn from HST and injected into BZ, m^3/s
$\gamma_{\text{hst,ch/dis}}$	Charging/discharging efficiency of HST
γ_{cp}	Electricity consumption coefficient of CP, kWh/m^3
$\tilde{q}_{\text{min / max}}$	Minimum/maximum pipeline flow rate, m^3/s
$\tilde{S}_{\text{min / max}}$	Minimum/maximum pipeline linepack, m^3

8.1 Introduction

With the continuous advancement of HCNG (hydrogen-enriched compressed natural gas) technology, centralized HCNG stations are increasingly becoming a key component at the upstream injection points of urban natural gas networks, such as city gate stations or pressure regulation stations, in the resilient planning of urban energy systems. As HCNG technology is further promoted and applied, a gradual transition toward distributed hydrogen blending is expected. Once the planning and design scheme for the electricity-HCNG-integrated energy system (E-HCNG-IES) is obtained, it is necessary to address the problem of operational optimization. This mainly includes the modeling and coordinated dispatch of the HCNG station with the downstream power network and HCNG network, as well as the impact of HCNG injection on the operational conditions of natural gas pipelines.

However, current studies on the dispatch of E-HCNG-IES rarely take into account the complex characteristics of the HCNG network. The gas pipeline flow models are typically derived from the fundamental equations of fluid dynamics under a series of assumptions [1]. These models, along with their simplified forms, are widely used for steady-state analysis and optimization dispatching of natural gas networks—such as the Weymouth equation. Some studies attempt to reflect the hydrogen volume fraction in HCNG by adjusting the transmission coefficient in the gas flow-pressure drop equation. However, whether this equation remains valid for HCNG networks with more complex gas compositions remains to be verified [2].

To achieve the low-carbon and efficient operation of the E-HCNG-IES, this chapter conducts a refined modeling study of the HCNG network under a centralized hydrogen blending mode. Based on this, an E-HCNG-IES operation optimization model and solution algorithm are proposed. First, considering the impact of hydrogen blending on gas dynamics, the pipeline flow-pressure drop equation and the linepack equation for the HCNG network are rederived and compared with the quasi-steady-state gas flow model based on the General Flow equation. Second, a dispatching model for the E-HCNG-IES is established, covering the "hydrogen production-blending-storage-transportation-utilization" links of the HCNG supply chain, along with the formulation of evaluation metrics for the HCNG network. Third, the original non-convex dispatching model is reformulated using second-order cone relaxation and piecewise linearization methods, and an iterative algorithm based on the convex-concave procedure and bound-tightening method is employed for solving the model. Finally, case studies are conducted on an E-HCNG-IES consisting of a 6-bus power grid+6-node gas network and a 24-bus power grid+20-node gas network to verify the effectiveness of the proposed methodology.

8.2 An improved Model of HCNG Network

This section considers the impact of gas composition variations on the parameters in the fundamental gas dynamics equations and derives the pipeline pressure drop-flow equation and linepack equation suitable for the HCNG network. Furthermore, the improved quasi-steady-state gas flow model for the HCNG network is compared with the traditional quasi-steady-state gas flow model based on the General Flow equation.

8.2.1 HCNG Network Modeling

Assuming the convective and inertial terms in the gas motion equation are omitted, Eq. (8.1) can be transformed into Eq. (8.2), which is used to describe a one-dimensional isothermal steady-state gas network [3]:

$$\frac{\partial(\rho v)}{\partial t}+\frac{\partial\left(\rho v^2\right)}{\partial x}=-\frac{\partial p}{\partial x}-g\rho\sin\theta-\frac{\lambda v^2}{2D}\rho \tag{8.1}$$

$$\frac{\partial p}{\partial x}+\frac{\lambda v^2}{2D}\rho=0 \tag{8.2}$$

Most existing studies on HCNG use volume fraction to describe the proportion of hydrogen in HCNG. Considering the real gas equation of state (8.3) and the density Eqs. (8.4), (8.2) can be transformed into Eq. (8.5).

$$\rho=\frac{pM}{ZRT} \tag{8.3}$$

$$m=\rho vA=\rho q \tag{8.4}$$

$$\frac{\mathrm{d}p}{\mathrm{d}x}+\frac{\lambda M}{2DRTA^2}\frac{p}{Z}\tilde{q}|\tilde{q}|=0 \tag{8.5}$$

For a uniformly mixed gas containing N_m gas components, its average molar mass M can be expressed as follows[4]:

$$M=\sum_{c=1}^{N_m}y_cM_c \tag{8.6}$$

Unlike conventional natural gas with a fixed composition, the compressibility factor Z of HCNG is no longer a constant. Under the isothermal steady-state assumption, its value depends on the hydrogen blending ratio y_{hcng} and the pressure level

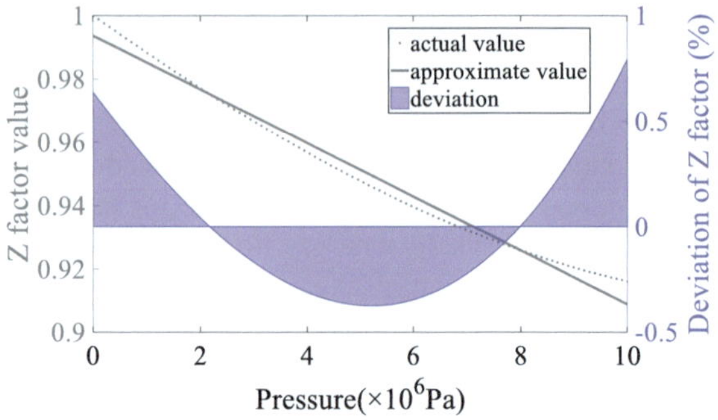

Fig. 8.1 Example of compressibility factor linear regression results for HCNG at a temperature of 288.15K and a hydrogen blending ratio of 20%

p. According to [5], when the hydrogen blending ratio is determined and remains below 50%, the relationship between *Z* and *p* can be approximated linearly using Eq. (8.7).

$$Z=\alpha p+\beta \tag{8.7}$$

where α and β are parameters related to the hydrogen blending ratio. The unit of α is Pa^{-1}, while β is dimensionless. The detailed calculation process can be seen in [5].

Figure 8.1 illustrates the linear approximation results of the compressibility factor *Z* for HCNG with a hydrogen blending ratio of 20% at a temperature of 288.15 K.

Specifically, in Fig. 8.1, the values of α and β are -8.44×10^{-9} and 0.99337, respectively. The gas pressure varies from 0.01×10^5 Pa to 100×10^5 Pa in increments of 1.01×10^5 Pa (a total of 100 data points). The average deviation between the approximated and actual values of *Z* is approximately 0.29%, which falls within an acceptable range.

Based on Eqs. (8.7) and (8.5) can be further rewritten as Eq. (8.8).

$$\frac{\mathrm{d}p}{\mathrm{d}x}+\frac{\lambda M}{2DRTA^2}\frac{p}{\alpha p+\beta}\tilde{q}|\tilde{q}|=0 \tag{8.8}$$

By integrating Eq. (8.8) over the pipeline length coordinate interval from the inlet $(x=0,\ p=p_i)$ to the outlet $(x=L_{ij},\ p=p_j)$, the relationship between the pressure drop at both ends of the pipeline and the average pipeline flow can be obtained. This relationship is referred to as the pipeline flow-pressure drop equation.

$$p_i^t-p_j^t+\frac{\beta}{\alpha}\left(\ln p_i^t-\ln p_j^t\right)=\frac{\lambda ML_{ij}}{2DRTA^2\alpha}\left(\tilde{q}_{ij}^t\right)^2, \forall ij, t \tag{8.9}$$

where $\tilde{q}_{ij}^t=\left(q_{ij,\mathrm{in}}^t+q_{ij,\mathrm{out}}^t\right)/2$.

To compare with the quasi-steady-state gas flow model based on the General Flow equation, the logarithmic term in Eq. (8.9) is incrementally linearized over the interval $[p_{\min}, p_{\max}]$. Meanwhile, the bilinear term in Eq. (8.9) is linearized using the special ordered sets 2 (SOS2) method over the interval $[\tilde{q}_{\min}, \tilde{q}_{\max}]$. The detailed procedures of the incremental linearization method and the SOS2 method can be found in [6].

The HCNG network must satisfy the node gas flow balance law:

$$LHV_{\text{hcng}}\left(\sum_{gs\in \text{up}(i)} q_{\text{gs}}^{t} + \sum_{j\in \text{up}(i)} q_{ij,\text{out}}^{t} - \sum_{j\in \text{down}(i)} q_{ij,\text{in}}^{t} - \sum_{gt\in \text{down}(i)} q_{\text{gt}}^{t}\right) \times 3600 \times 10^{-3} = L_{i,\text{gas}}^{t}, \forall i, t \tag{8.10}$$

where $LHV_{\text{hcng}} = y_{\text{hcng}}LHV_{h} + (1 - y_{\text{hcng}}LHV_{\text{ng}})$; q_{gs}^{t} and $q_{ij,\text{ out}}^{t}$ represent gas injection from source nodes and upstream pipelines connected to node i; $q_{ij,\text{ in}}^{t}$ and q_{gt}^{t} represent gas flowing out of node i to its downstream pipelines and gas turbines; $L_{i,\text{ gas}}^{t}$ denotes gas loads in the unit of MW.

The injection flow at gas source nodes and the gas pressure at each node must satisfy their respective upper and lower bound constraints:

$$q_{\text{gs,min}} \le q_{\text{gs}}^{t} \le q_{\text{gs,max}}, \forall gs, t \tag{8.11}$$

$$p_{i,\text{min}} \le p_{i}^{t} \le p_{i,\text{max}}, \forall i, t \tag{8.12}$$

The pipeline linepack state is determined by the physical parameters of the pipeline and the average pressure level, and it is expressed as follows [7, 8]:

$$\tilde{S}_{ij}^{t} = \frac{A_{ij}L_{ij}T_{n}}{p_{n}T}\frac{\tilde{p}_{ij}^{t}}{Z}, \forall ij, t \tag{8.13}$$

$$\tilde{S}_{ij}^{t} = \tilde{S}_{ij}^{t-1} + \left(q_{ij,\text{in}}^{t} - q_{ij,\text{out}}^{t}\right)\Delta t \times 3600, \forall ij, t \tag{8.14}$$

where $\tilde{p}_{ij}^{t} = \left(p_{i}^{t} + p_{j}^{t}\right)/2$.

Considering that the gas transmitted in the pipeline is hydrogen-enriched compressed natural gas (HCNG), the compressibility factor Z in Eq. (8.13) can be replaced by $\tilde{Z} = \alpha\tilde{p} + \beta$. Consequently, Eq. (8.13) can be transformed into Eq. (8.15).

$$\frac{\alpha p_{n}T}{A_{ij}L_{ij}T_{n}}\tilde{S}_{ij}^{t}\left(p_{i}^{t} + p_{j}^{t}\right) + \frac{2\beta p_{n}T}{A_{ij}L_{ij}T_{n}}\tilde{S}_{ij}^{t} = \left(p_{i}^{t} + p_{j}^{t}\right), \forall ij, t \tag{8.15}$$

Additionally, the total linepack of the HCNG network must be periodically restored to its initial level to ensure the repeatability of the dispatch instructions. The constraint is expressed as follows:

$$\sum_{t\in N_{t}} \sum_{ij\in \Omega_{\text{pipe}}} \tilde{S}_{ij}^{t} = \sum_{ij\in \Omega_{\text{pipe}}} \tilde{S}_{ij}^{1} \tag{8.16}$$

Equations (8.9)–(8.12) and (8.14)–(8.16) together constitute the improved quasi-steady-state flow model for the HCNG network.

8.2.2 Comparison with General Flow Model

In the traditional quasi-steady-state gas flow model based on the General Flow equation, the compressibility factor Z is regarded as a constant. The pipeline pressure drop-flow equation is expressed as follows:

$$\left(\frac{\pi}{4}\right)^2 \frac{D_{ij}^5 R T_n^2}{\lambda L_{ij} p_n^2 M Z T}\left[\left(p_i^t\right)^2 - \left(p_j^t\right)^2\right] = \left(\tilde{q}_{ij}^t\right)^2, \forall ij, t \tag{8.17}$$

The bilinear term in Eq. (8.17) is also linearized using the SOS2 method, while the hydrogen blending ratio in HCNG is reflected by adjusting the gas molar mass M. Additionally, the linepack equation follows Eq. (8.13).

To reveal the impact of hydrogen blending in HCNG on gas network dispatching under different pressure levels, this section conducts validation analysis using a 37-node gas network at 2-bar and a 6-node gas network at 60 bar, as shown in Fig. 8.2. In Fig. 8.2a, node 0 serves as the source node in the 37-node gas network. In Fig. 8.2b, nodes 4 and 6 act as source nodes in the 6-node gas network. The detailed system parameter configuration can be found in [9].

The gas network dispatching model based on the quasi-steady-state gas flow model using the General Flow equation is denoted as (M0), while the dispatching model based on the improved quasi-steady-state flow model for the HCNG network is denoted as (M1). The mathematical formulations of (M0) and (M1) are presented in Eqs. (8.18) and (8.19), respectively.

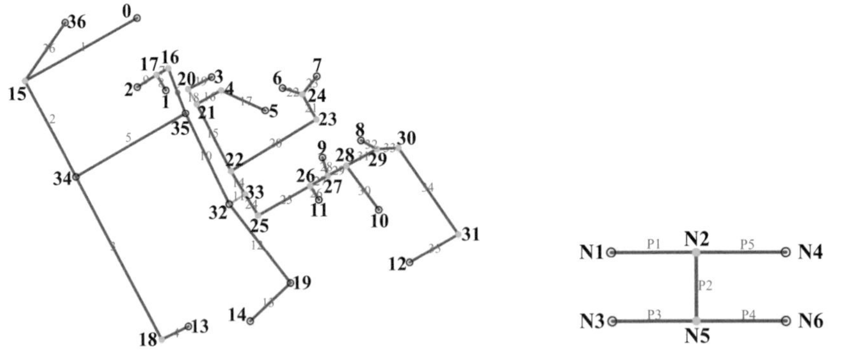

(a) 37-node gas network (b) 6-node gas network

Fig. 8.2 Gas network topologies of two examples.

$$\begin{cases} \min C_{\text{hcng}} = \sum\limits_{t \in N_t} \left(k^t_{\text{hcng,buy}} \sum\limits_{\text{gs} \in N_{\text{gs}}} q^t_{\text{gs}} \right) \Delta t \times 3600 \\ \text{s.t.}(8.10) - (8.14), (8.16), (8.17) \end{cases} \tag{8.18}$$

$$\begin{cases} \min C_{\text{hcng}} = \sum\limits_{t \in N_t} \left(k^t_{\text{hcng,buy}} \sum\limits_{\text{gs} \in N_{\text{gs}}} q^t_{\text{gs}} \right) \Delta t \times 3600 \\ \text{s.t.}(8.9) - (8.12), (8.14) - (8.16) \end{cases} \tag{8.19}$$

In Eqs. (8.18) and (8.19), N_{gs} represents the set of all gas source nodes; $k^t_{\text{hcng,buy}}$ denotes the price of HCNG at time t, ¥/m^3.It can be observed that the objective function in both models is the same, aiming to minimize the total daily cost of HCNG. The difference between the two models lies in the constraints they incorporate.

In the two examples shown in Fig. 8.2, the HCNG hydrogen blending ratio y_{hcng} varies from 1% to 20% in increments of 1%. Taking the optimized solution of dispatching model (M1) as the baseline, the deviation of the optimized solution of dispatching model (M0) relative to this baseline is calculated and illustrated in Fig. 8.3, where the gray line represents the scenario of the 37-node gas network at 2 bar, and the purple line represents the scenario of the 6-node gas network at 60 bar.

In the 37-node gas network at 2 bar, the maximum deviation in the optimization results is 0.0118%, with an average deviation of approximately 0.008%. In the 6-node gas network at 60 bar, the maximum deviation reaches 7.2359%, while the average deviation is around 4.81%. The average deviation in the 60-bar scenario is 601.25 times that of the 2-bar scenario, indicating that hydrogen blending has a significantly greater impact on high-pressure gas networks compared to low-pressure gas networks.

Taking the 6-node gas network at 60 bar as an example, when the HCNG hydrogen blending ratio y_{hcng} is 20%, the optimized results of node pressures and pipeline gas flows under the two dispatching models are shown in Fig. 8.4. The dashed

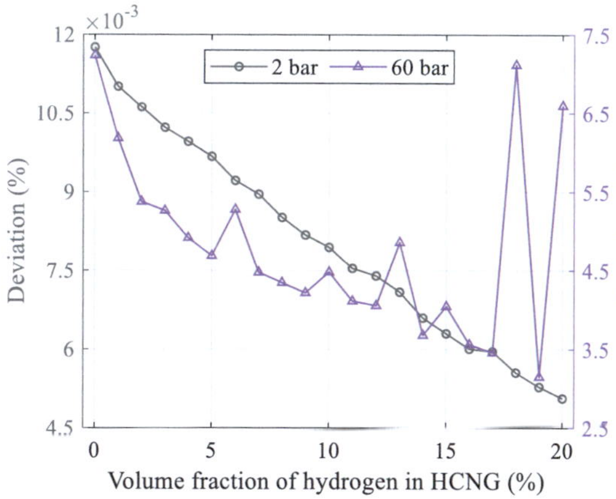

Fig. 8.3 Deviation of the optimization results of the objective function

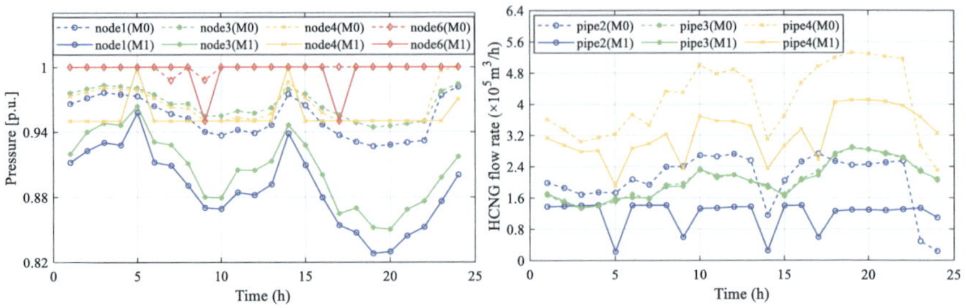

(a) Node pressure (b) HCNG flow rate

Fig. 8.4 Comparison of the optimization results of network operation conditions

lines represent the dispatching results of model (M0). The solid lines represent the dispatching results of model (M1).

It can be observed that the operating conditions of the gas network obtained using these two models differ significantly. In general, for high-pressure HCNG networks, assuming a fixed hydrogen blending ratio, the variation in the compressibility factor Z caused by pressure changes cannot be ignored. If the traditional gas flow model which assumes a constant compressibility factor is still used, it will lead to non-negligible deviations in the results. Both the improved HCNG network model and the traditional quasi-steady-state model based on the General Flow equation involve bilinear terms. However, the former contains more complex bilinear terms and additionally introduces logarithmic terms, which can be seen as the computational cost of considering the impact of pressure variations on the compressibility factor. To address these bilinear and logarithmic terms, a more effective solution will be presented in Sect. 8.4.1.

8.3 Optimal Dispatch Model for E-HCNG-IES

In this section, an operation optimization model is developed for the electricity-HCNG-integrated energy system (E-HCNG-IES) under the centralized hydrogen blending mode, as shown in Fig. 8.5. Next, the "production-storage-blending-transportation-utilization" link of the HCNG supply chain is introduced.

The centralized HCNG station integrates multiple functions, including renewable power generation, hydrogen production, compression, storage, methanation, and blending. In these stations, renewable energy resources are abundant, with power generation units consisting of wind turbines and photovoltaic panels. The generated electricity has two primary uses: supplying power to downstream electricity consumers via the public grid and meeting the internal electricity demand of equipment such as electrolyzers, compressors, and methanation reactors [10]. The hydrogen produced from electrolysis is compressed by compressors and can be stored in hydrogen storage tanks, injected into methanation reactors to convert into

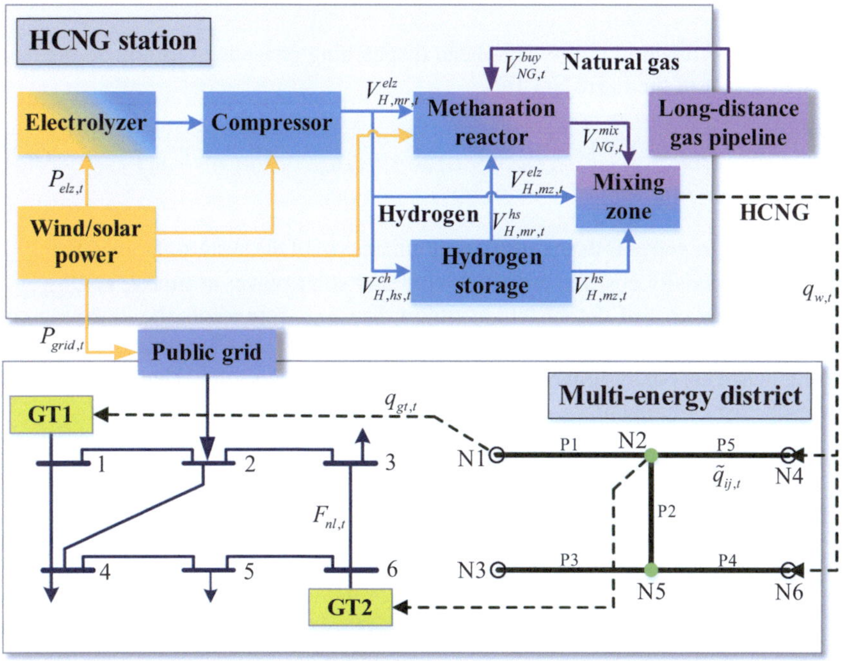

Fig. 8.5 Schematic diagram of E-HCNG-IES structure

synthetic natural gas, or directly blended with natural gas in the gas mixing zone to produce HCNG [11]. When hydrogen demand in the downstream region reaches peak levels, the HCNG station extracts hydrogen from storage tanks to supply both the methanation reactor and the gas blending zone. It is important to note that the primary gas source for the HCNG station remains the long-distance natural gas transmission pipeline. The HCNG output from the gas blending zone is injected into a downstream multi-energy load area, where it is transported via the HCNG network to meet the gas and electricity demands of industrial, commercial, and residential users. Furthermore, the downstream HCNG network is coupled with the electricity network through gas turbines or other gas-to-power units.

The centralized hydrogen blending mode shown in Fig. 8.5 can apply to the early stage of HCNG development and promotion. This mode can support the operation and management of HCNG demonstration projects and facilitate regional or industrial park-scale HCNG utilization.

8.3.1 Objective Function

The objective function in the day-ahead dispatching phase is to minimize the daily operating cost of the E-HCNG-IES:

$$\min T_{\text{cost}} = \sum_{t \in N_t} \left(C_{\text{ng,buy}}^t + C_{\text{res,curt}}^t + C_{\text{carb,tax}}^t + C_{\text{wheel}}^t + C_{\text{om}}^t \right) \tag{8.20}$$

where $C_{\text{ng,buy}}^t$ represents the natural gas purchase cost of the system at time t, ¥; $C_{\text{res,curt}}^t$ denotes the penalty cost for curtailed wind and solar power at time t, ¥; $C_{\text{carb,tax}}^t$ is the carbon tax cost of the system at time t, ¥; C_{wheel}^t represents the wheeling cost paid by the system to the public electricity grid at time t, ¥; C_{om}^t is the operation and maintenance cost of system equipment at time t, ¥. The specific calculation methods for each cost component are detailed as follows:

$$C_{\text{ng,buy}}^t = k_{\text{ng,buy}}^t V_{\text{ng,buy}}^t \Delta t \times 3600 \tag{8.21}$$

$$C_{\text{res,curt}}^t = \left[k_{\text{wt,curt}}^t \left(P_{\text{wt,fore}}^t - P_{\text{wt}}^t \right) + k_{\text{pv,curt}}^t \left(P_{\text{pv,fore}}^t - P_{\text{pv}}^t \right) \right] \Delta t \tag{8.22}$$

$$C_{\text{carb,tax}}^t = k_{\text{carb,tax}} \varepsilon_{\text{ng,carb}} V_{\text{ng,buy}}^t \Delta t \times 3600 \tag{8.23}$$

$$C_{\text{wheel}}^t = k_{\text{wheel}} P_{\text{grid}}^t \Delta t \tag{8.24}$$

$$C_{\text{om}}^t = \begin{bmatrix} k_{\text{om,wt}} P_{\text{wt}}^t + k_{\text{om,pv}} P_{\text{pv}}^t + k_{\text{om,elz}} P_{\text{elz}}^t \\ + k_{\text{om,gt}} \sum_{gt \in N_{gt}} P_{\text{gt}}^t + k_{\text{om,hst}} \left(V_{\text{hst,ch}}^t + V_{\text{hst,dis}}^t \right) \times 3600 \\ + k_{\text{om,mr}} \left(V_{\text{mr,elz,ng}}^t + V_{\text{mr,hst,ng}}^t \right) \times 3600 \\ + k_{\text{om,bz}} \left(V_{\text{bz,mix,ng}}^t + V_{\text{bz,mix,h}}^t \right) \times 3600 \end{bmatrix} \Delta t \tag{8.25}$$

Equation (8.21) provides the expression for natural gas purchase cost, where $V_{\text{ng,buy}}^t$ represents the natural gas purchase volume of the HCNG station at time t, m^3/s, and $k_{\text{ng,buy}}^t$ denotes the natural gas price at time t, ¥/m^3.

The penalty cost due to curtailed wind and solar power is given by Eq. (8.22), where $P_{\text{wt,fore}}^t$ and $P_{\text{pv,fore}}^t$ represent the forecasted output power of the wind turbines and photovoltaic panels at time t, respectively, MW; P_{wt}^t and P_{pv}^t denote the actual output power of the wind turbines and photovoltaic panels at time t, respectively, MW; $k_{\text{wt,curt}}^t$ and $k_{\text{pv,curt}}^t$ represent the penalty cost per unit of curtailed wind and solar power at time t, respectively, ¥/MWh.

The carbon tax cost resulting from natural gas consumption is calculated using Eq. (8.23), where $\varepsilon_{\text{ng,carb}}$ represents the carbon emission factor of natural gas, ton/m^3, and $k_{\text{carb,tax}}$ denotes the carbon tax price, ¥/ton.

In this chapter, the excess green electricity from the HCNG station is transmitted to the downstream power network via the public grid, and it is assumed that the system must pay a wheeling cost to the public grid, as expressed in Eq. (8.24), where k_{wheel} represents the unit electricity wheeling price, ¥/MWh, and P_{grid}^t denotes the electric power transmitted through the public grid at time t, MW.

Equation (8.25) provides the specific calculation method for the operation and maintenance costs of the equipment and facilities involved in both the HCNG station and the E-HCNG-IES network, where P_{elz}^t represents the input electric power of the electrolyzer at time t, MW; P_{gt}^t denotes the output electric power of the gas turbine at time t, MW; $V_{\mathrm{hst,ch}}^t$ and $V_{\mathrm{hst,dis}}^t$ represent the injection and withdrawal flow rates of the hydrogen storage tank at time t, respectively, m^3/s; $V_{\mathrm{mr,elz,ng}}^t$ and $V_{\mathrm{mr,hst,ng}}^t$ denote the synthetic natural gas flow rates produced in the methanation reactor from hydrogen from the electrolyzer and the hydrogen storage tank at time t, respectively, m^3/s; $V_{\mathrm{bz,mix,ng}}^t$ and $V_{\mathrm{bz,mix,h}}^t$ represent the total natural gas flow and total hydrogen flow injected into the gas blending zone at time t, respectively, m^3/s; N_{gt} represents the set of all gas turbine; $k_{\mathrm{om,wt}}$, $k_{\mathrm{om,pv}}$, $k_{\mathrm{om,elz}}$, $k_{\mathrm{om,gt}}$, $k_{\mathrm{om,hst}}$, $k_{\mathrm{om,mr}}$, $k_{\mathrm{om,bz}}$ denote the unit operation and maintenance cost costs for wind turbines, photovoltaic panels, electrolyzers, gas turbines, hydrogen storage tanks, methanation reactors, and the gas blending zone, respectively. Depending on the equipment type, the unit cost is given in either ¥/MWh or ¥/m^3.

8.3.2 *Constraints*

The operation optimization constraints of the E-HCNG-IES shown in Fig. 8.5 include HCNG network constraints, power network constraints, operational constraints of HCNG station equipment and facilities, and multi-energy flow supply-demand balance constraints. Among these, the HCNG network constraints have already been presented in Sect. 8.2.1.

Power Network Constraints: The power network in this chapter is modeled using the DC steady-state power flow model, with the relevant constraints expressed as follows [12]:

$$\sum_{l\rightarrow n} F_{\mathrm{nl}}^t + \sum_{\mathrm{gt}\rightarrow n} P_{\mathrm{gt}}^t = L_{n,\mathrm{elec}}^t, \forall n, t \tag{8.26}$$

$$F_{\mathrm{nl}}^t = \left(\theta_n^t - \theta_l^t\right)/X_{\mathrm{nl}}, \forall nl, t \tag{8.27}$$

$$-F_{\mathrm{nl,max}} \leq F_{\mathrm{nl}}^t \leq F_{\mathrm{nl,max}}, \forall nl, t \tag{8.28}$$

$$0 \leq P_{\mathrm{grid}}^t \leq P_{\mathrm{grid,max}}, \forall t \tag{8.29}$$

Operational Constraints of GT: In Fig. 8.5, the gas turbine is the key equipment that enables the coupling between the downstream power network and the HCNG network. In this chapter, heat load is not considered separately, so only the electrical output characteristics of the gas turbine are taken into account, as expressed in Eqs. (8.30)–(8.32).

$$P_{\mathrm{gt}}^{t} = \left[3.031\left(1 - y_{\mathrm{hcng}}\right)q_{\mathrm{gt}}^{t} + 1.019y_{\mathrm{hcng}}q_{\mathrm{gt}}^{t}\right] \times 3600 \times 10^{-3}, \forall t \tag{8.30}$$

$$\mu_{\mathrm{gt,min}}\mathrm{Cap}_{\mathrm{gt}} \leq P_{\mathrm{gt}}^{t} \leq \mu_{\mathrm{gt,min}}\mathrm{Cap}_{\mathrm{gt}}, \forall t \tag{8.31}$$

$$-\pi_{\mathrm{gt,max}}\mathrm{Cap}_{\mathrm{gt}}\Delta t \leq P_{\mathrm{gt}}^{t} - P_{\mathrm{gt}}^{(t-1)} \leq \pi_{\mathrm{gt,max}}\mathrm{Cap}_{\mathrm{gt}}\Delta t, \forall t \tag{8.32}$$

Renewable Energy Generation: The operational constraints for wind turbines are given in Eqs. (8.33) and (8.34), while the operational constraints for photovoltaic panels are provided in Eqs. (8.35) and (8.36).

$$P_{\mathrm{wt,fore}}^{t} = 0.5 \times 10^{-6} \times \pi \times \eta_{\mathrm{wt}}\rho_{\mathrm{air}}N_{\mathrm{wt}}R_{\mathrm{wt}}^{2}\left(u^{t}\right)^{3}, \forall t \tag{8.33}$$

$$0 \leq P_{\mathrm{wt}}^{t} \leq P_{\mathrm{wt,fore}}^{t}, \forall t \tag{8.34}$$

$$P_{\mathrm{pv,fore}}^{t} = \Re^{t} \cos\theta_{\mathrm{pv}}\eta_{\mathrm{MPPT}}A_{\mathrm{pv}}\eta_{\mathrm{pv}} \times 10^{-6}, \forall t \tag{8.35}$$

$$0 \leq P_{\mathrm{pv}}^{t} \leq P_{\mathrm{pv,fore}}^{t}, \forall t \tag{8.36}$$

Operational Constraints of ELZ: As expressed in Eq. (8.37), the hydrogen produced by the electrolyzer has three possible destinations: it can be directly injected into the gas blending zone, stored in the hydrogen storage tank, or supplied to the methanation reactor for the production of synthetic natural gas.

$$\begin{cases} \eta_{\mathrm{elz}}P_{\mathrm{elz}}^{t} = LHV_{h}V_{\mathrm{elz},h}^{t} \times 3600 \times 10^{-3} \\ V_{\mathrm{elz},h}^{t} = V_{\mathrm{bz,elz},h}^{t} + V_{\mathrm{hst,ch}}^{t} + V_{\mathrm{mr,elz},h}^{t} \end{cases}, \forall t \tag{8.37}$$

The electrical power input of the electrolyzer must comply with the following constraints:

$$\mu_{elz,\mathrm{min}}\mathrm{Cap}_{\mathrm{elz}} \leq P_{\mathrm{elz}}^{t} \leq \mu_{\mathrm{elz,max}}\mathrm{Cap}_{\mathrm{elz}}, \forall t \tag{8.38}$$

Operational Constraints of MR: The hydrogen supply for the methanation reactor originates from both the electrolyzer and the hydrogen storage tank. Its operational state must also comply with installed capacity constraints [13].

$$V_{\text{mr,elz,ng}}^t = \eta_{\text{mr}} V_{\text{mr,elz},h}^t LHV_h / LHV_{\text{ng}}, \forall t \tag{8.39}$$

$$V_{\text{mr,hst,ng}}^t = \eta_{\text{mr}} V_{\text{mr,hst},h}^t LHV_h / LHV_{\text{ng}}, \forall t \tag{8.40}$$

$$0 \le \gamma_{\text{mr}} \left(V_{\text{mr,elz,ng}}^t + V_{\text{mr,hst,ng}}^t \right) \times 3600 \times 10^{-3} \le \text{Cap}_{\text{mr}}, \forall t \tag{8.41}$$

Operational Constraints of HST: The HCNG station is equipped with hydrogen storage facilities to manage the intra-day variability and intermittency of renewable energy sources. In this chapter, compressed hydrogen storage tanks are considered, with their operational constraints defined as follows:

$$\mu_{\text{hst,min}} Cap_{\text{hst}} \le S_{\text{hst}}^t \le \mu_{\text{hst,max}} \text{Cap}_{\text{hst}}, \forall t \tag{8.42}$$

$$S_{\text{hst}}^1 = S_{\text{hst}}^{N_t} = \text{Cap}_{\text{hst}} (\mu_{\text{hst,min}} + \mu_{\text{hst,max}})/2 \tag{8.43}$$

$$S_{\text{hst}}^t = S_{\text{hst}}^{t-1} (1 - \sigma_{\text{hst}}) + \left(V_{\text{hst,ch}}^t \eta_{\text{hst,ch}} - V_{\text{hst,dis}}^t / \eta_{\text{hst,dis}} \right) \Delta t \times 3600, \forall 2 \le t \le N_t \tag{8.44}$$

$$V_{\text{hst,dis}}^t = V_{\text{mr,hst},h}^t + V_{\text{bz,hst},h}^t, \forall t \tag{8.45}$$

$$0 \le V_{\text{hst,ch}}^t \le \gamma_{\text{hst,ch}} \text{Cap}_{\text{hst}}, \forall t \tag{8.46}$$

$$0 \le V_{\text{hst,dis}}^t \le \gamma_{\text{hst,dis}} \text{Cap}_{\text{hst}}, \forall t \tag{8.47}$$

Multi-energy Flow Balance: The HCNG station must satisfy the supply-demand balance constraints for electricity, natural gas, and hydrogen. The electricity balance constraint is formulated as follows:

$$P_{\text{wt}}^t + P_{\text{pv}}^t = P_{\text{grid}}^t + P_{\text{elz}}^t (1 + \gamma_{\text{cp}}) + \gamma_{\text{mr}} \left(V_{\text{mr,elz,ng}}^t + V_{\text{mr,hst,ng}}^t \right) \times 3600 \times 10^{-3}, \forall t \tag{8.48}$$

In the gas blending zone, natural gas and hydrogen are mixed in a fixed proportion to produce HCNG. Therefore, the hydrogen blending ratio constraint and the law of energy conservation must be satisfied:

$$\left\{ \begin{array}{c} y_{\text{hcng}} = V_{\text{bz,mix},h}^t / \left(V_{\text{bz,mix},h}^t + V_{\text{bz,mix,ng}}^t \right) \\ V_{\text{bz,mix,ng}}^t = V_{\text{mr,elz,ng}}^t + V_{\text{mr,hst,ng}}^t + V_{\text{ng,buy}}^t \\ V_{\text{bz,mix},h}^t = V_{\text{bz,elz},h}^t + V_{\text{bz,hst},h}^t \end{array} \right., \forall t \tag{8.49}$$

$$\text{LHV}_h V_{\text{bz,mix},h}^t + \text{LHV}_{\text{ng}} V_{\text{bz,mix,ng}}^t = \text{LHV}_{\text{hcng}} \sum_{\text{gs} \in N_{\text{gs}}} q_{\text{gs}}^t, \forall t \tag{8.50}$$

Equation (8.49) describes the relationship between the hydrogen blending ratio of the HCNG output from the gas blending zone and the input flow rates of natural gas and hydrogen. Equation (8.50) represents the energy conservation law that the gas blending zone must adhere to, ensuring that the total energy contained in the input and output gases remains unchanged.

Additionally, the natural gas purchase volume of the HCNG station is constrained by the available gas extraction capacity of the upstream long-distance natural gas transmission pipeline:

$$0 \leq V_{\text{ng,buy}}^{t} \leq V_{\text{ng,buy,max}}, \forall t \tag{8.51}$$

8.3.3 Model Formulation

The optimal dispatching model for the E-HCNG-IES is formulated and denoted as Model (P0).

$$\begin{cases} \min T_{\text{cost}} \\ \text{s.t.}(8.9) - (8.12), (8.14) - (8.16), (8.26) - (8.51) \end{cases} \tag{8.52}$$

Model (P0) is a mixed-integer non-convex quadratically constrained optimization problem. The decision variables include: Outputs of energy production equipment, Inputs of energy conversion equipment, State of charge of energy storage devices, Electric power transmitted through the public grid, Natural gas purchase volume from the upstream long-distance pipeline, Gas flow injected by the HCNG station into the multi-energy load area, Pipeline gas flow, linepack, and node pressure in the HCNG network, Power flow in transmission lines and node voltage phase angles in the electricity network.

8.3.4 Evaluation indicators of HCNG network performance

Based on the theoretical derivation and calculations in [14], blending of hydrogen with natural gas will weaken the energy delivery capacity and linepack flexibility of the pipeline. This weakening effect becomes more pronounced as pressure levels and hydrogen blending ratios increase. Considering the key decision variables involved in Model (P0), this section establishes two metrics to evaluate the performance of the HCNG network and support the subsequent case study analysis.

(1) ***Relative Energy Delivery Capacity***

In the operation optimization problem formulated in this chapter, it is assumed that the hydrogen blending ratio of the HCNG produced at the HCNG station varies across different scenarios. However, in all scenarios, the multi-energy load area must meet the gas energy demand expressed in Eq. (8.10) as $L_{i,\text{gas}}^{t}$ in MW. Consequently, under the same pressure fluctuation range, scenarios with higher y_{hcng} require greater gas flow rates to supply the same level of gas energy demand. As a result, the average energy delivery capacity of the pipeline can be indirectly reflected through the concept of average gas flow. Specifically, for the same gas load, a higher average gas flow indicates a lower average energy delivery capacity of the pipeline.

The baseline scenario is denoted as S1, with the average gas flow of all pipelines in the HCNG network under S1 serving as the reference value. The variation in the average gas flow of all pipelines in other scenarios relative to the baseline value in S1 is defined as the relative pipeline energy delivery capacity, denoted as Γ_{S_k}.

$$\Gamma_{S_k} = \left[1 - \frac{\underset{S_k}{\text{mean}}\left(\tilde{q}_{ij,t}\right) - \underset{S_1}{\text{mean}}\left(\tilde{q}_{ij,t}\right)}{\underset{S_1}{\text{mean}}\left(\tilde{q}_{ij,t}\right)}\right] \times 100\%, \forall ij \in \Omega_{\text{pipe}}, \forall t \in N_t, \forall S_k \in \left\{S_2, \ldots, S_{N_k}\right\} \tag{8.53}$$

where "mean" refers to taking the average value; S_k denotes the k-th sub-scenario of total N_k scenarios.

(2) ***Relative Linepack Flexibility***

In this section, the average linepack flexibility of the HCNG network is defined as average peak-shaving capability of all pipelines. It is quantified as the ratio of the peak-to-valley difference in linepack over a dispatching day to the valley value, denoted as LP_{flexi,S_k}.

$$LP_{\text{flexi},S_k} = \underset{S_k}{\text{mean}}\left[\frac{\underset{S_k}{\max}\left(\tilde{S}_{ij,t}\right) - \underset{S_k}{\min}\left(\tilde{S}_{ij,t}\right)}{\underset{S_k}{\min}\left(\tilde{S}_{ij,t}\right)}\right], \forall ij \in \Omega_{\text{pipe}}, \forall t \in N_t, \forall S_k \in \left\{S_1, S_2, \ldots, S_{N_k}\right\} \tag{8.54}$$

where "max" and "min" represent the maximum and minimum values, respectively.

Similar to the relative pipeline energy delivery capacity, the relative linepack flexibility is also defined as a comparison against the baseline value, denoted as Ψ_{S_k}.

$$\Psi_{S_k} = \frac{LP_{\text{flexi},S_k}}{LP_{\text{flexi},S_1}} \times 100\%, \forall S_k \in \left\{S_2, \ldots, S_{N_k}\right\} \tag{8.55}$$

8.4 Solution Algorithm

This section first applies second-order cone (SOC) relaxation to the original non-convex model, then proposes a penalty function model based on the convex-concave procedure (CCP). Furthermore, a bound-tightening method is introduced to reduce errors caused by variable linearization processing. Finally, the overall algorithm framework is presented.

8.4.1 Second-Order Conic Relaxation

For the bilinear term $(\tilde{q}_{ij}^t)^2$ in Eq. (8.9), for simplicity, the subscripts and superscripts are omitted, and an auxiliary variable ϑ is introduced, setting $(\tilde{q})^2 = \vartheta$. For the bilinear terms $\tilde{S}_{ij}^t p_i^t$ and $\tilde{S}_{ij}^t p_j^t$ in Eq. (8.15), after omitting subscripts and superscripts, auxiliary variables ϖ_1 and ϖ_2 are introduced, setting $\tilde{S}_p = \varpi_1 - \varpi_2$. The relationship between $\tilde{S}_p$ and ϖ_1, ϖ_2 can be further reformulated as Eq. (8.56).

$$\begin{cases} 4\zeta\varpi_1 = \left(\tilde{S} + \zeta p\right)^2 \\ 4\zeta\varpi_2 = \left(\tilde{S} - \zeta p\right)^2 \end{cases} \tag{8.56}$$

where ζ is an introduced parameter that ensures $\tilde{S}$ and ζp remain on a similar order of magnitude, thereby preventing numerical issues that may arise from an ill-conditioned coefficient matrix.

Furthermore, the quadratic constraint $\vartheta = (\tilde{q})^2$ and the quadratic constraints in Eq. (8.56) can be equivalently reformulated as Eqs. (8.57) and (8.58).

$$\begin{cases} \vartheta \geq (\tilde{q})^2 \\ 4\zeta\varpi_1 \geq \left(\tilde{S} + \zeta p\right)^2 \\ 4\zeta\varpi_2 \geq \left(\tilde{S} - \zeta p\right)^2 \end{cases} \tag{8.57}$$

$$\begin{cases} \vartheta \leq (\tilde{q})^2 \\ 4\zeta\varpi_1 \leq \left(\tilde{S} + \zeta p\right)^2 \\ 4\zeta\varpi_2 \leq \left(\tilde{S} - \zeta p\right)^2 \end{cases} \tag{8.58}$$

The constraint in Eq. (8.57) is a convex constraint, while the constraint in Eq. (8.58) is non-convex. Based on the SOC relaxation method, the convex constraint (8.57) can be reformulated as follows:

$$\begin{cases} \left\| \begin{matrix} 2\tilde{q} \\ 1-\vartheta \end{matrix} \right\|_2 \le 1+\vartheta \\ \left\| \begin{matrix} \tilde{S}+\zeta p \\ 1-\zeta\varpi_1 \end{matrix} \right\|_2 \le 1+\zeta\varpi_1 \\ \left\| \begin{matrix} \tilde{S}-\zeta p \\ 1-\zeta\varpi_2 \end{matrix} \right\|_2 \le 1+\zeta\varpi_2 \end{cases} \tag{8.59}$$

By relaxing the quadratic constraint into Eq. (8.59) and discarding the non-convex constraint (8.58), Model (P0) is transformed into a SOCP model.

To reduce the error introduced by SOC relaxation, the McCormick envelope is applied to further tighten the feasible region boundaries of the relaxed problem. The approach is implemented as follows:

$$\begin{cases} \vartheta \le (\tilde{q}_{\min}+\tilde{q}_{\max})\tilde{q} - \tilde{q}_{\min}\tilde{q}_{\max} \\ \varpi_1 - \varpi_2 \ge \tilde{S}_{\min}p + \tilde{S}p_{\min} - \tilde{S}_{\min}p_{\min} \\ \varpi_1 - \varpi_2 \ge \tilde{S}_{\max}p + \tilde{S}p_{\max} - \tilde{S}_{\max}p_{\max} \\ \varpi_1 - \varpi_2 \le \tilde{S}_{\max}p + \tilde{S}p_{\min} - \tilde{S}_{\max}p_{\min} \\ \varpi_1 - \varpi_2 \le \tilde{S}_{\min}p + \tilde{S}p_{\max} - \tilde{S}_{\min}p_{\max} \end{cases} \tag{8.60}$$

In summary, by introducing the constraints in Eqs. (8.59) and (8.60), the MINCQCP Model (P0) is transformed into a MISOCP Model (P1).

8.4.2 Penalty Model

The SOC relaxation in Eq. (8.59) discards the non-convex constraint presented in Eq. (8.58), which may result in infeasibility of the solution to Model (P1) with respect to Model (P0). To address this issue, this section further applies the convex-concave procedure (CCP) to propose a penalty function model, which introduces penalty terms into the objective function to obtain a local minimum of Model (P0).

Next, an example is provided to illustrate the implementation of the convex-concave procedure. Taking the second non-convex constraint in Eq. (8.58) as an example, it can be expressed as the relationship between a linear function $f(\varpi_1) = 4\zeta\varpi_1$ and a convex function $g(\tilde{S}, p) = (\tilde{S}+\zeta p)^2$ [15].

$$f(\varpi_1) - g(\tilde{S}, p) \le 0 \tag{8.61}$$

Obviously, the convex function $g(\tilde{S}, p)$ satisfies the following constraint:

$$\begin{aligned} g(\tilde{S}, p) &\ge \hat{g}^{(k)}(\tilde{S}, p) \\ &= g'_{\tilde{S}}(\tilde{S}^{(k)}, p^{(k)})(\tilde{S}-\tilde{S}^{(k)}) + g'_p(\tilde{S}^{(k)}, p^{(k)})(p-p^{(k)}) + g(\tilde{S}^{(k)}, p^{(k)}) \end{aligned} \tag{8.62}$$

where $\hat{g}^{(k)}(\tilde{S},p)$ denotes the tangent plane of $g(\tilde{S},p)$ at $(\tilde{S}^{(k)},p^{(k)})$; $g'_{\tilde{S}}(\tilde{S}^{(k)},p^{(k)})$ and $g'_p(\tilde{S}^{(k)},p^{(k)})$ are partial derivatives of $g(\tilde{S},p)$ to $\tilde{S}$ and p at $(\tilde{S}^{(k)},p^{(k)})$, respectively.

Therefore, Eq. (8.61) can be relaxed to Eq. (8.63):

$$f(\varpi_1)-\hat{g}^{(k)}(\tilde{S},p)\leq s \tag{8.63}$$

where s is a slack variable that can be added as a penalty term to the objective function, thereby reducing the relaxation error.

The slack form of Eq. (8.58) can be written as Eq. (8.64):

$$\begin{cases} \vartheta-\left[2\cdot\tilde{q}^{(k)}\tilde{q}-\left(\tilde{q}^{(k)}\right)^2\right]\leq s_1 \\ 4\zeta\varpi_1-\left[2\cdot\left(\tilde{S}^{(k)}+\zeta p^{(k)}\right)\left(\tilde{S}+\zeta p\right)-\left(\tilde{S}^{(k)}+\zeta p^{(k)}\right)^2\right]\leq s_2 \\ 4\zeta\varpi_2-\left[2\cdot\left(\tilde{S}^{(k)}-\zeta p^{(k)}\right)\left(\tilde{S}-\zeta p\right)-\left(\tilde{S}^{(k)}-\zeta p^{(k)}\right)^2\right]\leq s_3 \\ s_1,s_2,s_3\geq 0 \end{cases} \tag{8.64}$$

Then, (P1) is modified into penalty model (P2), as shown in Eq. (8.65):

$$\begin{cases} \min\quad T_{cost}+\kappa^{(k)}s_{sum} \\ \text{s.t.}\begin{pmatrix} (8.9)-(8.12),(8.14)-(8.16), \\ (8.26)-(8.51),(8.59),(8.60),(8.64) \end{pmatrix} \end{cases} \tag{8.65}$$

where $\kappa^{(k)}$ represents the penalty coefficient at the k-th iteration, ensuring that the magnitudes of T_{cost} and s_{sum} remain comparable. s_{sum} is the penalty term composed of all slack variables, with its expression given in Eq. (8.66).

$$s_{\text{sum}}=\sum_{\Omega_{\text{pipe}}}s_1+\sum_{\Omega_{\text{node}},\Omega_{\text{pipe}}}(s_2+s_3) \tag{8.66}$$

Through the iterative solution, the objective function of (P2) will converge when stopping criteria (8.67) and (8.68) are satisfied.

$$\text{GAP}_1=1-\frac{T_{\text{cost}}^{(k)}+\kappa^{(k-1)}s_{\text{sum}}^{(k)}}{T_{\text{cost}}^{(k-1)}+\kappa^{(k-1)}s_{\text{sum}}^{(k-1)}}\leq\delta_1 \tag{8.67}$$

$$\text{GAP}_2=s_{\text{sum}}^{(k)}\leq\delta_2 \tag{8.68}$$

where GAP_1 and GAP_2 are convergence criteria related to the improvement in the objective function and the sum of slack variables, respectively. $T_{\text{cost}}^{(k)}$ represents the solution to the model obtained in the k-th iteration, while $s_{\text{sum}}^{(k)}$ denotes the sum of slack variables in the k-th iteration. δ_1 and δ_2 are the tolerance thresholds for GAP_1 and GAP_2, respectively.

8.4.3 Bound-tightening Method

When the number of segments is fixed, the linearization accuracy of the logarithmic term $\ln p$ in Eq. (8.9) depends on the segmentation interval settings. As illustrated in Fig. 8.6, assuming a single segment, for the point $(p^*,\ln(p^*))$, the linearization accuracy within the interval $[p_{\min}^{(1)}, p_{\max}^{(1)}]$ is significantly higher than that within $[p_{\min}^{(2)}, p_{\max}^{(2)}]$.

In light of this, the bound-tightening method is adopted in this section to improve the linearization accuracy of the logarithmic term. The bound-tightening rules are as follows [16]:

$$\begin{cases} p_{\min}^{(k+1)} \leftarrow \left(1 - \varepsilon^{(k)}\right)p^* \\ p_{\max}^{(k+1)} \leftarrow \left(1 + \varepsilon^{(k)}\right)p^* \end{cases} \tag{8.69}$$

where $\varepsilon^{(k)}$ is the control parameter of bound-tightening, and it is a positive constant less than 1 that monotonically decreases with the increase of iteration number k.

The termination condition for the bound-tightening iterations is set based on the deviation of the true value of p relative to its approximate value.

$$\mathrm{GAP}_3 = \max_{i \in \Omega_{\mathrm{node}}} \left| \frac{e^{\ln p_i} - p_i}{p_i} \right| \le \delta_3 \tag{8.70}$$

where GAP_3 is the convergence criterion associated with the deviation in the bound-tightening method, and δ_3 represents the tolerance threshold for GAP_3.

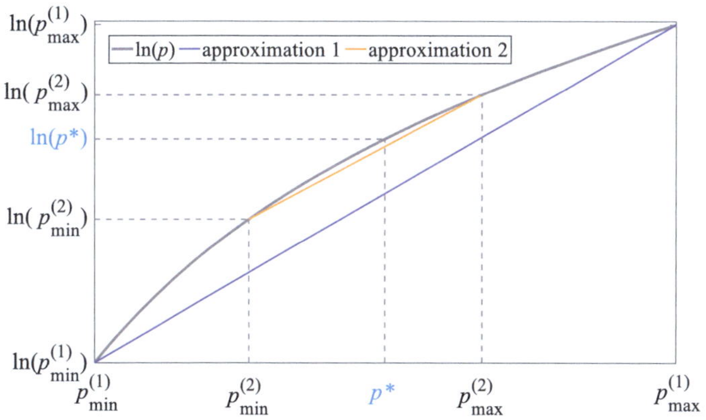

Fig. 8.6 Piecewise linearization of $\ln p$.

8.4.4 Algorithm Flow

Based on the above analysis, the operational process of the proposed iterative solution algorithm is illustrated in Fig. 8.7.

First, initialize parameters related to the iteration count, termination criteria, and variable bounds. Next, solve the second-order cone relaxation model (P1) to obtain the solution $(p^{(0)}, q^{(0)}, \tilde{S}^{(0)})$, which is then used as the initial input for the penalty function model (P2). During the iterative solving process of model (P2), the algorithm terminates once the iteration stopping criteria are satisfied. Otherwise, it updates the solution $(p^{(k+1)}, q^{(k+1)}, \tilde{S}^{(k+1)})$, the penalty coefficients, and the segmentation boundaries for piecewise linearization, before proceeding to the next iteration.

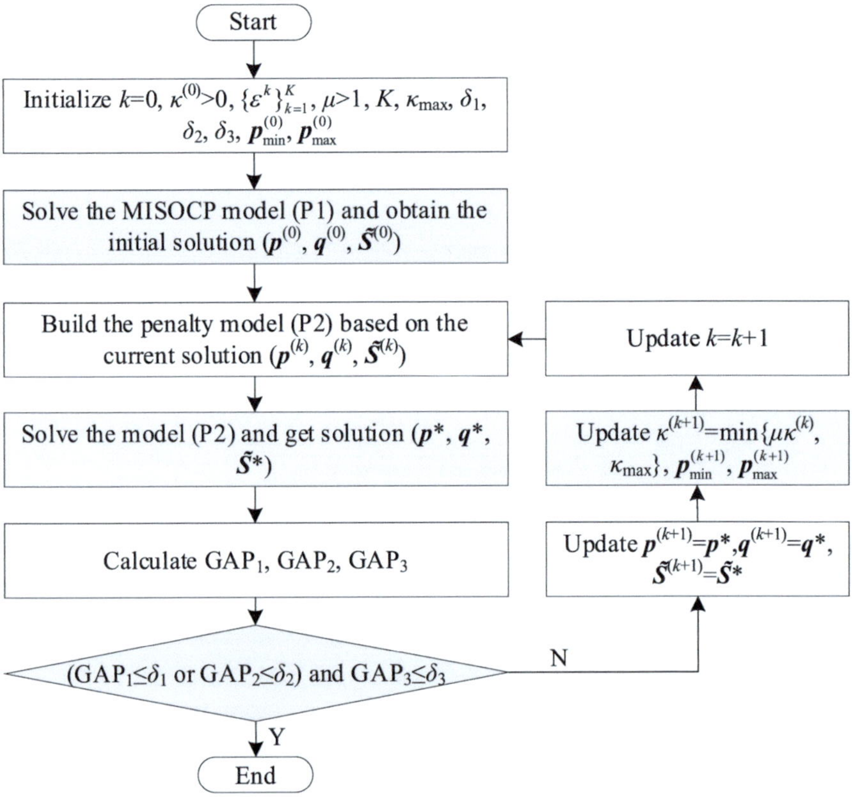

Fig. 8.7 Procedure of the proposed solving algorithm

8.5 Case Studies

To verify the effectiveness of the proposed operation optimization method, case studies are conducted on two E-HCNG-IESs of different scales. The system performance is analyzed under different hydrogen blending ratio scenarios.

8.5.1 System Description

According to the comparative results in Sect. 8.2.2, the proposed HCNG network model is more critical for application in high-pressure gas networks. Therefore, Case 1 conducts simulation analysis on the 6-bus+6-node E-HCNG-IES shown in Fig. 8.5, where the gas network operates at a pressure level of 60 bar, and the power and gas networks are coupled through two gas turbines. To analyze the impact of hydrogen blending ratio on system operation, five sub-scenarios are defined. S1 serves as the baseline scenario, where the hydrogen blending ratio y_{hcng} is set to 0, meaning no HCNG station is deployed. From S2 to S5, the hydrogen blending ratio y_{hcng} increases from 5% to 20%, with a step size of 5%. It is assumed that gas equipment and loads in the downstream E-HCNG network can normally operate with HCNG containing up to 20% hydrogen. Additionally, to ensure comparability of dispatch results, the gas load levels remain consistent across different hydrogen blending ratio scenarios and are uniformly measured in MW.

Case 2 examines a larger-scale 24-bus+20-node E-HCNG-IES, where the gas network operates at a pressure level of 80 bar. The scenario settings and assumptions remain identical to those in Case 1.

The system configurations for both case studies are provided in [9]. The input parameters for the iterative solution algorithm are set based on empirical values, as follows: $\kappa^{(0)} = 10^6$, $\kappa_{\max} = 2 \times 10^6$, $\mu = 1.05$, $\delta_1 = 10^{-4}$, $\delta_2 = \delta_3 = 10^{-3}$; $\varepsilon^{(k)} = 0.97^{4k}/2.8$, $K = 50$.

8.5.2 Case 1: 6-Bus Power Network + 6-Node Gas Network

(1) ***Performance of Convergence***

Table 8.1 compares the computational and convergence performance of scenarios S2–S5 in Case 1. It can be observed that for all four hydrogen blending scenarios, the proposed algorithm successfully converges within 3 to 5 iterations. The shortest solution time is 855.56 seconds, while the longest solution time is 2501.18 seconds.

(2) ***Economic Analysis***

Table 8.2 presents the overall dispatch results for the five scenarios, including economic costs, carbon emissions, and renewable energy utilization.

Table 8.1 Convergence performance of S2 to S5 in Case 1

Item		S2	S3	S4	S5
Number of iterations		5	4	3	5
Solver time (s)		2035.94	2501.18	855.56	2369.72
Stopping criteria	GAP_1	6.62×10^{-5}	5.13×10^{-5}	5.17×10^{-5}	5.50×10^{-5}
	GAP_2	1.46×10^{-2}	1.05×10^{-1}	2.30×10^{-2}	5.15×10^{-2}
	GAP_3	3.93×10^{-4}	3.73×10^{-4}	4.33×10^{-4}	4.01×10^{-4}

Table 8.2 Dispatch results of S1 to S5 in Case 1

Item	S1	S2	S3	S4	S5
Total cost ($\times 10^6$\$)	5.71	5.92	5.90	5.88	5.86
Daily cost of devices maintenance ($\times 10^4$\$)	8.22	90.38	91.39	92.27	93.20
Daily wheeling cost ($\times 10^4$\$)	–	1.31	1.31	1.31	1.31
Daily RES curtailment cost (\$)	–	768.42	0	0	0
Daily natural gas cost ($\times 10^6$\$)	5.16	4.59	4.56	4.54	4.51
Daily CO_2 emission cost ($\times 10^5$\$)	4.66	4.16	4.14	4.12	4.10
Daily CO_2 emission ($\times 10^3$t)	19.52	17.43	17.35	17.26	17.17
Average RES consumption (%)	–	99.67	100	100	100

First, the total daily equipment operation and maintenance cost in the baseline scenario (S1) is less than 10% of that in the other scenarios. This is because S1 does not consider renewable energy utilization or HCNG technology, meaning that the optimization model does not involve the operation and maintenance of renewable energy generators, electrolyzers, methanation reactors, or gas blending zones.

Second, the total daily natural gas purchase cost in S1 is 12.4%–14.4% higher than in the other scenarios. This is because, in S2–S5, a portion of the gas demand is met by green hydrogen produced from renewable energy sources, whereas in S1, the system must purchase more natural gas to meet the same level of gas energy demand. As a result, the higher natural gas consumption in S1 leads to higher carbon tax payments compared to the other scenarios.

Due to energy losses in the energy flow conversion processes within the HCNG station and the current technological limitations of renewable energy and hydrogen utilization, the total daily operating cost in S1, which does not consider green electricity and hydrogen utilization, remains lower than in the other four scenarios, amounting to 96.45% of S2 and 97.44% of S5.

In S2–S5, the economic performance of the system gradually improves as the hydrogen blending ratio in HCNG increases. Specifically, the O&M cost of station equipment and facilities increases with a higher hydrogen blending ratio, while the natural gas purchase cost and carbon tax cost decrease. From an overall perspective, the system continues to benefit economically from an increased hydrogen blending ratio. When the hydrogen blending ratio y_{hcng} reaches 20% in S5, the total daily operating cost of the system is reduced to 98.98% of that in S2, where $y_{hcng} = 5\%$. Additionally, as y_{hcng} increases, the system's renewable energy utilization capacity also improves. In S2, the system incurs a wind and solar curtailment penalty of ¥768.42, whereas when y_{hcng} increases to 10%, the system achieves 100% renewable energy utilization.

(3) *Operational analysis of HCNG station*

Taking S5 as an example, Fig. 8.8 illustrates operation conditions of devices in the HCNG station when y_{hcng} is 20%.

As shown in Fig. 8.8a, most wind and solar power are used for hydrogen production, compression, and methanation. Due to low cost and zero carbon emissions of renewable energy generation, the HCNG station transmits constant 20MW power to the multi-energy district during each dispatch period, despite wheeling fees that should be paid to the upper grid.

Figure 8.8b indicates that most natural gas in HCNG comes from the long-distance transmission line, and some natural gas is produced by hydrogen from 0:00 to

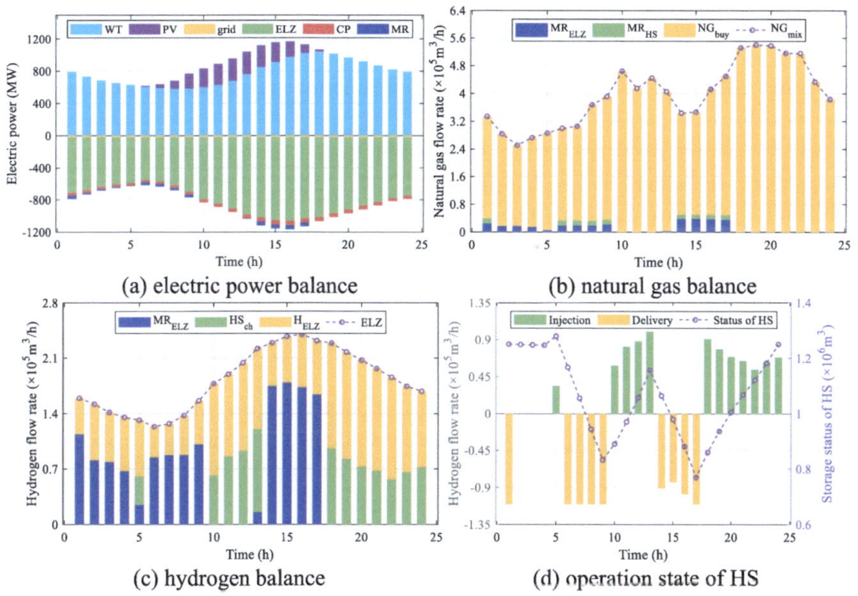

Fig. 8.8 Operation conditions of devices in the HCNG station in Case 1-S5.

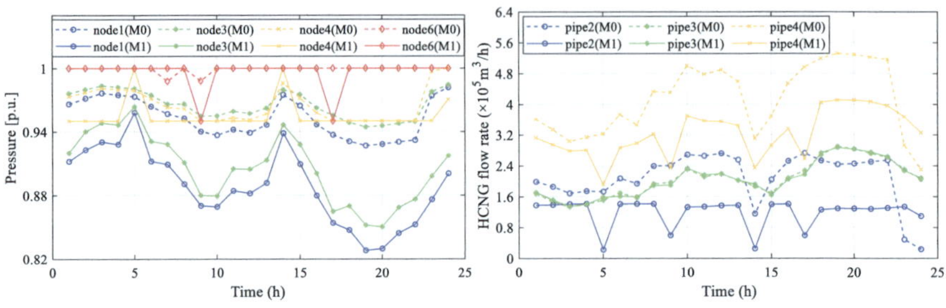

(a) node pressure (b) pipeline flow rate

Fig. 8.9 Operation conditions of the HCNG network in Case 1-S5.

9:00 and 12:00 to 17:00. In addition, some hydrogen is withdrawn from the HS for methanation during peak natural gas prices.

In each dispatch period, part of the hydrogen compressed by the CP is directly injected into the MZ to form HCNG, which can be seen in Fig. 8.8c. A considerable part of the compressed hydrogen will be used for methanation during peak natural gas prices to reduce purchasing gas from the long-distance transmission line. When natural gas price is low and renewable energy generation is abundant, some hydrogen will be injected into the HS as a reserve.

Operation status of HS is shown in Fig. 8.8d. Its hydrogen injection and withdrawal behavior is consistent with the above analysis and shows an effective response to changes in natural gas prices.

(4) ***Operational analysis of multi-energy district***

Figure 8.9 illustrates the node pressures and pipeline gas flow distribution in the HCNG network for Case 1-S5.

As shown in Fig. 8.9a, during a dispatching day, the gas pressures at source nodes 4 and 6 fluctuate between 95% and 100% of their baseline values, while the pressures at the other four nodes can drop to as low as 73.05% of their baseline values during peak load periods. The pressure variations across all nodes follow a similar trend. From Eq. (8.9), it can be observed that for a given pipeline, the gas flow rate is determined by the pressure difference between the pipeline's inlet and outlet. Therefore, the pressure trends in Fig. 8.9a exhibit a negative correlation with the pipeline gas flow trends in Fig. 8.9b.

The linepack variations of each pipeline throughout the dispatching day are illustrated in Fig. 8.10, where the gray-shaded areas indicate low natural gas price periods. It is evident that the linepack trends of all pipelines are highly consistent, with the system discharging linepack during high gas price periods and accumulating linepack during low gas price periods. The peak linepack occurs between 03:00–04:00, while the lowest linepack is observed between 19:00–20:00. The linepack variations in the HCNG network reflect the capability of the gas network to leverage its gas inertia in response to external energy price signals, which is further confirmed in Fig. 8.11.

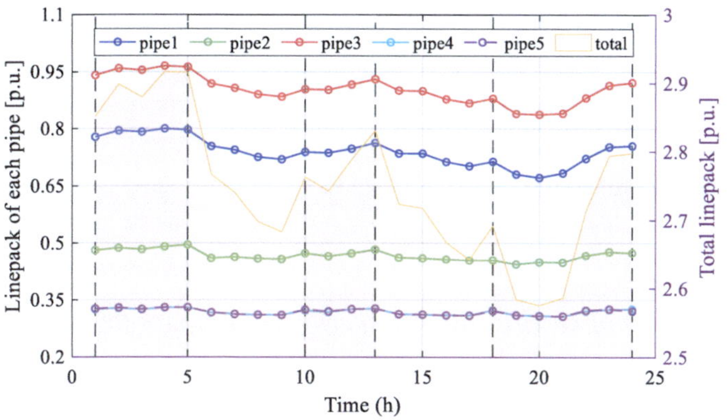

Fig. 8.10 Linepack of the HCNG network in Case 1-S5

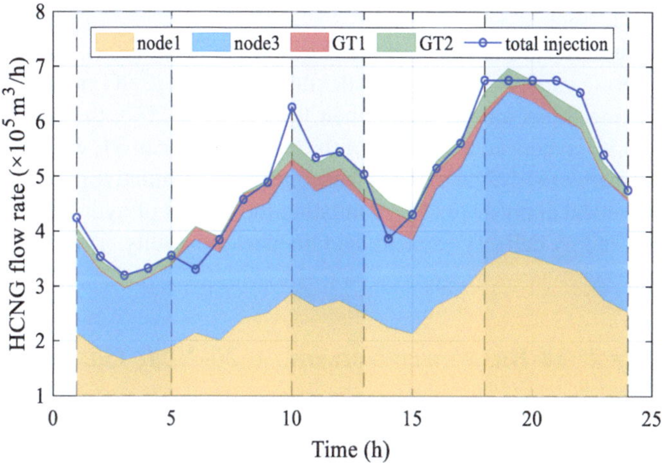

Fig. 8.11 HCNG flow rate in the HCNG network in Case 1-S5

As shown in Fig. 8.11, the gas injection flow from the HCNG station into gas source nodes 4 and 6 of the HCNG network does not fully align with the total gas load of the downstream E-HCNG network. The downstream gas load consists of the gas demands at HCNG network nodes 1 and 3 as well as the fuel requirements of gas turbines 1 and 2. It can be observed that during low natural gas price periods, the total HCNG injection flow exceeds the gas demand of the downstream E-HCNG network, leading to "gas injection" into the pipelines. Conversely, during other periods, the HCNG injection flow is lower than the gas demand, resulting in "gas withdrawal" from the pipelines. This behavior highlights the peak-shaving and valley-filling role of pipeline linepack, demonstrating its function as a special form of energy storage.

Table 8.3 Energy delivery capacity and linepack flexibility in Case 1

	S1	S2	S3	S4	S5
Average gas flow rate ($\times 10^4$m^3/h)	17.53	18.06	18.81	19.46	20.24
Relative energy delivery capacity, Γ_{S_k} (%)	–	96.98	92.71	88.98	84.54
Average linepack flexibility (%)	15.02	12.85	12.60	12.54	12.26
Relative linepack flexibility, Ψ_{S_k} (%)	-	85.51	83.89	83.43	81.59

(5) ***Comparison of energy delivery capacity and linepack flexibility in HCNG network***

Based on the dispatch results from Case 1, this section conducts a quantitative analysis of the impact of hydrogen blending on the HCNG network in terms of relative pipeline energy delivery capacity and relative linepack flexibility. Using the indicator calculation methods defined in Sect. 8.3.4, the computed values for each scenario are presented in Table 8.3.

It can be observed that as the hydrogen blending ratio increases, both the pipeline energy delivery capacity and linepack flexibility of the HCNG network gradually decrease. Taking S5 as an example, when y_{hcng} reaches 20%, the pipeline energy delivery capacity is reduced to 84.54% of the baseline value in S1, while the linepack flexibility decreases to 81.59% of the S1 baseline. This quantitative result is consistent with the theoretical analysis in [14], validating the impact of hydrogen blending on both pipeline energy delivery capacity and linepack flexibility.

8.5.3 *Case 2: 24-Bus Power Network + 20-Node Gas Network*

This section conducts a study on Case 2 to verify the effectiveness and scalability of the proposed method. The E-HCNG-IES in Case 2 is illustrated in Fig. 8.12, where the power network adopts the IEEE 24-bus system, and the gas network is based on the Belgian 20-node system. The power and gas networks are coupled through ten gas turbines. It is important to note that the HCNG network contains six gas source nodes, each assumed to have a centralized HCNG station nearby. The hydrogen blending ratio of the gas injected into these source nodes is assumed to be identical. The multi-energy flow conversion processes within the HCNG stations remain consistent with those in Case 1.

Consistent with Case 1, Case 2 also investigates five scenarios, where the hydrogen blending ratio y_{hcng} varies from 0% to 20% in 5% increments. The dispatch results are presented in Table 8.4.

In Case 2, the average solution time for S2–S5 is 14,399.76 seconds, with the iterative termination criteria being met after 6–9 iterations. It can be observed that the average solution time in Case 2 is 7.42 times that of Case 1, and the number of iterations is approximately twice as high. Compared to Case 1, the network topology

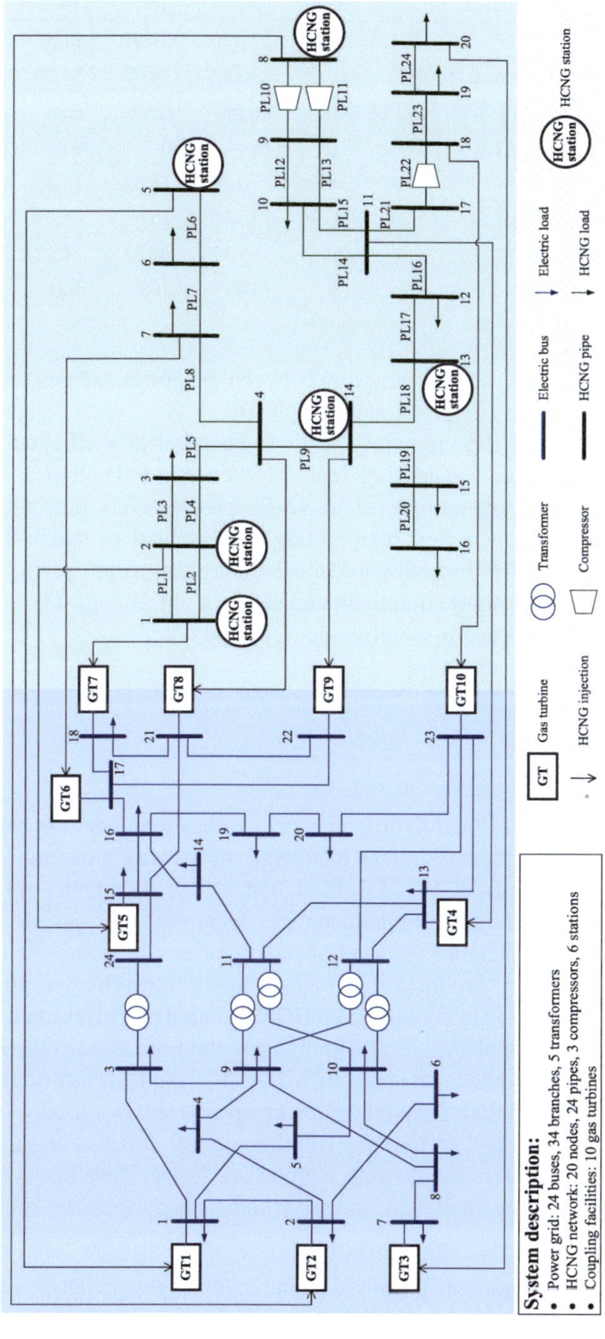

Fig. 8.12 Topology of E-HCNG-IES in Case 2.

Table 8.4 Dispatch results of S1 to S5 in Case 2

	S1	S2	S3	S4	S5
Total cost ($\times 10^6$\$)	21.41	21.76	21.67	21.57	21.47
Daily cost of devices maintenance ($\times 10^5$\$)	13.06	33.88	34.07	34.29	34.60
Daily wheeling cost ($\times 10^4$\$)	–	6.56	6.56	6.56	6.56
Daily RES curtailment cost (\$)	–	0	0	0	0
Daily natural gas cost ($\times 10^6$\$)	18.42	16.75	16.65	16.53	16.41
Daily CO_2 emission cost ($\times 10^5$\$)	16.94	15.55	15.48	15.41	15.33
Daily CO_2 emission ($\times 10^3$t)	70.94	65.12	64.83	64.52	64.19
Average RES consumption (%)	–	100	100	100	100

in Case 2 is more complex, and the coupling between the power and gas networks is stronger, leading to higher computational complexity.

The baseline scenario (S1) remains the most economically efficient scenario in Case 2. However, after considering renewable energy utilization and carbon tax costs, the total daily operating cost in S2–S5 progressively decreases as the hydrogen blending ratio increases, highlighting the potential of green electricity, green hydrogen, and HCNG technologies in enhancing the overall performance of energy systems. The analysis of equipment and network operational states in Case 2 is similar to that in Case 1 and is therefore not repeated here.

8.6 Conclusions

This chapter focuses on regional hydrogen energy utilization under the centralized hydrogen blending mode, addressing the challenges of network modeling and operation optimization for the E-HCNG-IES. First, considering the impact of hydrogen blending on key parameters in the fundamental gas dynamics equations, an improved HCNG network model suitable for centralized hydrogen blending is proposed. Second, incorporating the "production-storage-blending-transportation-utilization" link of the HCNG supply chain, a centralized HCNG station model and an E-HCNG-IES dispatch model are established. Third, applying second-order cone relaxation and piecewise linearization methods, the original MINCQCP model is reformulated into a MISOCP model. Finally, to reduce relaxation errors and obtain a feasible solution, a penalty function model and an iterative solution algorithm based on the convex-concave procedure and bound-tightening method are further developed. Based on case study analysis, the key qualitative and quantitative conclusions are summarized as follows:

1. The necessity of applying the proposed model in high-pressure HCNG networks is validated by comparing it with the quasi-steady-state gas flow model based on the General Flow equation. In the 60-bar gas network validation case, the

optimization result deviation of the traditional model from the improved model is 4.81%, which is 601.25 times that of the 2-bar gas network validation case.
2. Although non-hydrogen blending scenarios still hold an economic advantage under the current energy techno-economic environment, the comprehensive benefits of the E-HCNG-IES in terms of economic performance, carbon emissions, and renewable energy integration become more pronounced as the hydrogen blending ratio increases. In Case 1 (6-bus+6-node E-HCNG-IES), when the hydrogen blending ratio reaches 10%, the system achieves 100% renewable energy utilization. When the hydrogen blending ratio increases to 20%, the total daily operating cost is 1.02% lower than in the 5% blending scenario.
3. The impact of hydrogen blending on both pipeline energy delivery capacity and linepack flexibility is validated through the operation optimization problem of the E-HCNG-IES.

The proposed HCNG network modeling and system operation optimization methods are specifically designed for the centralized hydrogen blending mode in the E-HCNG-IES. However, the distributed nature of renewable energy sources suggests that a distributed hydrogen blending mode may hold long-term potential for widespread adoption. Therefore, future research will explore modeling and optimization methods for distributed hydrogen blending, providing theoretical support for transitional hydrogen energy utilization.

References

1. Menon, E. S. (2005). *Gas pipeline hydraulics*. Taylor & Francis Group.
2. Fu, C., Lin, J., Song, Y. H., Xu, Y., Zhang, Z., & Zhou, M. (2020). Optimal operation of an integrated energy system incorporated with HCNG distribution networks. *IEEE Transactions on Sustainable Energy, 11*(4), 2141–2151.
3. Wang, X. W. (2013). *Research and practice on hydraulic simulation of urban natural gas transmission and distribution network* [Master's thesis, Chongqing University].
4. Tabkhi, F. (2007). *Optimization of gas transmission networks* [Doctoral dissertation, National Polytechnic Institute of Toulouse].
5. Hafsi, Z., Elaoud, S., Akrout, M., & Mhimid, A. (2017). Numerical approach for steady state analysis of hydrogen-natural gas mixtures flows in looped network. *Arabian Journal for Science and Engineering, 42*(5), 1941–1950.
6. Hasan, M. M. F., & Karimi, I. A. (2010). Piecewise linear relaxation of bilinear programs using bivariate partitioning. *AIChE Journal, 56*(7), 1880–1893.
7. Lo Cascio, E., De Schutter, B., & Schenone, C. (2018). Flexible energy harvesting from natural gas distribution networks through line-bagging. *Applied Energy, 229*, 253–263.
8. Correa-Posada, C. M., & Sanchez-Martin, P. (2015). Integrated power and natural gas model for energy adequacy in short-term operation. *IEEE Transactions on Power Systems, 30*(6), 3347–3355.
9. Qiu, Y. (2022). *Test data of the E-HCNG-IES*. [Online]. Retrieved from: https://www.dropbox.com/scl/fi/ftcz64zxibz37fsk3skid/Test-data-of-the-E-HCNG-IES.xlsx?dl=0&rlkey=7d98o6eqvwsnlxqk71nnwwmqe
10. Li, B., Li, X., & Su, Q. Y. (2022). A system and game strategy for the isolated island electric-gas deeply coupled energy network. *Applied Energy, 306*, Article 117968.

11. Kouchachvili, L., & Entchev, E. (2018). Power to gas and H_2/NG blend in SMART energy networks concept. *Renewable Energy, 125*, 456–464.
12. Cai, Y., Lin, J., Wan, C., & Luo, Y. (2016). A bi-level optimization model for active distribution network operation and market trading based on stochastic programming under market environment. *Proceedings of the CSEE, 36*(20), 5391–5402.
13. Pan, G. S., Gu, W., Lu, Y. P., Chen, S., & Gao, S. (2021). Accurate modeling of a profit-driven power to hydrogen and methane plant toward strategic bidding within multi-type markets. *IEEE Transactions on Smart Grid, 12*(1), 338–349.
14. Quarton, C. J., & Samsatli, S. (2020). Should we inject hydrogen into gas grids? Practicalities and whole-system value chain optimisation. *Applied Energy, 275*, Article 115172.
15. Lu, S., Gu, W., Zhang, C., Wu, L., & Wu, T. (2020). Hydraulic-thermal cooperative optimization of integrated energy systems: A convex optimization approach. *IEEE Transactions on Smart Grid, 11*(6), 4818–4832.
16. Chen, S., Conejo, A. J., Sioshansi, R., & Plazas, C. R. (2019). Unit commitment with an enhanced natural gas-flow model. *IEEE Transactions on Power Systems, 34*(5), 3729–3738.

Chapter 9
Multi-stage Multi-energy Coordinated Restoration Method for IESs Under Extreme Disasters

Abstract With the increasing occurrence of extreme weather events (EWEs) in recent years, enhancing the resilience of multi-energy systems (MESs) has become a critical concern. This paper presents a comprehensive resilience enhancement strategy that accounts for the multi-stage recovery process and multi-energy coordination to strengthen MESs' ability to withstand and recover quickly from EWEs. The proposed approach systematically models the entire recovery process, spanning the pre-event stage, system disturbance stage, fault isolation stage, and remote controlled restoration stage, while considering the interactions between resilience measures at each phase. To minimize demand curtailment, an integrated energy flow model is developed to optimize coordination among different energy subsystems. The problem is formulated as a stochastic mixed-integer linear programming (MILP) model that accounts for the uncertainties in fault components. To address the computational challenges associated with multiple scenarios, a customized progressive hedging algorithm (PHA) is introduced. Case studies validate the effectiveness of the proposed framework, demonstrating that a combination of restoration strategies can mitigate the impact of EWEs, and that cross-energy carrier coordination can enhance the overall resilience of MESs.

Keywords Multi-energy system · Extreme weather event · Resilience · Multi-stage recovery process · Stochastic programming

Nomenclature

s	Index for fault scenarios
τ	Index for time stages
i/e	Index for nodes
Pre/Dis/Iso/Res	Pre-event/system disturbance/fault isolation/remote controlled restoration stage
$\mathcal{N}_{\mathrm{P}}/\mathcal{N}_{\mathrm{G}}/\varepsilon$	Set for nodes in PDS/NGS/EHs
$\mathcal{S}$	Set for fault scenarios

Z. Wu et al., *Integrated Energy System Resilience*,
https://doi.org/10.1007/978-981-95-2363-4_9

ω_i^{P}	Weight coefficients for power loads
ω_i^{G}	Weight coefficients for gas loads
ω_e^{EH}	Weight coefficients for energy hub loads
ρ_s	Probability occurrence of fault scenario
C_{g2e}	Gas-to-electricity conversion coefficient
$P_{\mathrm{Cut},i,s}^{\tau}$	Electrical power curtailment in PDS
$M_{\mathrm{Cut},i,s}^{\tau}$	Gas flow curtailment in NGS
$P_{\mathrm{Cut},e,s}^{\tau}$	Electrical power curtailment in EHs
$H_{\mathrm{Cut},e,s}^{\tau}$	Thermal power curtailment in EHs
ij	Index for power lines
$\mathcal{L}_P/\mathcal{L}_G$	Set for power lines/gas pipelines
$N_{\mathrm{RCS}}/N_{\mathrm{RCV}}$	Maximum allocation number of RCS/RCV
z_{ij}	Binary: equal to 1 if RCS/RCV is allocated
$\mathcal{N}_{\mathrm{Sub}}^{\mathrm{P}}$	Set for substation nodes
N_P	Number of nodes in PDS
f_{ij}^{Pre}	Virtual power of power line/pipeline
y_{ij}^{Pre}	Binary: equal to 1 if power line is operated
$\alpha_{ij}^{\mathrm{Pre}}/\alpha_{ij,s}^{\mathrm{Iso}}/\alpha_{ij,s}^{\mathrm{Res}}$	Binary: connection status of power line/pipeline (equal to 1 if connected, 0 otherwise)
$\beta_{ij}^{\mathrm{Pre}}$	Binary: fictitious connection status of power line(equal to 1 if connected, 0 otherwise)
$\mathcal{N}_{\mathrm{Sta}}^{\mathrm{G}}$	Set for natural gas station nodes
R_{ij}/X_{ij}	Resistance/reactance of power line
$P_{\mathrm{D},i}/Q_{\mathrm{D},i}$	Active/reactive demand at power node i
$S_{ij,\max}^{\mathrm{L}}$	Capacity limit of power line
$P_{\mathrm{L},ij}^{\mathrm{Pre}}/P_{\mathrm{L},ki}^{\mathrm{Pre}}$	Active power flow of power line
$Q_{\mathrm{L},ij}^{\mathrm{Pre}}/Q_{\mathrm{L},ki}^{\mathrm{Pre}}$	Reactive power flow of power line
$P_{\mathrm{Sub},i}^{\mathrm{Pre}}/Q_{\mathrm{Sub},i}^{\mathrm{Pre}}$	Active/reactive power of SSs
$P_{\mathrm{GT},i}^{\mathrm{Pre}}/Q_{\mathrm{GT},i}^{\mathrm{Pre}}$	Active/reactive power of GTs
$P_{\mathrm{EH},i}^{\mathrm{Pre}}/Q_{\mathrm{EH},i}^{\mathrm{Pre}}$	Active/reactive power of EHs
$V_i^{\mathrm{Pre}}/V_j^{\mathrm{Pre}}$	Square of node voltage
$\mathcal{N}_{\mathrm{GT}}^{\mathrm{P}}$	Set for gas turbine nodes in PDS
φ_{GT}	Power factor of GTs
φ_{Sub}	Power factor of SSs
$V_{i,\min}/V_{i,\max}$	Minimal/maximal node voltage
$P_{i,\max}^{\mathrm{GT}}$	Maximum output active power of GTs
$P_{i,\max}^{\mathrm{Sub}}$	Maximum output active power of SSs
$N_{\mathrm{Sta}}^{\mathrm{G}}$	Set for natural gas station nodes
A_{ij}^{G}	Cross-sectional area of natural gas pipelines
D_{ij}^{G}	Diameter of pipelines
L_{ij}^{G}	Length of pipelines
$\overline{\omega}_{ij}^{\mathrm{G}}$	Friction coefficient of pipelines
$\lambda_{ij}^{\mathrm{G}}$	Gas flow velocity of pipelines

Δt_{Pre}	Duration of the pre-event stage
$M_{D,i}$	Gas flow rate of gas loads
$Pr_{i,0}$	Node gas pressure of GSs
$\rho_{i,0}$	Gas density of natural gas station nodes
$M_{ij,\mathrm{out}}^{\mathrm{Pre}}/M_{ij,\mathrm{in}}^{\mathrm{Pre}}$	Outflow/inflow gas flow rate of natural gas pipelines in the pre-event stage
$M_{ij,\mathrm{out}}^{0}/M_{ij,\mathrm{in}}^{0}$	Outflow/inflow gas flow rate of natural gas pipelines in the initial stage
$M_{\mathrm{Sta},i}^{\mathrm{Pre}}$	Gas flow rate of GSs
$M_{\mathrm{GT},i}^{\mathrm{Pre}}$	Gas flow rate of GTs
$M_{\mathrm{EH},i}^{\mathrm{Pre}}$	Gas flow rate of EHs
$\mathrm{Pr}_i^{\mathrm{Pre}}/\mathrm{Pr}_j^{\mathrm{Pre}}$	Node gas pressure in the pre-event stage
$\mathrm{Pr}_i^0/\mathrm{Pr}_j^0$	Node gas pressure in the initial stage
$\rho_i^{\mathrm{Pre}}/\rho_j^{\mathrm{Pre}}$	Node gas density in the pre-event stage
ρ_i^0/ρ_j^0	Node gas density in the initial stage
$\mathrm{Pr}_{i,\min}/\mathrm{Pr}_{i,\max}$	Minimum/maximum node gas pressure
$M_{ij,\min}^{\mathrm{G}}/M_{ij,\max}^{\mathrm{G}}$	Minimum/maximum gas flow rate of pipelines
$M_{i,\max}^{\mathrm{Sta}}$	Maximum gas flow rate of GSs
$P_{\mathrm{D},e}/H_{\mathrm{D},e}$	Electric/thermal load power of EHs
$P_{\mathrm{PV},e}^{\mathrm{Pre}}$	Active power of photovoltaic
$P_{\mathrm{ES}+,e}^{\mathrm{Pre}}/P_{\mathrm{ES}-,e}^{\mathrm{Pre}}$	Charging/discharging power of electrical energy storage
$P_{\mathrm{PtG},e}^{\mathrm{Pre}}/M_{\mathrm{PtG},e}^{\mathrm{Pre}}$	Input electric power/output gas flow rate of power-to-gas equipment
$P_{\mathrm{HP},e}^{\mathrm{Pre}}/H_{\mathrm{HP}}^{\mathrm{Pre}}$	Input electric power/output thermal power of electric heat pumps
$M_{\mathrm{GT},e}^{\mathrm{Pre}}/P_{\mathrm{GT},e}^{\mathrm{Pre}}$	Input gas flow rate/output electric power of GTs
$M_{\mathrm{GB},e}^{\mathrm{Pre}}/H_{\mathrm{GB},e}^{\mathrm{Pre}}$	Input gas flow rate/output heat power of gas boilers
PtG/GT/HP	Power-to-gas equipment/gas turbines/electric heat pumps
$R_{x,e}$	Maximum ramping power of PtG/GT/HP
$R_{\mathrm{GB},e}$	Maximum ramping power of gas boilers
$P_{x,e}^{\mathrm{Pre}}/P_{x,e}^{0}$	Electric power of PtG/GT/HP in the pre-event stage/initial stage
$H_{\mathrm{GB},e}^{\mathrm{Pre}}/H_{\mathrm{GB},e}^{0}$	Heat power of gas boilers in the pre-event stage/initial stage
$E_{e,\min}^{\mathrm{ES}}/E_{\mathrm{S},\max}^{\mathrm{E}}$	Minimum/maximum remaining capacity of electrical energy storage
$\eta_{\mathrm{ES}+}/\eta_{\mathrm{ES}-}$	Charging/discharging efficiency of energy storage
$\eta_{\mathrm{PtG}}/\eta_{\mathrm{GB}}/\eta_{\mathrm{HP}}$	Conversion efficiency of PtG/GT/HP
$\eta_{\mathrm{GT,gp}}/\eta_{\mathrm{GT,gh}}$	Gas-to-electric and gas-to-heat efficiency of GTs
$P_{e,\max}^{x}$	Maximum electric power of PtG/GT/HP
$M_{e,\max}^{\mathrm{GB}}$	Maximum input gas flow rate of gas boilers
$P_{e,\max}^{\mathrm{EH}}$	Maximum active power transmitted from EHs to the power distribution system
$M_{e,\max}^{\mathrm{EH}}$	Maximum gas flow rate transmitted from EHs to the NGS
$E_{\mathrm{ES},e}^{\mathrm{Pre}}$	Remaining capacity of electrical energy storage
$x_{\mathrm{ES},e}^{\mathrm{Pre}}$	Binary: 1 for energy storage in charge, 0 in discharge

$\mathcal{N}_{\text{GT}}^{\text{G}}$	Set for gas turbine nodes in the NGS
$\varepsilon_{\text{P}}/\varepsilon_{\text{G}}$	Set for nodes in the PDS/NGS connected to EHs
$\xi_{ij,s}$	0-1 State parameters for damage status of power lines and pipelines post-disaster
$n_{i,s}^{\text{Deg}}/n_{j,s}^{\text{Deg}}$	0-1 State variables for power and gas nodes in the system disturbance stage
$n_{i,s}^{\text{Iso}}/n_{j,s}^{\text{Iso}}$	0-1 State variables for power and gas nodes in the fault isolation stage
$n_{i,s}^{\text{Res}}$	0-1 State variables for node status in the remote controlled restoration stage

9.1 Introduction

The resilience planning of integrated energy systems (IESs) can enhance the system's ability to withstand large-scale disturbances. However, under extreme conditions, energy supply interruptions may still occur. Implementing post-disaster emergency response measures and developing remote controlled restoration strategies to gradually recover power in affected areas are crucial steps to improving system resilience. The post-disaster emergency response and remote controlled restoration process in IESs involves multiple stages, including system disturbance, fault isolation, and remote controlled restoration. However, existing research often formulates strategies for these stages independently, failing to systematically model the deeply coupled multi-stage process under multi-energy interactions. This limitation reduces the overall efficiency of energy restoration. Furthermore, given the high uncertainty of disaster-induced failures, optimizing resource allocation across different scenarios to ensure a coordinated recovery remains an open challenge.

To address these issues, this chapter proposes a multi-stage and multi-energy coordinated remote controlled restoration method for IESs under extreme disasters. First, the interactions between multi-energy flow subsystems during emergency response and restoration are comprehensively considered, and a coordinated multi-stage process model is established. In the pre-event stage, the system undergoes proactive network reconfiguration to minimize potential damage to critical loads from extreme disasters. In the post-disaster stage, faults are rapidly isolated using remote controlled switches and valves, while coordinated multi-energy resource scheduling ensures swift power restoration in non-fault areas. Next, a unified modeling approach is developed to integrate multi-stage response and restoration processes, establishing an IES remote controlled restoration model that accounts for uncertainty in fault scenarios. Finally, a customized progressive hedging algorithm (C-PHA) is proposed to decompose the original large-scale model into multiple subproblems corresponding to different fault scenarios, enabling efficient iterative solving.

The remainder of this chapter is organized as follows: Sect. 9.2 introduces the conceptual framework of the multi-stage restoration model, Sect. 9.3 presents the mathematical formulation of the restoration model, Sect. 9.4 discusses the solution algorithm, Sect. 9.5 provides case studies and numerical analysis, Sect. 9.6 concludes the chapter.

9.2 Conceptual Framework of the Remote Controlled Restoration Model

In this section, the system architecture of integrated energy systems (IESs) and the multi-stage restoration process are introduced, followed by the establishment of an remote controlled restoration framework that incorporates multi-stage and multi-energy flow coordination.

9.2.1 Structure of IESs and Multi-stage Resilience Curve

This study focuses on distribution-level IESs, whose structure is illustrated in Fig. 9.1. The power distribution system (PDS), natural gas system (NGS), and energy hubs (EHs) are tightly coupled through energy conversion equipment [1, 2]. When extreme disasters cause widespread power and gas outages, failures within each energy subsystem can propagate through coupling links, triggering cascading failures across multiple energy domains. The coordinated dispatch of different energy subsystems is critical for enabling rapid post-disaster energy restoration [3]. By optimizing the scheduling of gas turbines in both PDS and NGS, the sequence of power and gas load restoration can be optimized to accelerate the recovery process. Additionally, EHs serve as key intermediaries between the PDS and NGS, enabling the transfer of surplus electricity and natural gas to support critical loads during restoration.

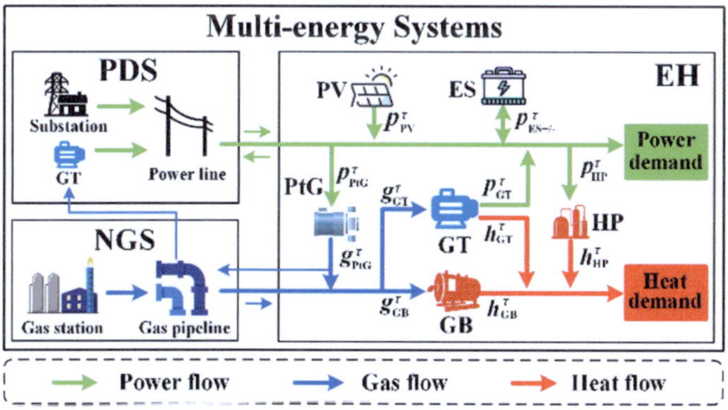

Fig. 9.1 Structure of the multi-energy system

To illustrate the multi-stage restoration process of IESs under EWEs, Fig. 9.2 presents the multi-stage resilience curve, which encompasses system disturbance, fault isolation, and remote controlled restoration [4]. As shown in the figure, at time t_1, an extreme event occurs, damaging power lines and gas pipelines, resulting in large-scale energy supply interruptions. During this period, system operators must quickly identify system disturbances and assess system damage. Between time t_2 and t_3, operators can remotely operate switches and valves within non-fault areas [1, 4, 5] to isolate faulted zones, reducing the impact of failures and partially restoring power supply to certain loads, leading to an initial recovery of energy supply. Between time t_3 and t_4, further remote operations enable additional load restoration in non-faulted areas, further improving energy supply levels. After time t_4, damaged power lines and pipelines undergo physical repairs, gradually restoring the system to normal operation.

The response and restoration measures in different stages are interdependent, with system states in earlier stages directly influencing subsequent decision-making. Therefore, coordinated multi-stage planning is necessary to enhance overall IES recovery performance. Given that physical repairs of damaged components typically require extended time, this chapter focuses on the rapid energy restoration phase facilitated by remote controlled devices before time t_4. The effectiveness of IES restoration is quantified by the cumulative energy supply level across multiple stages, represented by the gray area in Fig. 9.2.

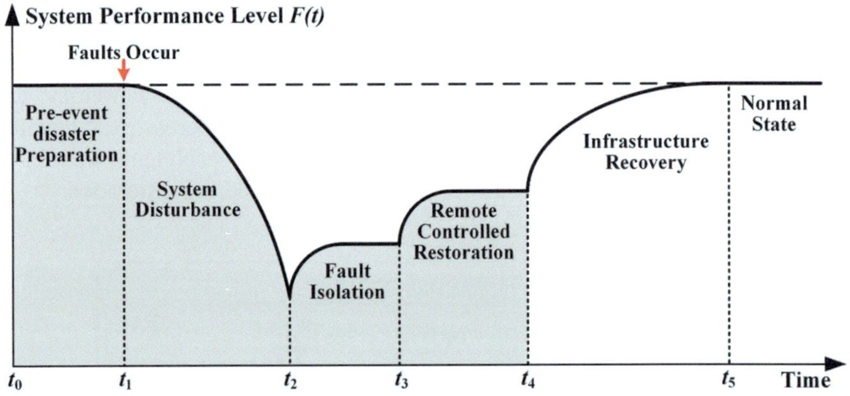

Fig. 9.2 Multi-stage resilience curve of IESs

9.2.2 *Framework for Multi-stage and Multi-energy Coordinated Remote Controlled Restoration*

This section provides a detailed explanation of the multi-energy coordination mechanisms between PDS, NGS, and EHs during the multi-stage restoration process, illustrated in Fig. 9.3.

Pre-event Stage: Given that remote controlled devices play a critical role in emergency response and energy restoration, their strategic deployment must be planned in advance. Additionally, preventive network reconfiguration is an effective approach to enhance the resilience of IESs by mitigating fault propagation effects [5, 6]. By segmenting the grid into several independent islanded subsystems, the risk of cascading failures is significantly reduced. In Fig. 9.3a, the PDS is reconfigured into four islanded microgrids, while the NGS is divided into three subsystems. This preventive reconfiguration, along with optimized deployment of remote controlled devices, provides greater flexibility and operability in subsequent restoration stages, thereby improving overall energy restoration efficiency.

System Disturbance Stage: After EWEs occur, multiple faults may appear within the IES. As shown in Fig. 9.3b, power lines and gas pipelines connected to fault points are affected by cascading failures, forming multiple fault zones. Additionally,

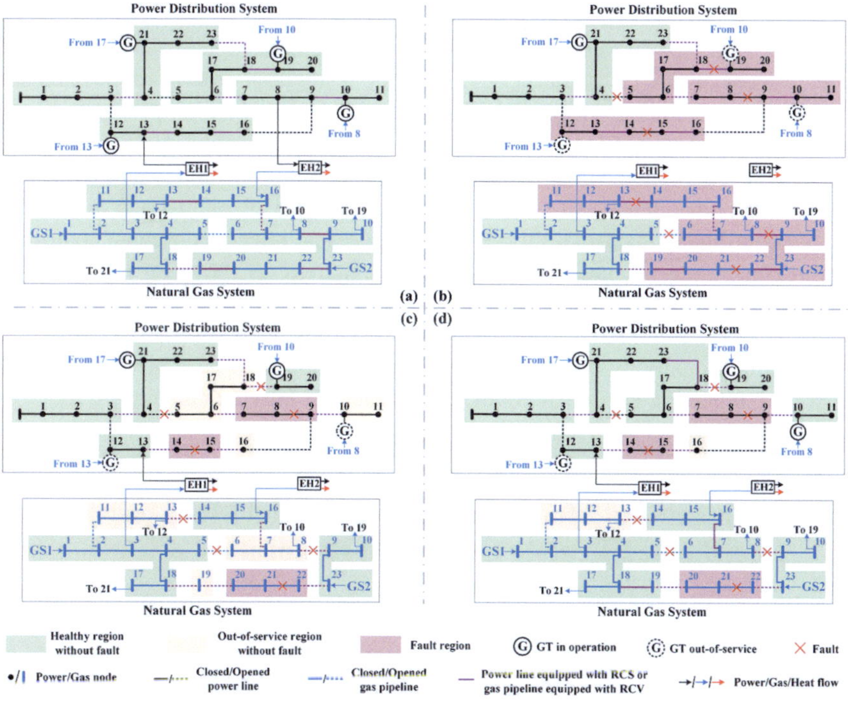

Fig. 9.3 Illustrative example for the multi-stage recovery process

energy exchanges between faulted areas and EHs are disrupted [1]. Some isolated microgrids formed during preventive reconfiguration remain unaffected and continue supplying energy. To facilitate subsequent restoration, system operators must quickly identify faulted areas using monitoring and information collection systems.

Fault Isolation Stage: Once faulted components are identified, system operators can remotely operate remote controlled switches (RCSs) and remote controlled valves (RCVs) to isolate faulted areas. As depicted in Fig. 9.3c, in the PDS, opening the RCS on power line 18–19 isolates nodes 19 and 20 from the faulted area, allowing power restoration via gas turbines. In the NGS, opening the RCVs on pipelines 8–9 and 22–23 isolates gas nodes 9, 10, and 23 from the fault zone. Despite fault isolation, some non-faulted nodes remain without energy supply, as indicated by the yellow areas in the figure. For example, power nodes 5, 6, and 16–18 and gas nodes 6–8, 11–13, and 19 lack energy supply due to insufficient energy resources. Additionally, power nodes 10 and 11 cannot be restored due to interrupted gas supply, leading to gas turbine shutdown.

Remote Controlled Restoration Stage: After fault isolation, system operators can use remote controlled devices to restore energy supply to some non-faulted areas. As shown in Fig. 9.3d, in the PDS, closing the RCS on power line 18–23 reconnects power nodes 5, 6, 17, and 18 to an energized microgrid, restoring power supply. In the NGS, switching RCVs on pipelines 7–16 and 18–19 restores gas supply to nodes 6–8 and 19. However, gas supply to nodes 11–13 remains unavailable as pipeline 2–11 requires manual valve operation.

The coordinated dispatch of different energy subsystems significantly impacts the effectiveness of IES restoration [7]. During the restoration process, power and gas loads can receive energy from EHs. For instance, in Fig. 9.3c, d, power nodes 12 and 13 are supplied by EH1, while gas nodes 14–16 receive supply from EH2. These interactions highlight the importance of multi-energy coordination in improving IES restoration performance. To fully leverage the benefits of multi-stage and multi-energy coordinated recovery, this chapter proposes an integrated remote controlled restoration framework under uncertain fault scenarios, as shown in Fig. 9.4. The unified framework systematically models the deep interdependencies between energy flows during the multi-stage restoration process to enhance overall recovery efficiency. Additionally, a stochastic approach is employed to generate diverse fault scenarios, reflecting the uncertainties associated with extreme disasters [2, 8].

9.3 Mathematical Model for Multi-stage and Multi-energy Coordinated Remote Controlled Restoration

This section presents the mathematical formulation of the proposed remote controlled restoration model. Before establishing the model, the following simplifications and assumptions are made: (1) The model assumes the existence of a central coordination agency (e.g., a disaster prevention committee) capable of effectively managing the resource dispatching of the PDS, NGS, and EHs [9]; (2) the model primarily focuses on the steady-state operation of IESs during the restoration process; (3) it is assumed that the communication systems of RCSs and RCVs remain functional during extreme events.

9.3.1 Objective Function

The objective of the restoration model is to minimize the expected energy supply interruption loss across the PDS, NGS, and EHs under multiple fault scenarios, as expressed in the following equation:

$$\begin{aligned} \min: \sum_{s \in \mathcal{S}} \rho_s \sum_{\tau = \{\text{ Deg, Iso, Res }\}} \Bigg[\sum_{i \in \mathcal{N}_{\text{P}}} \omega_i^{\text{P}} P_{\text{Cut},\, i,\, s}^{\tau} + \sum_{i \in \mathcal{N}_{\text{G}}} \omega_i^{\text{G}} C_{\text{g2e}} M_{\text{Cut},\, i,\, s}^{\tau} \\ + \sum_{e \in \mathcal{E}} \omega_e^{\text{EH}} \left(P_{\text{Cut},\, e,\, s}^{\tau} + H_{\text{Cut},\, e,\, s}^{\tau} \right) \Bigg] \end{aligned} \tag{9.1}$$

9.3.2 Constraints for the Pre-event Stage

Allocation Constraints for Remote Controlled Devicess: Since RCSs and RCVs are relatively expensive, their allocation is subject to investment budget constraints:

$$\sum_{ij \in \mathcal{L}_{\text{P}}} z_{ij} \le N_{\text{RCS}} \tag{9.2}$$

$$\sum_{ij \in \mathcal{L}_{\text{G}}} z_{ij} \le N_{\text{RCV}} \tag{9.3}$$

Preventive Network Reconfiguration Constraints: In the preventive network reconfiguration of the PDS, radial topology constraints must be satisfied. The "Spanning Tree" constraint (9.4)–(9.5) and virtual power constraints (9.6)–(9.10) are adopted to achieve more flexible network reconfiguration [10]. Specifically, Eqs. (9.4)–(9.5) ensure that each power node, except for substation nodes, has one and only one parent node. Equations (9.6)–(9.10) construct a virtual spanning tree topology for the power network under normal operating conditions, without considering fault scenarios. Equation (9.10) further defines the reconfigured power distribution network as a subgraph of the virtual tree topology, allowing islanded microgrids to designate gas turbines as root nodes.

$$\sum_{k:k\rightarrow i} y_{ki}^{\mathrm{Pre}} = 0\,; \qquad \forall i \in \mathcal{N}_{\mathrm{Sub}}^{\mathrm{P}} \tag{9.4}$$

$$\sum_{k:k\rightarrow i} y_{ki}^{\mathrm{Pre}} = 1\,; \qquad \forall i \in \mathcal{N}_{\mathrm{P}} \backslash \mathcal{N}_{\mathrm{Sub}}^{\mathrm{P}} \tag{9.5}$$

$$\sum_{k:k\rightarrow i} f_{ki}^{\mathrm{Pre}} - \sum_{j:i\rightarrow j} f_{ij}^{\mathrm{Pre}} = 1\,; \qquad \forall i \in \mathcal{N}_{\mathrm{P}} \backslash \mathcal{N}_{\mathrm{Sub}}^{\mathrm{P}} \tag{9.6}$$

$$\beta_{ij}^{\mathrm{Pre}} = y_{ij}^{\mathrm{Pre}} + y_{ji}^{\mathrm{Pre}}; \quad \forall ij \in \mathcal{L}_{\mathrm{P}} \tag{9.7}$$

$$-M\beta_{ij}^{\mathrm{Pre}} \le f_{ij}^{\mathrm{Pre}} \le M\beta_{ij}^{\mathrm{Pre}}; \quad \forall ij \in \mathcal{L}_{\mathrm{P}} \tag{9.8}$$

$$|f_{ij}^{\mathrm{Pre}}| \le \beta_{ij}^{\mathrm{Pre}} N_{\mathrm{P}}\,; \forall ij \in \mathcal{L}_{\mathrm{P}} \tag{9.9}$$

$$\alpha_{ij}^{\mathrm{Pre}} \le \beta_{ij}^{\mathrm{Pre}}; \quad \forall ij \in \mathcal{L}_{\mathrm{P}}. \tag{9.10}$$

Unlike the PDS, NGS does not require a radial topology. Instead, each gas load node must be directly connected to at least one upstream gas source:

$$\sum_{k:k\rightarrow i} f_{ki}^{\mathrm{Pre}} - \sum_{j:i\rightarrow j} f_{ij}^{\mathrm{Pre}} \ge 1\,; \quad \forall i \in \mathcal{N}_{\mathrm{G}} \backslash (\mathcal{N}_{\mathrm{Sta}}^{\mathrm{G}} \cup \mathcal{E}) \tag{9.11}$$

$$-M\alpha_{ij}^{\mathrm{Pre}} \le f_{ij}^{\mathrm{Pre}} \le M\alpha_{ij}^{\mathrm{Pre}}; \quad \forall ij \in \mathcal{L}_{\mathrm{G}}. \tag{9.12}$$

***Operational Constraints for PDS*:** The constraints for the PDS in the pre-event stage are given in Eqs. (9.13)–(9.22). A linearized DistFlow model is employed to represent power flow [11, 12]. Equations (9.13)–(9.14) define the active and reactive power balance constraints at nodes. Equation (9.15) represents the voltage drop constraint for power lines. The branch power flow constraint $P^2 + Q^2 \le S^2$ is linearized using two sets of polygonal approximations [13, 14], as shown in Eqs. (9.16)–(9.17).

$$\sum_{k:k\rightarrow i} P_{\mathrm{L},ki}^{\mathrm{Pre}} - \sum_{j:i\rightarrow j} P_{\mathrm{L},ij}^{\mathrm{Pre}} + P_{\mathrm{GT},i}^{\mathrm{Pre}} + P_{\mathrm{Sub},i}^{\mathrm{Pre}} + P_{\mathrm{EH},i}^{\mathrm{Pre}} = P_{\mathrm{D},i}\,; \; \forall i \in \mathcal{N}_{\mathrm{P}} \tag{9.13}$$

$$\sum_{k:k\rightarrow i} Q_{\mathrm{L},ki}^{\mathrm{Pre}} - \sum_{j:i\rightarrow j} Q_{\mathrm{L},ij}^{\mathrm{Pre}} + Q_{\mathrm{GT},i}^{\mathrm{Pre}} + Q_{\mathrm{Sub},i}^{\mathrm{Pre}} + Q_{\mathrm{EH},i}^{\mathrm{Pre}} = Q_{\mathrm{D},i}\,; \forall i \in \mathcal{N}_{\mathrm{P}} \tag{9.14}$$

$$-M(1-\alpha_{ij}^{\mathrm{Pre}}) \le V_i^{\mathrm{Pre}} - V_j^{\mathrm{Pre}} - 2(P_{\mathrm{L},ij}^{\mathrm{Pre}} R_{ij} + Q_{\mathrm{L},ij}^{\mathrm{Pre}} X_{ij}) \le M(1-\alpha_{ij}^{\mathrm{Pre}}); \;\; \forall ij \in \mathcal{L}_{\mathrm{P}} \tag{9.15}$$

$$-S_{ij,\max}^{\mathrm{L}}\alpha_{ij}^{\mathrm{Pre}} \le P_{\mathrm{L},ij}^{\mathrm{Pre}},\; Q_{\mathrm{L},ij}^{\mathrm{Pre}} \le S_{ij,\max}^{\mathrm{L}}\alpha_{ij}^{\mathrm{Pre}}; \forall ij \in \mathcal{L}_{\mathrm{P}} \tag{9.16}$$

$$-S_{ij,\max}^{\mathrm{L}}\alpha_{ij}^{\mathrm{Pre}} \leq P_{\mathrm{L},ij}^{\mathrm{Pre}} \pm Q_{\mathrm{L},ij}^{\mathrm{Pre}} \leq S_{ij,\max}^{\mathrm{L}}\alpha_{ij}^{\mathrm{Pre}};\quad \forall ij \in \mathcal{L}_{\mathrm{P}}. \tag{9.17}$$

Equation (9.18) is the node voltage upper and lower limit constraint. Equations (9.19)–(9.20) are the upper and lower limit constraints for the active and reactive power of gas turbines. Equations (9.21)–(9.22) are the upper and lower limit constraints for the active and reactive power of substations (SSs).

$$V_{i,\ \min}^{2} \leq V_{i}^{\mathrm{Pre}} \leq V_{i,\ \max}^{2};\quad \forall i \in \mathcal{N}_{\mathrm{P}} \tag{9.18}$$

$$0 \leq P_{\mathrm{GT},i}^{\mathrm{Pre}} \leq P_{i,\ \max}^{\mathrm{GT}};\quad \forall i \in \mathcal{N}_{\mathrm{GT}}^{\mathrm{P}} \tag{9.19}$$

$$|Q_{\mathrm{GT},i}^{\mathrm{Pre}}| \leq P_{\mathrm{GT},i}^{\mathrm{Pre}} \tan(\cos^{-1}\varphi_{\mathrm{GT}});\quad \forall i \in \mathcal{N}_{\mathrm{GT}}^{\mathrm{P}} \tag{9.20}$$

$$0 \leq P_{\mathrm{Sub},i}^{\mathrm{Pre}} \leq P_{i,\ \max}^{\mathrm{Sub}};\quad \forall i \in \mathcal{N}_{\mathrm{Sub}}^{\mathrm{P}} \tag{9.21}$$

$$|Q_{\mathrm{Sub},i}^{\mathrm{Pre}}| \leq P_{\mathrm{Sub},i}^{\mathrm{Pre}} \tan(\cos^{-1}\varphi_{\mathrm{Sub}});\quad \forall i \in \mathcal{N}_{\mathrm{Sub}}^{\mathrm{P}}. \tag{9.22}$$

Operational Constraints for NGS: The operational constraints for NGS in the pre-event stage are given in Equations (9.23)–(9.30). The gas flow in pipelines is modeled using the quasi-dynamic gas flow model based on the Wendroff difference scheme [15, 16]. Equations (9.23)–(9.25) represent the Navier–Stokes momentum equation, mass conservation equation, and natural gas state equation, respectively. Equation (9.26) is the gas flow balance constraint at nodes. Equation (9.27) constrains the gas pressure and gas density at natural gas station nodes to remain constant [17].

$$\begin{aligned}
&\Big|\frac{1}{A_{ij}^{\mathrm{G}}}\left(M_{ij,\mathrm{out}}^{\mathrm{Pre}} + M_{ij,\mathrm{in}}^{\mathrm{Pre}} - M_{ij,\mathrm{out}}^{0} - M_{ij,\mathrm{in}}^{0}\right) \\
&+ \frac{\Delta t_{\mathrm{Pre}}}{L_{ij}^{\mathrm{G}}}(\mathrm{Pr}_{j}^{\mathrm{Pre}} - \mathrm{Pr}_{i}^{\mathrm{Pre}} + \mathrm{Pr}_{j}^{0} - \mathrm{Pr}_{i}^{0}) + \frac{\lambda_{ij}^{\mathrm{G}}\overline{\omega}_{ij}^{\mathrm{G}}\Delta t_{\mathrm{Pre}}}{4D_{ij}^{\mathrm{G}}A_{ij}^{\mathrm{G}}} \\
&(M_{ij,\mathrm{out}}^{\mathrm{Pre}} + M_{ij,\mathrm{in}}^{\mathrm{Pre}} + M_{ij,\mathrm{out}}^{0} + M_{ij,\mathrm{in}}^{0})\Big| \leq M(1-\alpha_{ij}^{\mathrm{Pre}});\quad \forall ij \in \mathcal{L}_{\mathrm{G}}
\end{aligned} \tag{9.23}$$

$$\begin{aligned}
&\Big|\frac{\Delta t_{\mathrm{Pre}}}{L_{ij}^{\mathrm{G}}A_{ij}^{\mathrm{G}}}(M_{ij,\mathrm{out}}^{\mathrm{Pre}} - M_{ij,\mathrm{in}}^{\mathrm{Pre}} + M_{ij,\mathrm{out}}^{0} - M_{ij,\mathrm{in}}^{0}) \\
&+ \rho_{j}^{\mathrm{Pre}} + \rho_{i}^{\mathrm{Pre}} - \rho_{j}^{0} - \rho_{i}^{0}\Big| \leq M(1-\alpha_{ij}^{\mathrm{Pre}});\quad \forall ij \in \mathcal{L}_{\mathrm{G}}
\end{aligned} \tag{9.24}$$

$$\mathrm{Pr}_{i}^{\mathrm{Pre}} = c^{2}\rho_{i}^{\mathrm{Pre}};\ \forall i \in \mathcal{N}_{\mathrm{G}} \tag{9.25}$$

$$\sum_{k:k\to i}\frac{M_{ki,\ \mathrm{out}}^{\mathrm{Pre}}}{A_{ki}} - \sum_{j:i\to j}\frac{M_{ij,\ \mathrm{in}}^{\mathrm{Pre}}}{A_{ij}} + M_{\mathrm{Sta},i}^{\mathrm{Pre}} - M_{\mathrm{GT},i}^{\mathrm{Pre}} + M_{\mathrm{EH},i}^{\mathrm{Pre}} = M_{\mathrm{D},i};\quad \forall i \in \mathcal{N}_{\mathrm{G}} \tag{9.26}$$

$$\begin{cases} \mathrm{Pr}_i^{\mathrm{Pre}} = \mathrm{Pr}_{i,0}; \\ \rho_i^{\mathrm{Pre}} = \rho_{i,0}; \end{cases} \quad \forall i \in \mathcal{N}_{\mathrm{Sta}}^{\mathrm{G}}. \tag{9.27}$$

Equation (9.28) is the node gas pressure upper and lower limit constraint. Equation (9.29) is the node gas flow upper and lower limit constraint. Equation (9.30) is the upper and lower limit constraint for the gas flow rate at natural gas stations (GSs).

$$\mathrm{Pr}_{i,\min} \leq \mathrm{Pr}_i^{\mathrm{Pre}} \leq \mathrm{Pr}_{i,\max}; \forall i \in \mathcal{N}_{\mathrm{G}} \tag{9.28}$$

$$M_{ij,\min}^{\mathrm{G}} \alpha_{ij}^{\mathrm{Pre}} \leq M_{ij,\mathrm{in}}^{\mathrm{Pre}}, M_{ij,\mathrm{out}}^{\mathrm{Pre}} \leq M_{ij,\max}^{\mathrm{G}} \alpha_{ij}^{\mathrm{Pre}}; \ \forall ij \in \mathcal{L}_{\mathrm{G}} \tag{9.29}$$

$$0 \leq M_{\mathrm{Sta},i}^{\mathrm{Pre}} \leq M_{i,\max}^{\mathrm{Sta}}; \forall ij \in \mathcal{N}_{\mathrm{Sta}}^{\mathrm{G}}. \tag{9.30}$$

Operational Constraints for EHs: The operational constraints for EHs in the pre-event stage are given in Eqs. (9.31)–(9.42) [18].

$$P_{\mathrm{PV},e}^{\mathrm{Pre}} - P_{\mathrm{ES+},e}^{\mathrm{Pre}} + P_{\mathrm{ES-},e}^{\mathrm{Pre}} - P_{\mathrm{PtG},e}^{\mathrm{Pre}} + P_{\mathrm{GT},e}^{\mathrm{Pre}} - P_{\mathrm{HP},e}^{\mathrm{Pre}} - P_{\mathrm{EH},e}^{\mathrm{Pre}} = P_{\mathrm{D},e}; \quad \forall e \in \mathcal{E} \tag{9.31}$$

$$M_{\mathrm{PtG},e}^{\mathrm{Pre}} - M_{\mathrm{GT},e}^{\mathrm{Pre}} - M_{\mathrm{GB},e}^{\mathrm{Pre}} - M_{\mathrm{EH},e}^{\mathrm{Pre}} = 0; \quad \forall e \in \mathcal{E} \tag{9.32}$$

$$H_{\mathrm{GT},e}^{\mathrm{Pre}} + H_{\mathrm{HP},e}^{\mathrm{Pre}} + H_{\mathrm{GB},e}^{\mathrm{Pre}} = H_{\mathrm{D},e}; \quad \forall e \in \mathcal{E} \tag{9.33}$$

$$[0,\ 0] \leq [P_{\mathrm{ES+},e}^{\mathrm{Pre}},\ P_{\mathrm{ES-},e}^{\mathrm{Pre}}] \leq [x_{\mathrm{ES},e}^{\mathrm{Pre}} P_{e,\max}^{\mathrm{ES}},\ (1 - x_{\mathrm{ES},e}^{\mathrm{Pre}}) P_{e,\max}^{\mathrm{ES}}]; \quad \forall e \in \mathcal{E} \tag{9.34}$$

$$E_{\mathrm{ES},e}^{\mathrm{Pre}} = E_{\mathrm{ES},e}^{0} + P_{\mathrm{ES+},e}^{\mathrm{Pre}} \eta_{\mathrm{ES+}} \Delta t_{\mathrm{Pre}} - P_{\mathrm{ES-},e}^{\mathrm{Pre}} \Delta t_{\mathrm{Pre}} / \eta_{\mathrm{ES-}}; \quad \forall e \in \mathcal{E} \tag{9.35}$$

$$E_{e,\min}^{\mathrm{ES}} \leq E_{\mathrm{ES},e}^{\mathrm{Pre}} \leq E_{e,\max}^{\mathrm{ES}}; \quad \forall e \in \mathcal{E} \tag{9.36}$$

$$0 \leq P_{x,e}^{\mathrm{Pre}} \leq P_{e,\max}^{x}, x = \{\mathrm{PtG}, \mathrm{GT}, \mathrm{HP}\}; \quad \forall e \in \mathcal{E} \tag{9.37}$$

$$0 \leq M_{\mathrm{GB},e}^{\mathrm{Pre}} \leq M_{e,\max}^{\mathrm{GB}}; \quad \forall e \in \mathcal{E} \tag{9.38}$$

$$|H_{\mathrm{GB},e}^{\mathrm{Pre}} - H_{\mathrm{GB},e}^{0}| \leq R_{\mathrm{GB},e} \Delta t_{\mathrm{Pre}}; \quad \forall e \in \mathcal{E} \tag{9.39}$$

$$|P_{x,e}^{\mathrm{Pre}} - P_{x,e}^{0}| \leq R_{x,e} \Delta t_{\mathrm{Pre}}, x = \{\mathrm{PtG}, \mathrm{GT}, \mathrm{HP}\}; \quad \forall e \in \mathcal{E} \tag{9.40}$$

$$\begin{cases} M_{\mathrm{PtG},e}^{\mathrm{Pre}} = P_{\mathrm{PtG},e}^{\mathrm{Pre}} \eta_{\mathrm{PtG}}; \\ P_{\mathrm{GT},e}^{\mathrm{Pre}} = M_{\mathrm{GT},e}^{\mathrm{Pre}} \eta_{\mathrm{GT,gp}}, \quad H_{\mathrm{GT},e}^{\mathrm{Pre}} = M_{\mathrm{GT},e}^{\mathrm{Pre}} \eta_{\mathrm{GT,gh}}; \forall e \in \mathcal{E} \\ H_{\mathrm{GB},e}^{\mathrm{Pre}} = M_{\mathrm{GB},e}^{\mathrm{Pre}} \eta_{\mathrm{GB}}, \quad H_{\mathrm{HP},e}^{\mathrm{Pre}} = P_{\mathrm{HP},e}^{\mathrm{Pre}} \eta_{\mathrm{HP}}; \end{cases} \tag{9.41}$$

$$[-P_{e,\max}^{\mathrm{EH}}, \ -M_{e,\max}^{\mathrm{EH}}] \le [P_{\mathrm{EH},e}^{\mathrm{Pre}}, \ M_{\mathrm{EH},e}^{\mathrm{Pre}}] \le [P_{e,\max}^{\mathrm{EH}}, \ M_{e,\max}^{\mathrm{EH}}]; \quad \forall e \in \mathcal{E}. \tag{9.42}$$

Multi-energy Coupling Constraints: Equation (9.43) represents the energy conversion constraint between PDS and NGS through gas turbines (GTs). Equation (9.44) represents the energy coupling constraint among PDS, NGS, and EHs.

$$M_{\mathrm{GT},n}^{\mathrm{Pre}} = P_{\mathrm{GT},i}^{\mathrm{Pre}} / \eta_{\mathrm{GT,gp}}; \quad \forall i \in \mathcal{N}_{\mathrm{GT}}^{\mathrm{P}}, n \in \mathcal{N}_{\mathrm{GT}}^{\mathrm{G}} \tag{9.43}$$

$$P_{\mathrm{EH},e}^{\mathrm{Pre}} = P_{\mathrm{EH},i}^{\mathrm{Pre}}, \quad M_{\mathrm{EH},e}^{\mathrm{Pre}} = M_{\mathrm{EH},i}^{\mathrm{Pre}}; \quad \forall e \in \mathcal{E}, \quad i \in \mathcal{E}_{\mathrm{P}}, \quad n \in \mathcal{E}_{\mathrm{G}}. \tag{9.44}$$

9.3.3 Emergency Response and Remote Controlled Restoration Constraints

After EWEs occur, the post-disaster response and remote controlled restoration strategies of the IES vary depending on the specific system disturbances. As illustrated in Fig. 9.4, the post-disaster recovery process consists of interdependent stages. This section models three key stages: system disturbance, fault isolation, and remote controlled restoration based on remote controlled devices. The constraints include how to identify faulted areas, how to operate remote controlled devices for fault isolation, and how to restore power supply to non-faulted areas. The detailed constraints are as follows.

Network Constraints in System Disturbance Stage: When an extreme event causes power line failures, the fault propagates through connected circuits if the faulted line is not disconnected. Similarly, failures in NGS spread through connected pipelines until an isolation valve is closed. The post-disaster state of islanded subsystems formed through preventive reconfiguration depends on the network topology and system disturbances, as described in Eqs. (9.45)–(9.46). Equation (9.45) ensures that nodes on both sides of a faulted power line or gas pipeline are included in the faulted area. Equation (9.46) guarantees that both ends of an operational power line or gas pipeline remain in the same state, either both faulted or both non-faulted. These equations model the fault propagation mechanism within the network while also capturing the coupling relationship between the pre-event stage and system disturbance stage [6, 19].

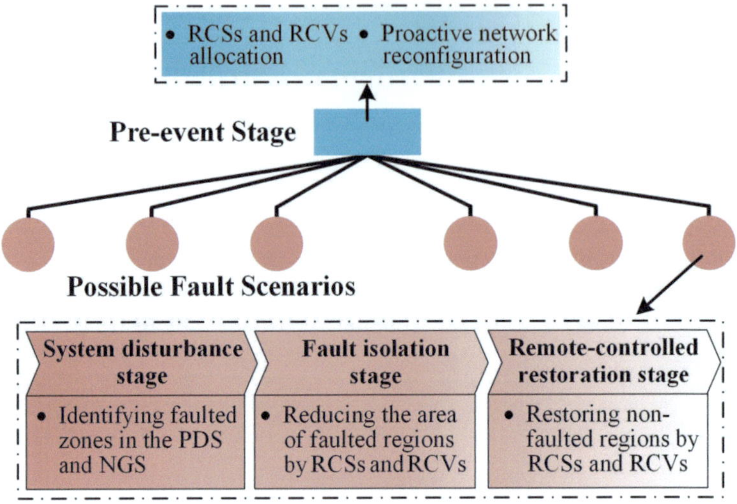

Fig. 9.4 Framework of the remote controlled restoration model under diverse fault scenarios

$$n_{i,s}^{\text{Deg}},\ n_{j,s}^{\text{Deg}} \geq \xi_{ij,s}\alpha_{ij}^{\text{Pre}};\ \forall ij \in \mathcal{L}_{\text{P}} \cup \mathcal{L}_{\text{G}},\quad s \in \mathcal{S} \tag{9.45}$$

$$\begin{cases} n_{i,s}^{\text{Deg}} + \alpha_{ij}^{\text{Pre}} - 1 \leq n_{j,s}^{\text{Deg}}; \\ n_{j,s}^{\text{Deg}} + \alpha_{ij}^{\text{Pre}} - 1 \leq n_{i,s}^{\text{Deg}}; \end{cases} \forall ij \in \mathcal{L}_{\text{P}} \cup \mathcal{L}_{\text{G}},\quad s \in \mathcal{S}. \tag{9.46}$$

Network Constraints in the Fault Isolation Stage: After system disturbance is completed, RCSs and RCVs can be controlled remotely to reduce the faulted area. The network constraints in the fault isolation stage are formulated in Eqs. (9.47)–(9.49). Equation (9.47) ensures that faulted power lines and gas pipelines are removed from operation, while non-faulted lines and pipelines can be isolated using RCSs and RCVs to prevent fault propagation. Equation (9.48) imposes constraints on the operation of RCSs and RCVs to effectively block fault propagation. Equation (9.49) ensures that nodes at both ends of an operational power line or gas pipeline remain in the same state, consistent with Eq. (9.46).

$$(1 - \xi_{ij,s})(\alpha_{ij}^{\text{Pre}} - z_{ij}) \leq \alpha_{ij,s}^{\text{Iso}} \leq (1 - \xi_{ij,s})\alpha_{ij}^{\text{Pre}};\ \ \forall ij \in \mathcal{L}_{\text{P}} \cup \mathcal{L}_{\text{G}},\quad s \in \mathcal{S} \tag{9.47}$$

$$n_{i,s}^{\text{Iso}},\ n_{j,s}^{\text{Iso}} \geq \xi_{ij,s}(1 - z_{ij}) + \alpha_{ij}^{\text{Pre}} - 1;\ \ \forall ij \in \mathcal{L}_{\text{P}} \cup \mathcal{L}_{\text{G}},\quad s \in \mathcal{S} \tag{9.48}$$

$$\begin{cases} n_{i,s}^{\text{Iso}} + \alpha_{ij,s}^{\text{Iso}} - 1 \leq n_{j,s}^{\text{Iso}}; \\ n_{j,s}^{\text{Iso}} + \alpha_{ij,s}^{\text{Iso}} - 1 \leq n_{i,s}^{\text{Iso}}; \end{cases} \forall ij \in \mathcal{L}_{\text{P}} \cup \mathcal{L}_{\text{G}},\quad s \in \mathcal{S}. \tag{9.49}$$

Network Constraints in the Remote Controlled Restoration Stage: After fault isolation is completed, RCSs and RCVs can be controlled remotely to transfer loads from non-faulted areas to adjacent islanded subsystems, enabling partial remote controlled restoration. The network constraints for the remote controlled restoration stage are formulated in Eqs. (9.50)–(9.52). Equation (9.50) ensures that the node states in the restoration stage remain the same as in the fault isolation stage. Equation (9.51) allows reconfiguration of the network topology through RCSs and RCVs. Equation (9.52) ensures that non-faulted areas do not reconnect to faulted areas. Additionally, the PDS must satisfy the radial topology constraints defined in Eqs. (9.4)–(9.10), and each recoverable gas node must have at least one upstream gas source, as described in Eqs. (9.11)–(9.12).

$$n_{i,s}^{\mathrm{Res}} = n_{i,s}^{\mathrm{Iso}};\ \forall ij \in \mathcal{L}_{\mathrm{P}} \cup \mathcal{L}_{\mathrm{G}},\quad s \in \mathcal{S} \tag{9.50}$$

$$\begin{aligned}&(1-\xi_{ij,s})(\alpha_{ij,s}^{\mathrm{Iso}} - z_{ij}) \le \alpha_{ij,s}^{\mathrm{Res}} \le (1-\xi_{ij,s})(\alpha_{ij,s}^{\mathrm{Iso}} + z_{ij});\\ &\forall ij \in \mathcal{L}_{\mathrm{P}} \cup \mathcal{L}_{\mathrm{G}},\quad s \in \mathcal{S}\end{aligned} \tag{9.51}$$

$$\begin{cases} n_{i,s}^{\mathrm{Res}} + \alpha_{ij,s}^{\mathrm{Res}} - 1 \le n_{j,s}^{\mathrm{Res}}; \\ n_{j,s}^{\mathrm{Res}} + \alpha_{ij,s}^{\mathrm{Res}} - 1 \le n_{i,s}^{\mathrm{Res}}; \end{cases} \forall ij \in \mathcal{L}_{\mathrm{P}} \cup \mathcal{L}_{\mathrm{G}},\quad s \in \mathcal{S}. \tag{9.52}$$

Operational Constraints for the System Disturbance, Fault Isolation, and Remote Controlled Restoration Stages: The operational constraints for the IES in the system disturbance, fault isolation, and remote controlled restoration stages are similar to Eqs. (9.13)–(9.44). However, due to the consideration of load-shedding variables and node fault state variables, additional constraints must be satisfied, as formulated in Eqs. (9.53)–(9.64). Equation (9.53) defines the set of three recovery stages: system disturbance, fault isolation, and remote controlled restoration. Equations (9.54)–(9.55) enforce the active and reactive power balance constraints for nodes in PDS across all three stages. Equation (9.56) ensures the gas flow balance at nodes in NGS, while Eqs. (9.57)–(9.58) impose power and heat balance constraints in EHs during all three stages.

$$\mathcal{T}' = \{\ \mathrm{Deg},\ \mathrm{Iso},\ \mathrm{Res}\ \}; \tag{9.53}$$

$$\begin{aligned}&\sum_{k:k\to i} P_{\mathrm{L},ki,s}^{\tau} - \sum_{j:i\to j} P_{\mathrm{L},ij,s}^{\tau} + P_{\mathrm{GT},i,s}^{\tau} + P_{\mathrm{Sub},i,s}^{\tau}\\ &+ P_{\mathrm{EH},i,s}^{\tau} = P_{\mathrm{D},i} - P_{\mathrm{Cut},i,s}^{\tau},\quad \tau = \mathcal{T}';\ \ \forall i \in \mathcal{N}_{\mathrm{P}},\ s \in \mathcal{S}\end{aligned} \tag{9.54}$$

$$\begin{aligned}&\sum_{k:k\to i} Q_{\mathrm{L},ki,s}^{\tau} - \sum_{j:i\to j} Q_{\mathrm{L},ij,s}^{\tau} + Q_{\mathrm{GT},i,s}^{\tau} + Q_{\mathrm{Sub},i,s}^{\tau}\\ &+ Q_{\mathrm{EH},i,s}^{\tau} = Q_{\mathrm{D},i} - Q_{\mathrm{Cut},i,s}^{\tau},\quad \tau = \mathcal{T}';\quad \forall i \in \mathcal{N}_{\mathrm{P}},\ s \in \mathcal{S}\end{aligned} \tag{9.55}$$

$$\sum_{k:k\to i}\frac{M_{ki,\,\mathrm{out}}^{\tau}}{A_{ki}}-\sum_{j:i\to j}\frac{M_{ij,\,\mathrm{in}}^{\tau}}{A_{ij}}+M_{\mathrm{Sta},i,s}^{\tau}-M_{\mathrm{GT},i,s}^{\tau}$$
$$+M_{\mathrm{EH},i,s}^{\tau}=M_{\mathrm{D},i}-M_{\mathrm{Cut},i,s}^{\tau},\ \tau=\mathcal{T}';\quad \forall i\in\mathcal{N}_{\mathrm{G}},\ s\in\mathcal{S} \tag{9.56}$$

$$P_{\mathrm{PV},e,s}^{\tau}-P_{\mathrm{ES}+,e,s}^{\tau}+P_{\mathrm{ES}-,e,s}^{\tau}-P_{\mathrm{PtG},e,s}^{\tau}+P_{\mathrm{GT},e,s}^{\tau}-P_{\mathrm{HP},e,s}^{\tau}$$
$$-P_{\mathrm{EH},e,s}^{\tau}=P_{\mathrm{D},e}-P_{\mathrm{Cut},e,s}^{\{\cdot\}},\ \tau=\mathcal{T}';\quad \forall e\in\mathcal{E},\ s\in\mathcal{S} \tag{9.57}$$

$$H_{\mathrm{GT},e,s}^{\tau}+H_{\mathrm{HP},e,s}^{\tau}+H_{\mathrm{GB},e,s}^{\tau}=H_{\mathrm{D},e}-H_{\mathrm{Cut},e,s}^{\tau},\quad \tau=\mathcal{T}';\quad \forall e\in\mathcal{E},\ s\in\mathcal{S}. \tag{9.58}$$

Equations (9.59)–(9.61) ensure that SSs, GSs, and GTs within the faulted area remain non-operational. Equations (9.62)–(9.63) impose constraints ensuring that power and gas loads within the faulted areas of PDS and NGS remain interrupted. Equation (9.64) defines the upper and lower limits for the curtailed power and heat loads in EHs.

$$0\le P_{\mathrm{Sub},i,s}^{\tau}\le P_{i,\max}^{\mathrm{Sub}}(1-n_{i,s}^{\tau}),\ \tau=\mathcal{T}';\ \forall i\in\mathcal{N}_{\mathrm{Sub}}^{\mathrm{P}},\ s\in\mathcal{S} \tag{9.59}$$

$$0\le M_{\mathrm{Sta},i,s}^{\tau}\le M_{i,\max}^{\mathrm{Sta}}(1-n_{i,s}^{\tau}),\ \tau=\mathcal{T}';\ \forall i\in\mathcal{N}_{\mathrm{Sta}}^{\mathrm{G}},\ s\in\mathcal{S} \tag{9.60}$$

$$0\le P_{\mathrm{GT},i,s}^{\tau}\le P_{i,\max}^{\mathrm{GT}}(1-n_{i,s}^{\tau}),\ \tau=\mathcal{T}';\ \forall i\in\mathcal{N}_{\mathrm{GT}}^{\mathrm{P}},\ s\in\mathcal{S} \tag{9.61}$$

$$n_{i,s}^{\tau}P_{\mathrm{D},i}\le P_{\mathrm{Cut},i,s}^{\tau}\le P_{\mathrm{D},i},\ \tau=\mathcal{T}';\quad \forall i\in\mathcal{N}_{\mathrm{P}},\ s\in\mathcal{S} \tag{9.62}$$

$$n_{i,s}^{\tau}M_{\mathrm{D},i}\le M_{\mathrm{Cut},i,s}^{\tau}\le M_{\mathrm{D},i},\ \tau=\mathcal{T}';\quad \forall i\in\mathcal{N}_{\mathrm{G}},\ s\in\mathcal{S} \tag{9.63}$$

$$[0,\ 0]\ \le\ [P_{\mathrm{Cut},e,s}^{\tau},\ H_{\mathrm{Cut},e,s}^{\tau}]\ \le\ [P_{\mathrm{D},e},\ H_{\mathrm{D},e}],\ \tau=\mathcal{T}';\forall e\in\mathcal{E},\ s\in\mathcal{S}. \tag{9.64}$$

9.3.4 Energy Supply Restoration Metrics

This section introduces a set of energy supply restoration metrics to quantify the performance of the IES in restoring energy supply across multiple stages. Equation (9.65) defines $\mathrm{SPL}_{\tau,s}$, which represents the energy supply level of the system at a single stage τ under a specific fault scenario s. Equation (9.66) defines SPL_{s}, which represents the cumulative energy supply level of the system across the pre-event, post-disaster response, and remote controlled restoration stages for a given fault scenario s. Equation (9.67) defines SPL, which represents the overall energy supply level of the system considering all fault scenarios in the scenario set.

$$\mathrm{SPL}_{\tau,s} = 1 - \frac{\sum_{i\in\mathcal{N}_\mathrm{P}}\omega_i^\mathrm{P}P_{\mathrm{Cut},i,s}^{\tau} + \sum_{i\in\mathcal{N}_\mathrm{G}}\omega_i^\mathrm{G}C_{\mathrm{g2e}}M_{\mathrm{Cut},i,s}^{\tau} + \sum_{e\in\mathcal{E}}\omega_e^\mathrm{EH}(P_{\mathrm{Cut},e,s}^{\tau} + H_{\mathrm{Cut},e,s}^{\tau})}{\sum_{i\in\mathcal{N}_\mathrm{P}}\omega_i^\mathrm{P}P_{\mathrm{D},i} + \sum_{i\in\mathcal{N}_\mathrm{G}}\omega_i^\mathrm{G}C_{\mathrm{g2e}}M_{\mathrm{D},i} + \sum_{e\in\mathcal{E}}\omega_e^\mathrm{EH}(P_{\mathrm{D},e} + H_{\mathrm{D},e})}$$
$$\tau\in\{\,\mathrm{Deg},\mathrm{Iso},\mathrm{Res}\,\};\ \forall s\in\mathcal{S} \tag{9.65}$$

$$\mathrm{SPL}_s = 1 - \frac{\sum_{\tau=\{\,\mathrm{Deg},\,\mathrm{Iso},\,\mathrm{Res}\,\}}\Delta t_\tau \left[\begin{array}{l}\sum_{i\in\mathcal{N}_\mathrm{P}}\omega_i^\mathrm{P}P_{\mathrm{Cut},i,s}^{\tau} + \sum_{i\in\mathcal{N}_\mathrm{G}}\omega_i^\mathrm{G}C_{\mathrm{g2e}}M_{\mathrm{Cut},i,s}^{\tau}\\ +\sum_{e\in\mathcal{E}}\omega_e^\mathrm{EH}(P_{\mathrm{Cut},e,s}^{\tau} + H_{\mathrm{Cut},e,s}^{\tau})]\end{array}\right]}{\sum_{\tau=\{\,\mathrm{Deg},\,\mathrm{Iso},\,\mathrm{Res}\,\}}\Delta t_\tau \left[\begin{array}{l}\sum_{i\in\mathcal{N}_\mathrm{P}}\omega_i^\mathrm{P}P_{\mathrm{D},i} + \sum_{i\in\mathcal{N}_\mathrm{G}}\omega_i^\mathrm{G}C_{\mathrm{g2e}}M_{\mathrm{D},i}\\ +\sum_{e\in\mathcal{E}}\omega_e^\mathrm{EH}(P_{\mathrm{D},e} + H_{\mathrm{D},e})\end{array}\right]};\ \forall s\in\mathcal{S} \tag{9.66}$$

$$\mathrm{SPL} = 1 - \frac{\sum_{s\in\mathcal{S}}\rho_s\sum_{\tau=\{\,\mathrm{Deg},\,\mathrm{Iso},\,\mathrm{Res}\,\}}\Delta t_\tau \left[\begin{array}{l}\sum_{i\in\mathcal{N}_\mathrm{P}}\omega_i^\mathrm{P}P_{\mathrm{Cut},i,s}^{\tau} + \sum_{i\in\mathcal{N}_\mathrm{G}}\omega_i^\mathrm{G}C_{\mathrm{g2e}}M_{\mathrm{Cut},i,s}^{\tau}\\ +\sum_{e\in\mathcal{E}}\omega_e^\mathrm{EH}(P_{\mathrm{Cut},e,s}^{\tau} + H_{\mathrm{Cut},e,s}^{\tau})\end{array}\right]}{\sum_{\tau=\{\,\mathrm{Deg},\,\mathrm{Iso},\,\mathrm{Res}\,\}}\Delta t_\tau \left[\begin{array}{l}\sum_{i\in\mathcal{N}_\mathrm{P}}\omega_i^\mathrm{P}P_{\mathrm{D},i} + \sum_{i\in\mathcal{N}_\mathrm{G}}\omega_i^\mathrm{G}C_{\mathrm{g2e}}M_{\mathrm{D},i}\\ +\sum_{e\in\mathcal{E}}\omega_e^\mathrm{EH}(P_{\mathrm{D},e} + H_{\mathrm{D},e})\end{array}\right]}. \tag{9.67}$$

9.4 Solution Algorithm for the Energy Supply Restoration Model

The proposed multi-stage and multi-energy coordinated energy supply restoration model is formulated as a large-scale two-stage stochastic mixed-integer linear programming (MILP) problem. While commercial solvers can be used for direct solving, the computational burden becomes significantly high as the number of fault scenarios and system scale increase. This section introduces the customized progressive hedging algorithm (C-PHA), which decomposes the original model into smaller subproblems associated with different fault scenarios, allowing for iterative solving to reduce computational complexity.

9.4.1 Compact Form of the Model

The compact form of the energy supply restoration model is expressed in Eqs. (9.68)–(9.69). Equation (9.68) represents the objective function, which corresponds to the original formulation presented earlier. Equation (9.69) defines the operational constraints, encompassing all constraints from the original model in a generalized form.

$$\min_{\boldsymbol{x},\boldsymbol{y}_s} \sum_{s\in\mathcal{S}} \rho_s \cdot \boldsymbol{b}_s^{\mathrm{T}}\boldsymbol{y}_s \tag{9.68}$$

$$\text{s.t.} \quad (\boldsymbol{x}, \boldsymbol{y}_s) \in \mathcal{Q}_s; \quad \forall s \in \mathcal{S}, \tag{9.69}$$

where $\boldsymbol{x}$ represents the first-stage decision variables, while $\boldsymbol{y}_s$ represents the second-stage operational variables under fault scenario s. The set $\mathcal{Q}_s$ defines the feasible region for $\boldsymbol{x}$ and $\boldsymbol{y}_s$. $\boldsymbol{b}_s$ denotes the coefficient matrix related to system parameters. Due to the presence of the first-stage variables $\boldsymbol{x}$, the decision variables $\boldsymbol{y}_s$ across multiple scenarios are interdependent, creating a coupling effect between different fault scenarios.

9.4.2 Customized Progressive Hedging Algorithm

Based on the traditional progressive hedging algorithm (PHA) framework, the proposed customized progressive hedging algorithm (C-PHA) decomposes the original model into multiple fault scenario subproblems, as formulated in Eqs. (9.70)–(9.72). The total number of fault scenarios is denoted as N_s, and each scenario subproblem is a mixed-integer quadratic programming (MIQP) problem.

$$\min_{x_s,\boldsymbol{y}_s} \; \boldsymbol{b}_s^{\mathrm{T}}\boldsymbol{y}_s + \boldsymbol{\eta}_s^{k*\mathrm{T}}\boldsymbol{x} + \upsilon/2||\boldsymbol{x}_s^{k*} - \overline{\boldsymbol{x}}_s^{k-1*}||^2 \tag{9.70}$$

$$\text{s.t.} \quad (\boldsymbol{x}, \boldsymbol{y}_s) \in \mathcal{Q}_s, \tag{9.71}$$

where k represents the iteration index, $\boldsymbol{\eta}_s^{k*}$ is the Lagrange multiplier vector at iteration k, v is the penalty coefficient, $\boldsymbol{x}_s^{k*}$ and $\overline{\boldsymbol{x}}_s^{k-1*}$ represent the solution values of variable x in iteration k and iteration $k-1$, respectively, $||\cdot||^2$ denotes the L2-norm.

PHA is a heuristic algorithm designed to accelerate the solution of two-stage stochastic programming problems. However, when integer variables are included in the first stage, the algorithm may exhibit slow convergence [20]. In the proposed remote controlled restoration model, both the first-stage and second-stage formulations contain binary (0–1) variables, which further increases computational complexity. To improve convergence performance, this section introduces a

Table 9.1 Algorithm workflow

Algorithm 1: Customized Progressive Hedging Algorithm
1: Initialization: Set k=0, initialize the fixed variable index $\sigma_k = 0$, $\boldsymbol{\eta}_s^k = 0$
2: Solve all scenario subproblems: $\boldsymbol{x}_s^k$=arg min $_{x,y_s}$ $\{\boldsymbol{b}_s^{\mathrm{T}}\boldsymbol{y}_s: (\boldsymbol{x}, \boldsymbol{y}_s) \in \mathcal{Q}_s\}$
3: $\overline{\boldsymbol{x}}^k := \Sigma_{s\in\mathcal{S}}\, \rho_s \boldsymbol{x}_s^k$
4: For all $s \in \mathcal{S}$, $\boldsymbol{\eta}_s^k := \boldsymbol{\eta}_s^k + \upsilon(\boldsymbol{x}_s^k - \overline{\boldsymbol{x}}^k)$
5: Set k=k+1
6: For all $s \in \mathcal{S}$, $\overline{\boldsymbol{x}}^k :=$ arg $\min_{\boldsymbol{x},\boldsymbol{y}_s}\{\boldsymbol{b}_s^{\mathrm{T}}\boldsymbol{y}_s + \boldsymbol{\eta}_s^{k-1}\boldsymbol{x} + \upsilon/2\Vert\boldsymbol{x}_s^k - \overline{\boldsymbol{x}}_s^{k-1}\Vert^2: (\boldsymbol{x}, \boldsymbol{y}_s) \in \mathcal{Q}_s\}$
7: $\overline{\boldsymbol{x}}^k := \Sigma_{s\in\mathcal{S}}\, \rho_s \boldsymbol{x}_s^k$
8: For all $s \in \mathcal{S}$, $\boldsymbol{\eta}_s^k := \boldsymbol{\eta}_s^{k-1} + \upsilon(\boldsymbol{x}_s^k - \overline{\boldsymbol{x}}^k)$
9: If $\Sigma_{s\in\mathcal{S}}\, \rho_s \Vert\boldsymbol{x}_s^k - \overline{\boldsymbol{x}}_s^{k-1}\Vert^2 \le \varepsilon$, then terminate; otherwise, go to Step 10
10: If $k \leqslant K_3$ or $\sigma_{k+1} - \sigma_k \ge 1$, go to Step 11; otherwise, go to Step 16
11: If $k \geqslant K_1$, go to Step 12; otherwise, go to Step 13
12: If $z_{ij,1}^k = z_{ij,s}^k$, $\forall ij \in \mathcal{L}^{\mathrm{p}} \cup \mathcal{L}^{\mathrm{g}}, s \in \mathcal{S}$, fix $z_{ij} = z_{ij,s}^k$, set $\sigma_{k+1} = \sigma_k + 1$
13: If $k \geqslant K_2$, go to Step 14; otherwise, go to Step 15
14: If $\alpha_{ij,1}^k = \alpha_{ij,s}^k$, $\forall ij \in \mathcal{L}^{\mathrm{p}} \cup \mathcal{L}^{\mathrm{g}}, s \in \mathcal{S}$, fix $\alpha_{ij} = \alpha_{ij,s}^k$, set $\sigma_{k+1} = \sigma_k + 1$
15: Return to Step 5
16: Solve $\min_{x,\, y_s} \Sigma_{s\in\mathcal{S}}\, \rho_s b_s^{\mathrm{T}} y_s$: $(x, y_s) \in \mathcal{Q}_s$ $\forall s \in S$, end the iterative process

customized progressive hedging algorithm (C-PHA). The detailed algorithmic steps are outlined in Table 9.1, and the solution framework is illustrated in Fig. 9.5.

The main improvements of the algorithm are implemented in Steps 10–16. First, after several iterations, if certain first-stage binary variables yield the same solution across different scenario subproblems, they are fixed in the next iteration [20]. In the proposed algorithm, the placement variables of remote controlled switches and valves are fixed after K_1 iterations, while network reconfiguration variables are fixed after K_2 iterations ($K_2 \ge K_1$). Once variables are fixed, they are treated as parameters in subsequent iterations. If the algorithm fails to converge after K_3 iterations, or if the number of fixed variables remains unchanged ($K_3 \ge K_2$), the process directly solves the deterministic model in Step 16. Compared to the original model (9.68)–(9.69), this reduced-scale model has fewer binary variables, significantly improving computational efficiency.

9.5 Case Studies

To verify the effectiveness of the proposed multi-stage and multi-energy flow coordinated IES energy supply restoration method, this section conducts case studies on the IES E123-G40 system, as shown in Fig. 9.6. The PDS and NGS are coupled through five GTs and three EHs. The total power and natural gas loads are 3.49 MW + 1.92 MVar and 4.10 MW (corresponding to a gas flow of 7.8975 kg/s), respectively [1].

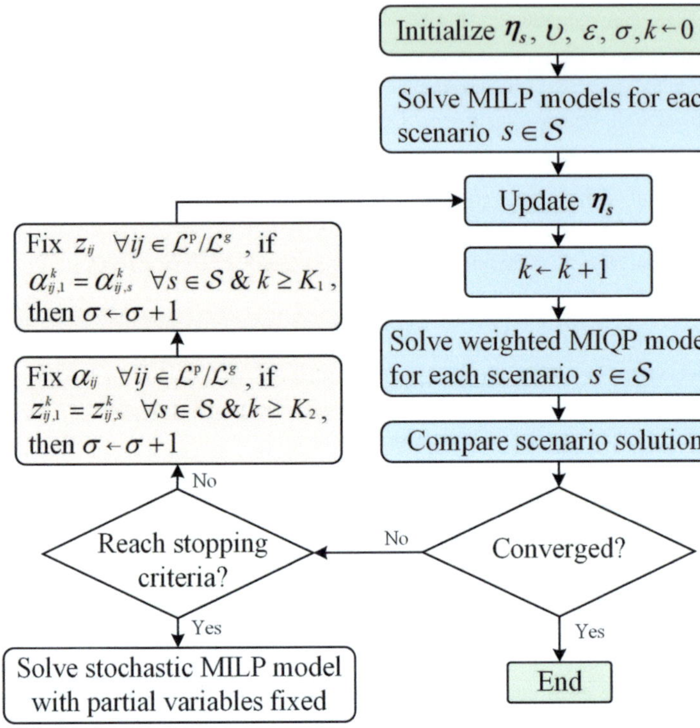

Fig. 9.5 Solution flowchart

Each energy hub has specific power and heat loads: 300 kW and 450 kW, 250 kW and 400 kW, and 200 kW and 350 kW, respectively. The parameters of relevant equipment are listed in Table 9.2. The key loads are assigned at power nodes 9, 18, 28, 37, 47, 50, 62, 73, 82, 103 and gas nodes 6, 14, 22, 29, 37, with a weight coefficient of 5, while loads at other nodes have a weight coefficient of 1. Detailed system parameters are available in [21]. A Monte Carlo method is employed to generate the fault scenario set [22], with the number of scenarios set to 20, where each scenario contains eight faulted power lines and four faulted natural gas pipelines, as depicted in Fig. 9.7, with an occurrence probability of 1/20. The durations of each stage in the restoration process are set as follows [23]: 0.20 h for the pre-event stage, 0.46 h for the system disturbance stage, 0.23 h for the fault isolation stage, and 0.22 h for the remote controlled restoration stage based on remote controlled devices. The maximum number of remote controlled switches and valves is set to 30 and 14, respectively. All simulations are implemented in MATLAB and solved using Gurobi.

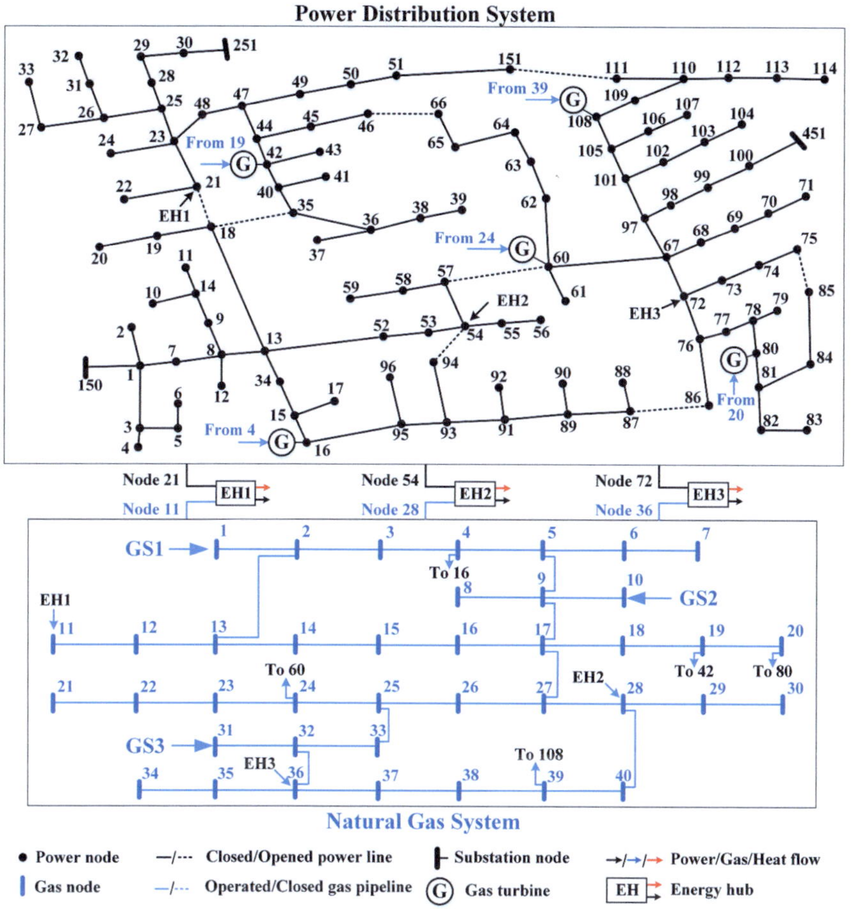

Fig. 9.6 IES E123-G40 test case

Table 9.2 Equipment parameters of EHs

Parameter	Value	Parameter	Value	Parameter	Value
$Soc_{\mathrm{Es},e}^{\min/\max}$ (kW)	0.1/0.9	$G_{e\max}^{\mathrm{EH}}$ (kW)	650	$R_{\mathrm{PtG},e}$ (kW)	60/120
$P_{e,\max}^{\mathrm{ES+/ES-}}$ (kW)	400/400	$\eta_{\mathrm{ES+/ES}}$	0.95/0.95	$R_{\mathrm{GT},e}$ (kW)	60/150
$P_{e,\max}^{\mathrm{PtG/HP}}$ (kW)	1000/80	η_{PtG}	0.65	$R_{\mathrm{HP},e}$ (kW)	60/80
$P_{e,\max}^{\mathrm{PV/GT}}$ (kW)	800/400	η_{GB}	0.85	$R_{\mathrm{GB},e}$ (kW)	30/50
$P_{e,\max}^{\mathrm{GB}}$ (kW)	300	η_{HP}	3.0	ς (kW/kg/s)	519.2
$P_{e,\max}^{\mathrm{EH}}$ (kW)	1000	$\eta_{\mathrm{GT},gp/gh}$	0.36/0.68		

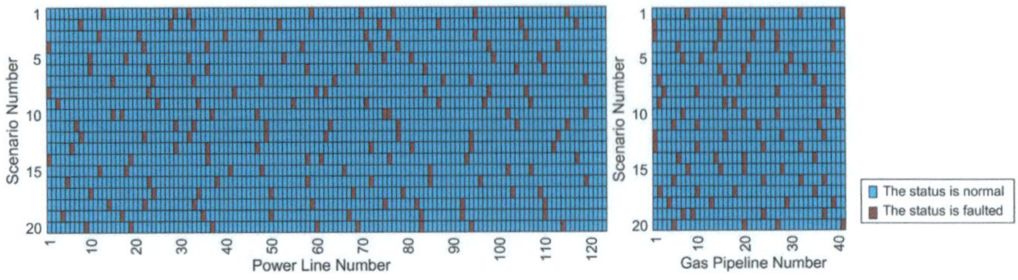

Fig. 9.7 Scenario set

9.5.1 Numerical Simulation Results Under Baseline Settings

Under the baseline settings, the proposed method yields a multi-stage response and energy supply restoration strategy for the IES. The objective function value for the fault scenario set is 5.75×10^5 kW, and the corresponding value of SPL is 64.26%.

Taking scenario 6 as an example, Fig. 9.8 illustrates the pre-event, emergency response, and energy supply restoration process of the IES. As shown in Fig. 9.8a, during preventive network reconfiguration, the PDS and NGS are divided into eight and six subsystems, respectively. In Fig. 9.8b, three power subsystems are affected by the disaster, but due to proactive topological reconfiguration, some subsystems remain operational post-disaster. In Fig. 9.8c, during the fault isolation stage, partial load restoration is achieved, and the faulted area is reduced. Finally, in Fig. 9.8d, during the remote controlled restoration stage, certain non-faulted loads are reconnected to operational areas, enabling further energy restoration. These results confirm the effectiveness of the proposed multi-stage restoration model, demonstrating its ability to enhance system resilience and ensure efficient energy recovery.

To further analyze the remote controlled restoration process across multiple stages, Fig. 9.9 illustrates the load restoration ratio of power and gas nodes under a fault scenario. Given the large number of nodes in the PDS, only the restoration ratios for nodes 1–60 are displayed in the figure. As shown, the node load restoration results align with those in Fig. 9.8, confirming the effectiveness of the proposed approach. It is noteworthy that not all nodes within the operational areas achieve full energy restoration. For instance, due to capacity limitations of the gas turbine at node 16, several power nodes (nodes 8, 12, 13, 15, and 34) are not fully restored during the fault isolation stage. However, in the NGS, all nodes within the operational area reach their full gas supply capacity, indicating an effective restoration process for gas loads.

Figure 9.10 illustrates the variation trends of the IES energy supply level $SPL_{\tau,s}$ across multiple stages under different fault scenarios, where each curve corresponds to a fault scenario. As shown in the figure, in each fault scenario, the IES energy supply level increases during the system disturbance, fault isolation, and remote controlled restoration stages. Significant differences in system energy supply levels are observed across different fault scenarios. Preventive reconfiguration and remote controlled device deployment decisions are not optimal for all fault scenarios but

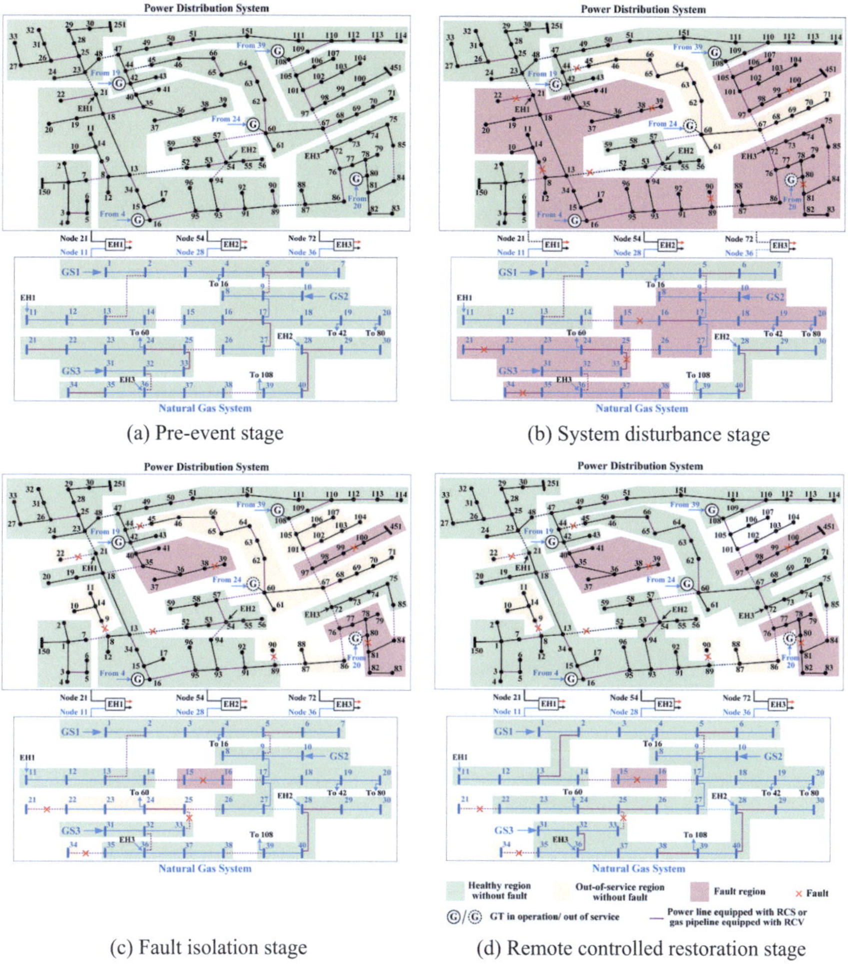

(a) Pre-event stage

(b) System disturbance stage

(c) Fault isolation stage

(d) Remote controlled restoration stage

Fig. 9.8 Multi-stage remote controlled restoration process of IES

represent a comprehensive optimal decision for differentiated scenarios. Figure 9.11 presents the comprehensive energy supply level SPL_s across multiple stages under different fault scenarios. The results indicate that the IES performs worst in scenarios 3, 17, and 5. To further enhance IES resilience, targeted reinforcement measures can be developed for these fault scenarios.

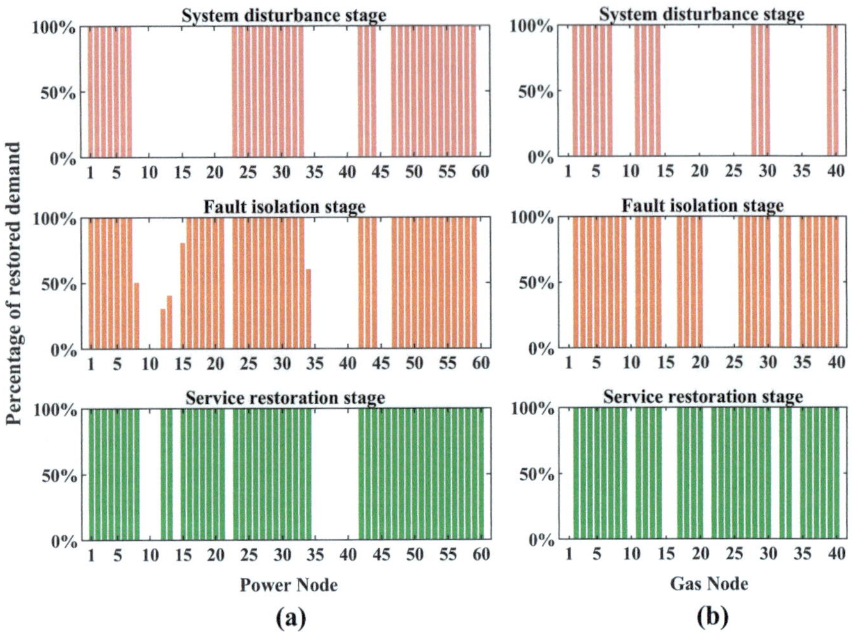

Fig. 9.9 Multi-stage load restoration of power and gas nodes

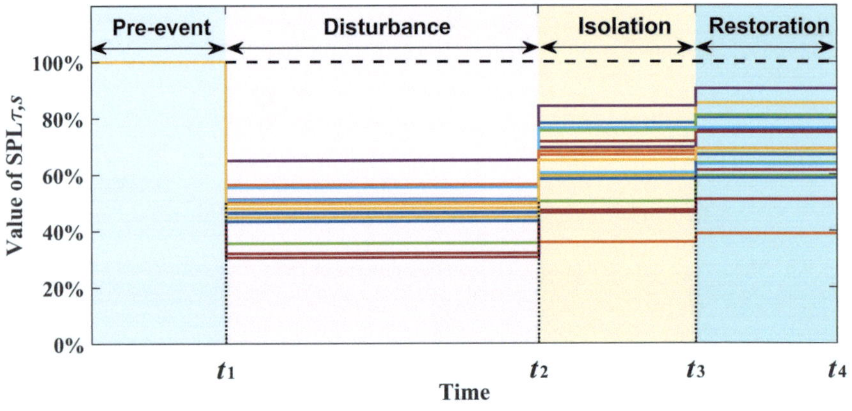

Fig. 9.10 Comparison of $SPL_{\tau,s}$ under different fault scenarios

9.5.2 *Impact of Multi-stage and Multi-energy Flow Coordination on Results*

To evaluate the impact of multi-stage coordination on IES remote controlled restoration, this section compares three cases with different configurations. Case 1 is the

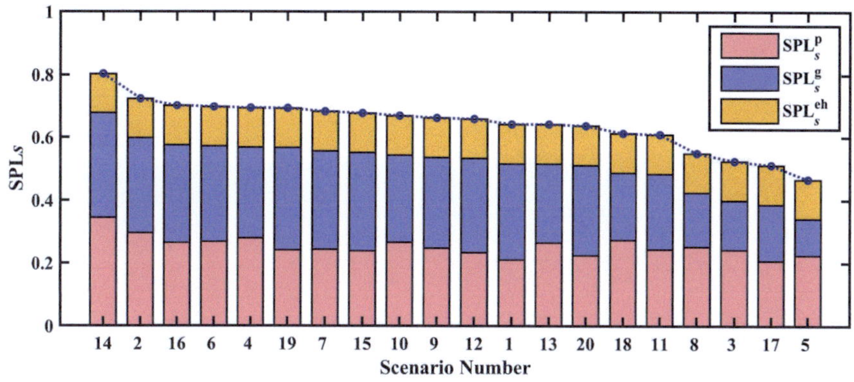

Fig. 9.11 Comparison of SPL*s* under different fault scenarios

proposed method; Case 2 only considers pre-event measures; Case 3 only considers fault isolation and remote controlled restoration using remote controlled devices.

The energy supply levels (SPL) in Case 1, Case 2, and Case 3 across multiple stages are 64.26%, 43.22%, and 30.19%, respectively. Figure 9.12 illustrates the trends of $SPL_{\tau,s}$ over multiple stages for the three cases. At the initial fault stage, the $SPL_{\tau,s}$ of Case 2 is slightly higher than that of Case 1. However, since Case 2 does not include emergency response and energy restoration measures, its energy supply level cannot be further improved in later stages. Case 3 accounts for fault isolation and energy restoration using remote controlled devices, but since preventive reconfiguration is not considered, the IES experiences a complete energy supply interruption during the system disturbance stage. The results indicate that compared to methods that only consider partial restoration processes, the proposed method effectively enhances the IES remote controlled restoration level. Additionally, the energy supply level in each stage of Case 1 is not a simple summation of Case 2 and Case 3, demonstrating that the measures taken in different stages are not independent but highly coupled. These findings verify the importance of multi-stage coordination in IES remote controlled restoration.

To evaluate the impact of multi-energy flow coordination on IES remote controlled restoration, this section compares three cases with different configurations. Case 1 is the proposed method, where PDS, NGS, and EHs collaboratively restore energy supply; in Case 4, each subsystem independently restores energy supply; in Case 5, PDS and NGS restore energy supply in coordination, but EHs are not considered. Figure 9.13 presents the energy supply levels under different fault scenarios for the three cases. The results indicate that in most scenarios, SPL_s for Case 4 is slightly lower than that of Case 1, demonstrating that independent energy restoration by heterogeneous energy flow subsystems negatively impacts the overall IES energy supply level. Case 5 consistently shows the lowest SPL across all fault scenarios, verifying the critical role of EHs in supporting PDS and NGS during EWEs.

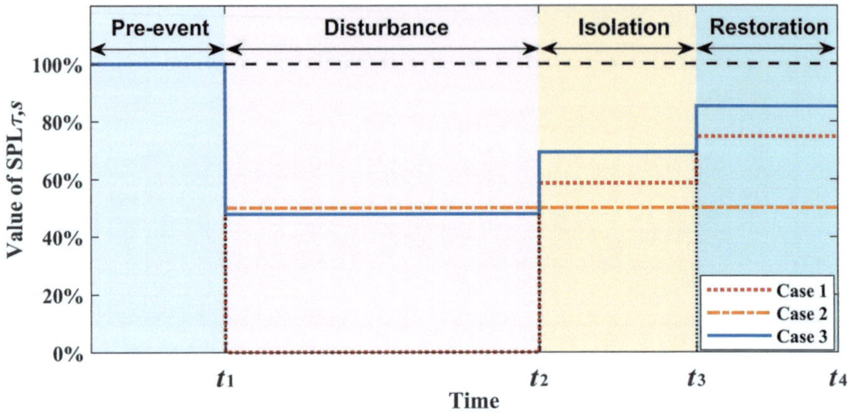

Fig. 9.12 Comparison of $SPL_{\tau,s}$ for Case 1–Case 3

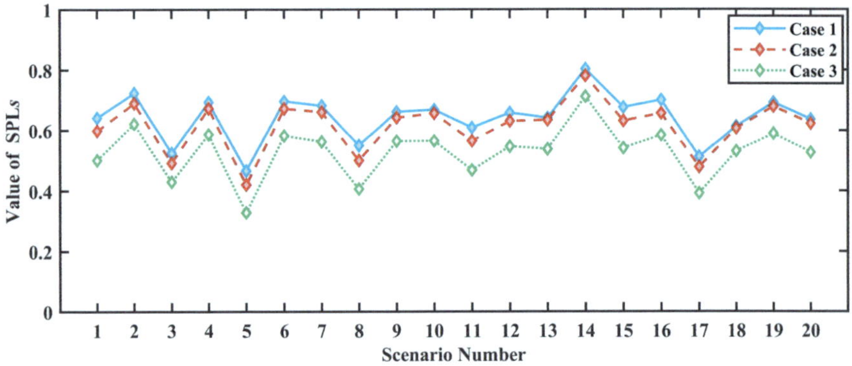

Fig. 9.13 Comparison of SPLs between different cases

To further illustrate the role of EHs in the multi-stage remote controlled restoration process, Fig. 9.14 presents the dispatch results of electrical and thermal power within EHs under fault scenario 6. As shown in Fig. 9.14a, the electrical load within the EHs is fully supplied, and surplus electricity is transferred to the PDS to support the restoration of electric loads. As depicted in Fig. 9.14b, EH1 experiences partial load shedding during the remote controlled restoration stage, indicating that the dispatch strategy of internal energy hub devices prioritizes optimizing the overall IES energy supply level. These results confirm the supportive role of EHs in the remote controlled restoration process.

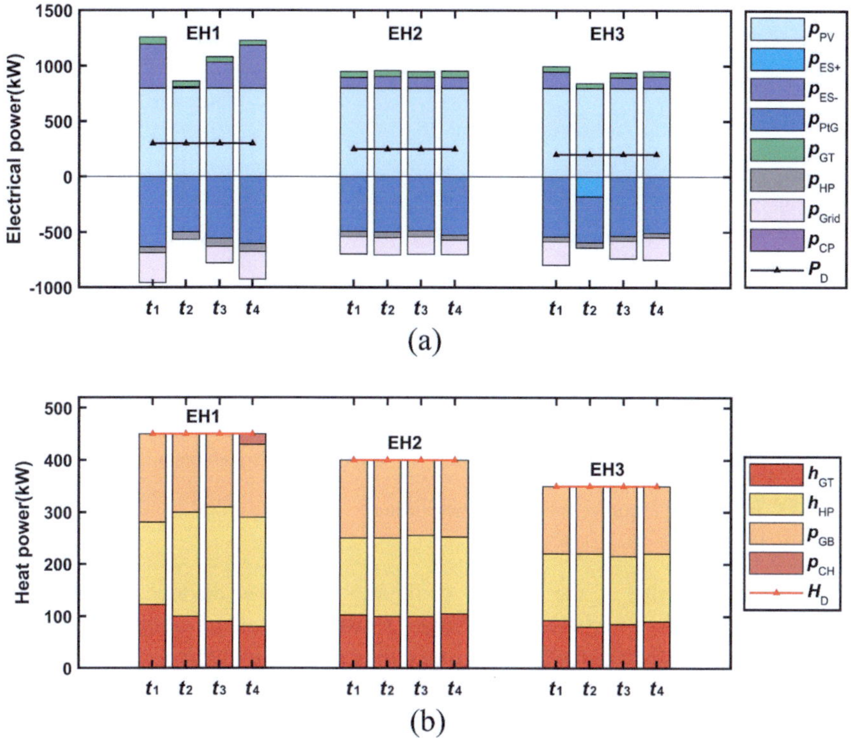

Fig. 9.14 Electrical and heat power balance in EHs

9.5.3 Sensitivity Analysis of Key Parameters

This section conducts a sensitivity analysis on key parameters that impact the IES remote controlled restoration performance in the proposed model. First, an analysis is performed on the number of faulted components in each fault scenario. Figure 9.15 illustrates the impact of the number of faulted components on the IES remote controlled restoration level. The results show that as the number of faulted components increases, the remote controlled restoration level of the IES declines at an accelerated rate throughout the multi-stage process. Therefore, for high-intensity disaster events, it is essential to implement resilience-enhancing protective measures to mitigate the impact on the system. Additionally, the results indicate that an increase in the number of faulted gas pipelines has a more significant impact on IES remote controlled restoration than power line failures, demonstrating that gas pipeline disruptions have a greater effect on the overall restoration process.

RCSs and RCVs play a crucial role in enhancing the flexibility of network reconfiguration and optimizing remote controlled restoration paths during the multi-stage recovery process of IES. Figure 9.16 further analyzes the impact of the number of remote controlled devices on the remote controlled restoration level of the IES.

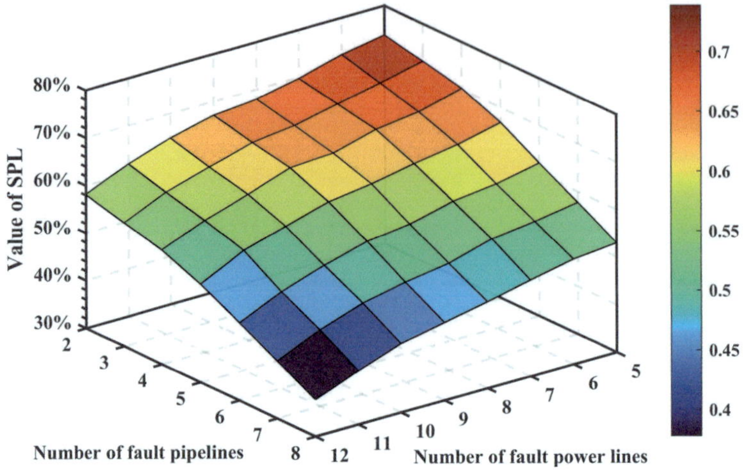

Fig. 9.15 Impact of the number of faulted components

The results show that as the number of remote controlled devices increases, the remote controlled restoration level improves. However, when the number of installed RCSs and RCVs continues to increase, the rate of improvement in remote controlled restoration exhibits a diminishing marginal effect.

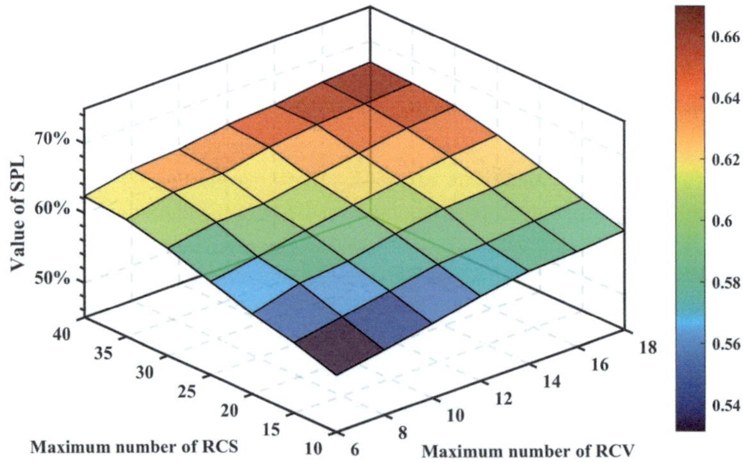

Fig. 9.16 Impact of the number of remote controlled devices

9.5.4 Computational Performance Testing of the Solution Algorithm

This section evaluates the computational performance of the proposed C-PHA algorithm. Table 9.3 first compares the computational efficiency of the proposed algorithm with the non-decomposition method (NDM). The results indicate that as the number of fault scenarios increases, the computational time of the NDM grows exponentially, and when the number of scenarios exceeds 20, it fails to obtain a solution within the maximum time limit of 10 h. In contrast, the proposed C-PHA algorithm consistently finds a high-quality solution within the maximum solving time, with a relative error of no more than 5%. Moreover, when the number of fault scenarios exceeds 10, the proposed algorithm significantly reduces computation time compared to the NDM. Additionally, the results show that the impact of the number of fault scenarios on computational time exhibits a nonlinear relationship.

Table 9.4 further compares the computational performance of the proposed C-PHA algorithm with the parallelized C-PHA algorithm, where the scenario subproblems in Steps 2 and 6 of Table 9.1 are solved in parallel. As shown in the second row of Table 9.4, the number of iterations required for the C-PHA algorithm increases as the number of fault scenarios grows. Meanwhile, with an increasing number of iterations, the average solving time per scenario subproblem gradually decreases. This is because, during the iterative process, the proposed algorithm progressively fixes more integer variables, reducing the computational burden of subproblems in later iterations. By incorporating parallelized subproblem solving, the computational efficiency of the C-PHA algorithm improves by more than 30%. These results verify the effectiveness of the proposed algorithm in handling large-scale IES remote controlled restoration problems efficiently.

Table 9.3 Comparison of NDM and C-PHA

Number of scenarios	CPU time (min)		Objectives (10^5 kW)		
	NDM	C-PHA	NDM	C-PHA	Relative errors
5	5.7	9.2	4.99	5.11	2.40%
10	25.7	18.1	5.30	5.54	4.53%
15	165.7	32.6	5.42	5.61	3.51%
20	N/A	49.3	N/A	5.75	N/A
25	N/A	78.5	N/A	5.86	N/A
30	N/A	113.1	N/A	6.01	N/A

Table 9.4 Comparison of C-PHA and PC-PHA

Number of scenarios	5	10	15	20	25	30
Iteration number	14	18	29	41	55	69
Average CPU time for each scenario (s)	7.88	6.03	4.50	3.61	3.42	3.28
CPU time of C-PHA (min)	9.2	18.1	32.6	49.3	78.5	113.1
CPU time of PC-PHA	5.2	11.0	20.6	32.7	49.8	67.6
Reduced CPU time (%)	43.5	39.2	36.8	33.7	36.6	40.2

9.6 Conclusions

This chapter proposes a multi-stage and multi-energy flow coordinated remote controlled restoration method for IES in the post-disaster phase. First, a refined set of operational constraints is developed, comprehensively considering the multi-stage emergency response and remote controlled restoration process, which includes pre-event, system disturbance, fault isolation, and remote controlled restoration stages. Next, based on unified multi-stage modeling, a mathematical model is formulated to optimize multi-energy flow coordination in IES energy restoration. Finally, a customized progressive hedging algorithm is introduced, which decomposes the large-scale optimization model into multiple smaller subproblems related to different fault scenarios, enabling iterative solving. From the simulation results, the following conclusions can be drawn:

1. The proposed model effectively leverages multi-energy flow resources to enhance synergy in multi-stage response and restoration, significantly improving post-disaster resilience and energy recovery rates, and accelerating the remote controlled restoration process.
2. Allocating additional flexible resources, such as RCSs and RCVs, enhances the network reconfiguration flexibility during the multi-stage restoration process, thereby improving the overall efficiency of IES energy supply recovery.
3. The proposed solution algorithm efficiently decomposes the large-scale model, significantly reducing the computational burden of the multi-stage IES remote controlled restoration problem, achieving efficient optimization.

The proposed multi-stage and multi-energy flow coordinated restoration method provides a practical solution for rapid post-disaster IES energy recovery. Future work will further explore the impact of communication and transportation system disruptions on IES energy restoration, study the collaborative restoration of IES with other critical infrastructures, and investigate IES remote controlled restoration strategies incorporating damaged equipment repair processes.

References

1. Bao, M. L., Ding, Y., Sang, M. S., Li, D. Q., Shao, C. Z., & Yan, J. Y. (2020). Modeling and evaluating nodal resilience of multi-energy systems under windstorms. *Applied Energy, 270*, Article 115136.
2. Sayed, A. R., Wang, C., & Bi, T. (2019). Resilient operational strategies for power systems considering the interactions with natural gas systems. *Applied Energy, 241*, 548–566.
3. Mehrjerdi, H., Mahdavi, S., & Hemmati, R. (2021). Resilience maximization through mobile battery storage and diesel DG in integrated electrical and heating networks. *Energy, 237*, Article 121195.
4. Lei, S., Wang, J., & Hou, Y. (2017). Remote-controlled switch allocation enabling prompt restoration of distribution systems. *IEEE Transactions on Power Systems, 33*(3), 3129–3142.
5. Yan, M. Q. (2010). Faulted pipeline segment isolation decision algorithm for gas distribution networks. *Gas & Heat, 30*(10), 35–39.
6. Liu, J. C., Qin, C., & Yu, Y. X. (2020). Enhancing distribution system resilience with proactive islanding and RCS-based fast fault isolation and service restoration. *IEEE Transactions on Smart Grid, 11*(3), 2381–2395.
7. Kou, Y., Bie, Z., Li, G., Liu, F., & Jiang, J. (2021). Reliability evaluation of multi-agent integrated energy systems with fully distributed communication. *Energy, 224*, Article 120123.
8. Shao, C. C., Shahidehpour, M., Wang, X. F., Wang, X. L., & Wang, B. Y. (2017). Integrated planning of electricity and natural gas transportation systems for enhancing the power grid resilience. *IEEE Transactions on Power Systems, 32*(6), 4418–4429.
9. Lin, Y. L., Chen, B., Wang, J. H., & Bie, Z. H. (2019). A combined repair crew dispatch problem for resilient electric and natural gas system considering reconfiguration and DG islanding. *IEEE Transactions on Power Systems, 34*(4), 2755–2767.
10. Lei, S., Chen, C., Song, Y., & Hou, Y. (2020). Radiality constraints for resilient reconfiguration of distribution systems: Formulation and application to microgrid formation. *IEEE Transactions on Smart Grid, 11*(5), 3944–3956.
11. Cao, X. Y., Cao, T. X., Gao, F., & Guan, X. H. (2021). Risk-averse storage planning for improving RES hosting capacity under uncertain siting choices. *IEEE Transactions on Sustainable Energy, 12*(4), 1984–1995.
12. Qiu, H. F., Wang, L., Gu, W., Pan, G. S., Ning, C., Wu, Z., & Sun, Q. R. (2022). Multistage scheduling of regional power grids against sequential outage and power uncertainties. *IEEE Transactions on Smart Grid, 13*(6), 4624–4637.
13. Zheng, W. Y., Huang, W. J., Hill, D. J., & Hou, Y. H. (2021). An adaptive distributionally robust model for three-phase distribution network reconfiguration. *IEEE Transactions on Smart Grid, 12*(2), 1224–1237.
14. Zou, Y., Xu, Y., Feng, X., & Nguyen, H. D. (2023). Peer-to-peer transactive energy trading of a reconfigurable multi-energy network. *IEEE Transactions on Smart Grid, 14*(3), 2236–2249.
15. Tao, R., Zhao, D., Xu, C., Wang, H., & Xia, X. (2023). Resilience enhancement of integrated electricity-gas-heat urban energy system with data centres considering waste heat reuse. *IEEE Transactions on Smart Grid, 14*(1), 183–198.
16. Fang, J., Zeng, Q., Ai, X., Chen, Z., & Wen, J. (2018). Dynamic optimal energy flow in the integrated natural gas and electrical power systems. *IEEE Transactions on Sustainable Energy, 9*(1), 188–198.
17. Yan, M. Y., He, Y. B., Shahidehpour, M., Ai, X. M., Li, Z. Y., & Wen, J. Y. (2019). Coordinated regional-district operation of integrated energy systems for resilience enhancement in natural disasters. *IEEE Transactions on Smart Grid, 10*(5), 4881–4892.
18. Shams, M. H., Shahabi, M., MansourLakouraj, M., Shafie-khah, M., & Catalão, J. P. (2021). Adjustable robust optimization approach for two-stage operation of energy hub-based microgrids. *Energy, 222*, Article 119894.
19. Liu, J., Qin, C., & Yu, Y. (2020). A comprehensive resilience-oriented FLISR method for distribution systems. *IEEE Transactions on Smart Grid, 12*(3), 2136–2152.

20. Arif, A., Wang, Z. Y., Chen, C., & Chen, B. (2020). A stochastic multi-commodity logistic model for disaster preparation in distribution systems. *IEEE Transactions on Smart Grid, 11*(1), 565–576.
21. Sun, Q. Test data of 123-node power distribution system and 40-node natural gas system [Online]. Retrieved from: https://drive.google.com/file/d/1Bb9dXcVb016xLuRmXLnNlTK5qJ7t9C3z/view?usp=sharing
22. Ma, S. S., Li, S. Y., Wang, Z. Y., & Qiu, F. (2019). Resilience-oriented design of distribution systems. *IEEE Transactions on Power Systems, 34*(4), 2880–2891.
23. de Assis, L. S., Vizcaíno, J. F., Usberti, F. L., Lyra, C., Cavellucci, C., & Von Zuben, F. J. (2014). Switch allocation problems in power distribution systems. *IEEE Transactions on Power Systems, 30*(1), 246–253.

MIX
Papier aus verantwortungsvollen Quellen
Paper from responsible sources
FSC® C105338
FSC
www.fsc.org

If you have any concerns about our products,
you can contact us on
ProductSafety@springernature.com

In case Publisher is established outside the EU,
the EU authorized representative is:
Springer Nature Customer Service Center GmbH
Europaplatz 3, 69115 Heidelberg, Germany

Printed by Libri Plureos GmbH
in Hamburg, Germany